1 MONTH OF FREE READING

at
www.ForgottenBooks.com

By purchasing this book you are eligible for one month membership to ForgottenBooks.com, giving you unlimited access to our entire collection of over 1,000,000 titles via our web site and mobile apps.

To claim your free month visit:
www.forgottenbooks.com/free849809

* Offer is valid for 45 days from date of purchase. Terms and conditions apply.

ISBN 978-0-666-77370-8
PIBN 10849809

This book is a reproduction of an important historical work. Forgotten Books uses state-of-the-art technology to digitally reconstruct the work, preserving the original format whilst repairing imperfections present in the aged copy. In rare cases, an imperfection in the original, such as a blemish or missing page, may be replicated in our edition. We do, however, repair the vast majority of imperfections successfully; any imperfections that remain are intentionally left to preserve the state of such historical works.

Forgotten Books is a registered trademark of FB &c Ltd.
Copyright © 2018 FB &c Ltd.
FB &c Ltd, Dalton House, 60 Windsor Avenue, London, SW19 2RR.
Company number 08720141. Registered in England and Wales.

For support please visit www.forgottenbooks.com

Griffin & Co., Publishers, The Hard, Portsea.

THE RIGGER'S GUIDE

By CHARLES BUSHELL. Fully Illustrated. Being the best and only complete book on the Rigging of Ships.

Sixth Edition, Revised and enlarged, 3s. cloth.

CAPT. ALSTON'S SEAMANSHIP

Crown 8vo. cloth, price 12s. 6d. Contains 200 Illustrations of Rigging, Sails, Masts, &c.; with Instructions for Officers of the Merchant Service, by W. H. ROSSER; forming a complete Manual of Practical Seamanship.—*Second Edition.*

THE SAILOR'S POCKET BOOK

By Commander F. G. D. BEDFORD, R.N. A Collection of Practical Rules, Notes, and Tables, for the use of the Royal Navy, the Mercantile Marine, and Yacht Squadrons: With Coloured Signal Flags, Charts, and Illustrations. Bound in leather, 400 pages, and carefully compiled Index. *Second Edition.*
Price 7s. 6d.

THE DEFENCE OF ENGLAND

Demy 8vo, with Map; cloth, price 7s. 6d. By Lieutenant-General SYNGE, Royal Engineers.

THE
SAILOR'S POCKET BOOK.

Stewart R.E.

THE OUTER fixed card represents a Compass not affected by the iron of the ship:—
or the Horizon with the magnetic points and degrees marked on it.

The INNER moveable card represents the Compass as affected by the iron of the ship;
and the Deviation produced is called East or West, according as the North point of
his Compass i drawn to the East or West of the Correct Magnetic North, as shewn
n the fixed card.

THE
SAILOR'S POCKET BOOK:

A COLLECTION OF
PRACTICAL RULES, NOTES, AND TABLES,
FOR THE USE OF
THE ROYAL NAVY, THE MERCANTILE MARINE, AND YACHT SQUADRONS.

By COMMANDER F. G. D. BEDFORD, R.N.,
H.M.S. "Agincourt."

Second Edition, Revised and Enlarged.

WITH CHARTS, ILLUSTRATIONS, AND INDEX.

J. GRIFFIN & Co.,
(Publishers to H.R.H. The Duke of Edinburgh,)
15, COCKSPUR STREET, | AND 2, THE HARD,
PALL MALL, LONDON. | PORTSEA, PORTSMOUTH.

1875.
All rights reserved.

790138

Transportation Lib.

PRINTED AT THE OFFICE OF THE PUBLISHERS.

PREFACE TO SECOND EDITION.

ON issuing this, the Second Edition of the SAILOR'S POCKET BOOK, it seems unnecessary to say more than has previously been said as to the object and scope of the publication.

The fact of another Edition being called for so soon is sufficient evidence that such a Work was wanted, and the kindly criticisms of my brother Officers and others, have shown me that the endeavour to call attention to the necessity of a thorough acquaintance with the important simplicities of the Naval profession, has been appreciated.

The book has not only been carefully revised and corrected, but considerable additions have been made to several of its sections; Wind and Current charts have been introduced, the Passage Tables enlarged, other Tables added, and additional Hydrographic information given. The Foreign sea terms and phrases have been kindly corrected by Admiral LOBO, of the Spanish Navy, and Captain Count CANEVARO, Naval Attaché to the Italian Legation.

My absence from England and the nature of my duties, have prevented me from doing as much of the revision as I should have wished: but the care of the Second Edition has fallen into good hands, and I can never be sufficiently grateful to my old messmate, Staff-Commander Thomas A. Hull, R.N., for the unselfish manner in which he has devoted so much of his scant leisure to insure its correctness.

F.G.D.B.

Gibraltar, January 30th, 1875.

PREFACE TO FIRST EDITION.

THE SAILOR'S POCKET BOOK is an attempt to collect in a compact form, a series of those small practical facts, a thorough and ready knowledge of which tend to secure the success of the mariner and the safety of life and property confided to his charge. Some of the Notes are taken from the excellent works on Seamanship now extant, but the greater number are extracted from the small professional pamphlets, published by or with the approval of the Admiralty, the Board of Trade, and the Meteorological Office.

These useful little works are liable, from their very form, to be either frequently mislaid or seldom at hand for reference when required; and it was more especially as an endeavour to concentrate the information contained in these pamphlets, that this work was commenced.

In this undertaking I have been much encouraged by the sound practical advice of Captain W. H. CUMING, to whom, in 1869, the idea was first communicated; the interval of half-pay following promotion afforded time to begin in earnest; helping hands and brains soon introduced new features, and the work grew so rapidly, that it threatened to assume the proportions of a nautical cyclopædia rather than a Pocket book—it became a difficulty to know where to stop, what to omit or condense, and what to enlarge upon.

Fortunately, from the outset the work commended itself to my old friend Staff-Commander Thomas A. HULL, the Superintendent of Admiralty Charts, who at once undertook to assist me in the compilation, wrote portions of Sections IV. and V., and pointed out the vast amount of valuable information that was to be gleaned from the Charts and books published by the Hydrographic Office of the Admiralty.

Many thanks are also due to Admiral A. P. RYDER for many important suggestions and contributions, particularly in Sections VII. and VIII.; and to Admiral Sir C. F. A. SHADWELL for permission to make extracts from his useful work on Chronometers. Sections II. and IV. owe much to Staff-Commander E. H. HILLS, whose help has been most willingly rendered; Section V. to Admiral G. A. BEDFORD, and Staff-Commander V. F. JOHNSON, Instructor of Surveying at the Royal Naval College; Sections VI. to Captain J. C. WILSON, and Mr. George TUCK, Chief Engineer, R.N., Instructor in Steam at the Royal Naval College; and to the Secretary of the Royal National Life-boat Institution, who courteously furnished me with copies of the instructions issued by the Institution, and also allowed me the use of the blocks for illustrating Section VIII. Lieutenant H. H. GRENFELL kindly undertook Section VII., while Section VIII. was revised, and partly re-written by Staff-Surgeon D. J. DUIGAN, M.D. of this ship; Mr. W. WESTON, of the Chemical Department, Portsmouth, supplied the Article on Disinfectants.

Mr. R. C. CARRINGTON, of the Hydrographic Office, kindly superintended the production of the last Section during my absence from England, and also furnished the book with its comprehensive Index.

PREFACE. ix.

Thanks to the hearty co-operation of my brother officers the book has thus reached its present form. I trust it may be of service to the sailor, as a handy work of reference, especially to those who are not gifted with retentive memories ; and that after passing through the necessary criticism, pruned down where diffuse, and enlarged where too brief, it may at some future time find favour with those who from motives of duty, or pleasure, take an interest in the art of navigation.

F.G.D.B.

H.M.S. *Agincourt.*
Madeira, November, 1873.

ERRATA.

Readers are earnestly requested to make the following corrections :—

Page 20, note under Dip Table—*insert* "198" after "page."

Page 39, 10th line from top—for "di" *read* "dip."

Page 77, 12th line from top—*insert* "November" after "October."

Page 80, 4th line from bottom—for "advance somewhat faster than, and in the same direction as, an approaching storm, having the wind also in same direction (ship running)" *read* "continue her course."

Page 82, 10th line from top—for "veering" *read* "shifting."

Page 125, Panama to Honolulu—for "5,500" *read* "5,200."

Page 232, 14th line from top—for "Ormani" *read* "Omani."

CONTENTS.

SECTION I.
COMPASS SIGNALS. MASTHEAD ANGLES. HORIZON TABLE. FLEET SAILING. FLASHING SIGNALS. INTERNATIONAL SIGNALS. 1

PAGE

SECTION II.
THE COMPASS 37

SECTION III.
RULE OF THE ROAD AT SEA. MERCHANT SHIPPING ACTS . . 57

SECTION IV.
WIND. REVOLVING STORMS. WEATHER. BAROMETER. THERMOMETER. CURRENTS. ICE. PASSAGE TABLES 73

SECTION V.
LIGHTS. BUOYS. HYDROGRAPHICAL ABBREVIATIONS. ADMIRALTY CHARTS. HYDROGRAPHICAL INFORMATION. SOUNDING. TIDES. DETERMINING POSITIONS. MEASURING DISTANCES AND HEIGHTS. CHRONOMETERS. MERIDIAN DISTANCES. USEFUL TABLES 131

SECTION VI.
BOATS :—WEIGHTS AND DIMENSIONS. MANAGEMENT OF STEAMBOATS. GENERAL HINTS ON MANAGEMENT. BOARDING. BOAT CRUISING. BOAT RACING . . . . , ... 209

CONTENTS.

SECTION VII.

OPERATIONS ON SHORE. LANDING A BATTALION. ORDER OF MARCH. BIVOUAC. ATTACK OF A POSITION. DISEMBARKING TROOPS. 241

SECTION VIII.

ROCKET AND MORTAR APPARATUS FOR SAVING LIFE FROM SHIPWRECK. LIFE BUOYS. LIFE BELTS AND CORK MATTRESSES. HINTS TO BATHERS. INSTRUCTIONS FOR RESTORING THE APPARENTLY DROWNED. FIRST HELP IN ACCIDENTS, &C. DISINFECTANTS, &C. 281

SECTION IX.

MONEY, WEIGHTS AND MEASURES, OF ALL NATIONS. USEFUL RULES IN MENSURATION. RULES FOR CALCULATING THE FLOATING POWER OF SPARS; TANKS AND CASKS; TONNAGE OF VESSELS; NUMBER OF SQUARE YARDS IN SAILS, &C. . . . 313

SECTION X.

MISCELLANEOUS.—PARTICULARS OF DOCKS ABROAD. FOREIGN SEA TERMS AND PHRASES. ROPE MAKING. ANCHORS AND CABLES. PAINTING SHIP. PRACTICAL RECIPES FOR MIXING COLOURS, STAINING AND DYEING, &C. COOKERY, &C. DAYLIGHT GUN. USEFUL NOTES ON THE MARINE STEAM ENGINE; COALS, &C. TABLES OF PROVISIONING AND CLOTHING; LOGARITHMS AND NATURAL SINES, &C. INFORMATION AND INSTRUCTIONS FOR DIVERS. 343

GENERAL REMARKS ON WIND, WEATHER, AND CURRENTS OF THE MALAY ARCHIPELAGO. . . . . APP. No. 1

CHINA,—MONEY, WEIGHTS AND MEASURES . . . . . APP. No. 2

LIST OF ILLUSTRATIONS AND DIAGRAMS.

	PAGE
Compass signals	3, 5
Flashing signals, with flags . . .	27
Semaphore signs	29
Signal flags used by British men-of-war and merchant ships	31 and 32
Archibald Smith's (straight line) Course table.	46 and 47
Wind charts	74 and 76
Barometer scales.	92
Current chart	108
Sounding on shoals out of sight of land	156
To explain the terms "Spring rise," "Neap rise," &c. . . .	176
To determine positions. .	177, 178, 179
To determine the distance from a lighthouse, or other object . .	180

	PAGE
The danger angle -	180
To find the distance of a target at sea . . . . , ...	181
To measure heights	185
To measure distances on shore. .	186
Order of marching	249
Bivouac for the night	251
To save life from shipwreck—Fig. 1, 3 4.—Life belts .	286, 287
„ 2.—Cork mattresses . .	287
„ 5, 6, 7.—Hammocks	288, 289
To restore the apparently drowned—Inspiration . , ...	295, 297
Expiration	296, 298
To illustrate the new measurement of the tonnage of ships. . .	337
To find the area of sails	339

Section 1.

**SIGNALS. MASTHEAD ANGLES.
HORIZON METHOD OF ASCERTAINING DISTANCES.
FLASHING SIGNALS; ETC.**

When Signals are hoisted in different positions, that at the Main Truck is to be read first, and considered as the first hoist of the Signal, and the others in the following order:—

Fore Truck, Mizen Truck, Peak, Starboard Main Topsail Yard, Port Main Topsail Yard, Starboard Mizen Topsail Yard, Port Mizen Topsail Yard.

Compass  Signals.

COMPASS PENDANT OVER.

No.	Letters.	Points.	No.	Letters.	Points.
42	A B	North.	18	B P	East.
421	A C	N. ¼ E.	181	B Q	E. ¼ S.
422	A D	N. ½ E.	182	B R	E. ½ S.
423	A E	N. ¾ E	183	B S	E. ¾ S
11	A F	N. by E.	19	B T	E. by S.
111	A G	N. by E. ¼ E.	191	B V	E. by S. ¼ S.
112	A H	N. by E. ½ E.	192	B Y	E. by S. ½ S.
113	A I	N. by E. ¾ E.	193	C D	E. by S. ¾ S.
12	A K	N. N. E	20	C E	E. S. E.
121	A L	N. N. E. ¼ E.	201	C F	S. E. by E. ¾ E.
122	A M	N. N. E. ½ E.	202	C G	S. E. by E. ½ E.
123	A N	N. N. E. ¾ E.	203	C H	S. E. by E. ¼ E.
13	A O	N. E. by N.	21	C I	S. E. by E.
131	A P	N. E. ¾ N.	211	C K	S. E. ¾ E.
132	A Q	N. E. ½ N.	212	C L	S. E. ½ E.
133	A R	N. E. ¼ N.	213	C M	S. E. ¼ E.
14	A S	N. E.	22	C N	S. E.
141	A T	N. E. ¼ E.	221	C O	S. E. ¼ S.
142	A V	N. E. ½ E.	222	C P	S. E. ½ S.
143	A Y	N. E. ¾ E.	223	C Q	S. E. ¾ S.
15	B C	N. E. by E.	23	C R	S. E. by S.
151	B D	N. E. by E. ¼ E.	231	C S	S. S. E. ¾ E.
152	B E	N. E. by E. ½ E.	232	C T	S. S. E. ½ E.
153	B F	N. E. by E. ¾ E.	233	C V	S. S. E. ¼ E.
16	B G	E. N. E.	24	C Y	S. S. E.
161	B H	E. by N. ¾ N.	241	D E	S. by E. ¾ E.
162	B I	E. by N. ½ N.	242	D F	S. by E. ½ E.
163	B K	E. by N. ¼ N.	243	D G	S. by E. ¼ E.
17	B L	E. by N.	25	D H	S. by E.
171	B M	E. ¾ N.	251	D I	S. ¾ E.
172	B N	E. ½ N.	252	D K	S. ½ E.
173	B O	E. ¼ N.	253	D L	S. ¼ E.

The Letters are used with Flags, and numbers with distant or Flashing Signals.

COMPASS SIGNALS.—*(continued.)*

COMPASS PENDANT OVER.

No.	Letters.	Points.	No.	Letters.	Points.
26	D M	South.	34	F N	West.
261	D N	S. ¼ W.	341	F O	W. ¼ N.
262	D O	S. ½ W.	342	F P	W. ½ N.
263	D P	S. ¾ W.	343	F Q	W. ¾ N.
27	D Q	S. by W.	35	F R	W. by N.
271	D R	S. by W. ¼ W.	351	F S	W. by N. ¼ N.
272	D S	S. by W. ½ W.	352	F T	W. by N. ½ N.
273	D T	S. by W. ¾ W.	353	F V	W. by N. ¾ N.
28	D V	S. S. W.	36	F Y	W. N. W.
281	D Y	S. S. W. ¼ W.	361	G H	N. W. by W. ¾ W.
282	E F	S. S. W. ½ W.	362	G I	N. W. by W. ½ W.
283	E G	S. S. W. ¾ W.	363	G K	N. W. by W. ¼ W.
29	E H	S. W. by S.	37	G L	N. W. by W.
291	E I	S. W. ¾ S.	371	G M	N. W. ¾ W.
292	E K	S. W. ½ S.	372	G N	N. W. ½ W.
293	E L	S. W. ¼ S.	373	G O	N. W. ¼ W.
30	E M	S. W.	38	G P	N. W.
301	E N	S. W. ¼ W.	381	G Q	N. W. ¼ N.
302	E O	S. W. ½ W.	382	G R	N. W. ½ N.
303	E P	S. W. ¾ W.	383	G S	N. W. ¾ W.
31	E Q	S. W. by W.	39	G T	N. W. by N.
311	E R	S. W. by W. ¼ W.	391	G V	N. N. W. ¾ W.
312	E S	S. W. by W. ½ W.	392	G Y	N. N. W. ½ W.
313	E T	S. W. by W. ¾ W.	393	H I	N. N. W. ¼ W.
32	E V	W. S. W.	40	H K	N. N. W.
321	E Y	W. by S. ¾ S.	401	H L	N. by W. ¾ W.
322	F G	W. by S. ½ S.	402	H M	N. by W. ½ W.
323	F H	W. by S. ¼ S.	403	H N	N. by W. ¼ W.
33	F I	W. by S.	41	H O	N. by W.
331	F K	W. ¾ S.	411	H P	N. ¾ W.
332	F L	W. ½ S.	412	H Q	N. ½ W.
333	F M	W. ¼ S.	413	H R	N. ¼ W.

The Letters are used with Flags, and numbers with distant or Flashing Signals.

PENDANT BOARD.

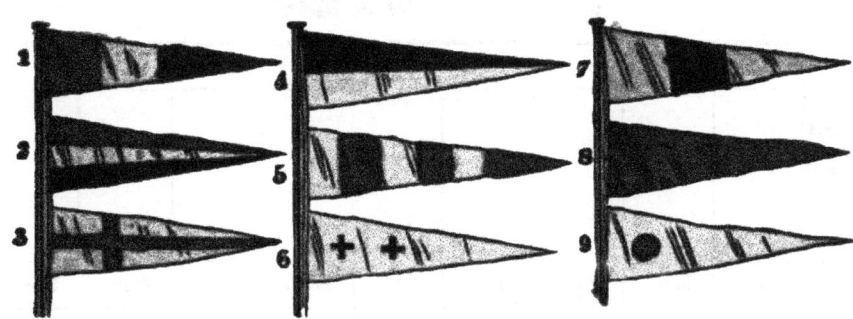

Pendants.	NAMES.	Pendants.	NAMES.	Pendants.	NAMES.
12		41		71	
13		42		72	
14		43		73	
15		45		74	
16		46		75	
17		47		76	
18		48		78	
19		49		79	
21		51		81	
23		52		82	
24		53		83	
25		54		84	
26		56		85	
27		57		86	
28		58		87	
29		59		89	
31		61		91	
32		62		92	
34		63		93	
35		64		94	
36		65		95	
37		67		96	
38		68		97	
39		69		98	

SPEED TABLE.

Table to be filled up showing the No. of Revolutions required to be made by each Ship of Squadron to obtain a given speed.

Speed required	NAME OF SHIP.								
Knots	Rev.	Rev.	Rev.	Rev.	Rev.	Rev.	Rev.	Rev.	Rev.
3									
4									
5									
6									
7									
8									
9									
10									
11									
12									
13									

SEC. I. HEIGHT OF MASTHEAD. 7

Table to be filled up with Heights of Mastheads of Ships comprising Squadron, or on Station.

HEIGHT OF MASTHEAD.

NAME.	Main Truck to Hammock Netting.	Main Topmast Crosstrees to Hammock Netting.	NAME.	Main Truck to Hammock Netting.	Main Topmast Crosstrees to Hammock Netting.

HEIGHT OF MASTHEAD.

The following are the Heights from Main Truck to Hammock Netting, of Ships representing the different classes of Vessels in the Navy.

Name.	Class.	Height.
		Feet.
Minotaur ..	Iron-clad.	149
Achilles		166
Hercules ..		173
Invincible ..		162
Lord Warden ..		155
Royal Oak ..	,,	152
Inconstant ..	Frigate (Iron, Wood-cased).	168
Immortalité ..	Frigate (Wood).	163
Volage	Corvette (Iron, Wood-cased).	130
Sirius	Corvette (Wood).	107
Plover ..	Sloop (Wood).	94
Cracker	Gun Vessel (Wood).	82

TABLE OF MASTHEAD ANGLES.

Distance in		HEIGHT OF MASTHEAD.					
Cables	Yards.	80 Feet.	85 Feet.	90 Feet.	95 Feet.	100 Feet.	105 Feet.
		° ′ ″	° ′ ″	° ′ ″	° ′ ″	° ′ ″	° ′ ″
½	100	14 55 53	15 49 10	16 41 57	17 34 17	18 26 6	19 17 23
¾	150	10 4 51	10 41 48	11 18 36	11 55 15	12 31 44	13 8 2
1·0	200	7 35 40	8 3 48	8 31 51	8 59 50	9 27 44	9 55 34
1¼	250	6 5 19	6 27 57	6 50 34	7 13 8	7 35 40	7 58 16
½	300	5 4 46	5 23 43	5 42 38	6 1 32	6 20 25	6 39 16
¾	350	4 21 25	4 37 42	4 53 57	5 10 12	5 26 25	5 42 39
2·0	400	3 48 51	4 3 6	4 17 21	4 31 35	4 45 49	5 0 2
¼	450	3 23 28	3 36 10	3 48 51	4 1 31	4 14 11	4 26 51
½	500	3 3 11	3 14 36	3 26 1	3 37 26	3 48 50	4 0 15
¾	550	2 46 33	2 56 57	3 7 20	3 17 43	3 28 6	3 38 28
3·0	600	2 32 41	2 42 13	2 51 45	3 1 16	3 10 47	3 20 19
¼	650	2 20 57	2 29 45	2 38 33	2 47 21	2 56 8	3 4 56
½	700	2 10 54	2 19 5	2 27 15	2 35 25	2 43 35	2 51 44
¾	750	2 2 11	2 9 49	2 17 26	2 24 4	2 32 41	2 40 19
4·0	800	1 54 33	2 1 43	2 8 51	2 25 0	2 23 9	2 30 18
¼	850	1 47 49	1 54 33	2 1 17	2 8 0	2 14 45	2 21 28
½	900	1 41 50	1 48 12	1 54 33	2 0 54	2 7 16	2 13 37
¾	950	1 36 28	1 42 30	1 48 32	1 54 33	2 0 34	2 6 36
5·0	1000	1 31 39	1 37 23	1 43 6	1 48 50	1 54 33	2 0 16
¼	1050	1 27 17	1 32 45	1 38 12	1 43 39	1 49 6	1 54 33
½	1100	1 23 19	1 28 32	1 33 44	1 38 57	1 44 8	1 49 21
¾	1150	1 19 42	1 24 41	1 29 40	1 34 39	1 39 37	1 44 35

NOTE.—When observed Masthead Angles are not more than 5° they should be taken both on and off the arc, and the mean will give angle free of index error.

Table of Masthead Angles.—(CONTINUED.)

Distance in		HEIGHT OF MASTHEAD.					
Cables.	Yards.	80 Feet.	85 Feet.	90 Feet.	95 Feet.	100 Feet.	105 Feet.
		° ′ ″	° ′ ″	° ′ ″	° ′ ″	° ′ ″	° ′ ″
6·0	1200	1 16 23	1 21 10	1 25 56	1 30 42	1 35 28	1 40 14
¼	1250	1 13 20	1 17 54	1 22 29	1 27 3	1 31 39	1 36 14
½	1300	1 10 31	1 14 55	1 19 19	1 23 43	1 28 8	1 32 32
¾	1350	1 7 54	1 12 8	1 16 23	1 20 37	1 24 52	1 29 6
7·0	1400	1 5 28	1 9 34	1 13 39	1 17 45	1 21 50	1 25 56
¼	1450	1 3 13	1 7 10	1 11 7	1 15 4	1 19 1	1 22 58
½	1500	1 1 7	1 4 56	1 8 45	1 12 34	1 16 23	1 20 12
¾	1550	0 59 8	1 2 50	1 6 32	1 10 13	1 13 55	1 17 37
8·0	1600	0 57 18	1 0 52	1 4 27	1 8 2	1 11 37	1 15 11
¼	1650	0 55 33	0 59 2	1 2 30	1 5 58	1 9 26	1 12 55
½	1700	0 53 55	0 57 17	1 0 39	1 4 2	1 7 24	1 10 46
¾	1750	0 52 23	0 55 39	0 58 55	1 2 12	1 5 28	1 8 45
9·0	1800	0 50 56	0 54 7	0 57 18	1 0 28	1 3 39	1 6 50
¼	1850	0 49 33	0 52 39	0 55 44	0 58 50	1 1 56	1 5 1
½	1900	0 48 15	0 51 16	0 54 16	0 57 17	1 0 18	1 3 19
¾	1950	0 47 1	0 49 57	0 52 53	0 55 49	0 58 46	1 1 42
1 Mile	2000	0 45 50	0 48 42	0 51 34	0 54 26	0 57 17	1 0 9
¼	2500	0 36 40	0 38 57	0 41 15	0 43 32	0 45 50	0 48 7
½	3000	0 30 34	0 32 28	0 34 23	0 36 17	0 38 12	0 40 6
¾	3500	0 26 12	0 27 50	0 29 28	0 31 6	0 32 44	0 34 23
2·0	4000	0 22 55	0 24 21	0 25 47	0 27 13	0 28 39	0 30 5
¼	4500	0 20 22	0 21 39	0 22 55	0 24 11	0 25 28	0 26 44
½	5000	0 18 20	0 19 29	0 20 38	0 21 46	0 22 55	0 24 4
¾	5500	0 16 40	0 17 43	0 18 45	0 19 48	0 20 50	0 21 52
3·0	6000	0 15 16	0 16 14	0 17 11	0 18 8	0 19 6	0 20 3

NOTE.—When observed Masthead Angles are not more than 5° they should be taken both on and off the arc, and the mean will give angle free of index error.

Table of Masthead Angles.—(Continued.)

Distance in		HEIGHT OF MASTHEAD.					
Cables.	Yards.	110 Feet.	115 Feet.	120 Feet.	125 Feet.	130 Feet.	135 Feet.
		° ′ ″	° ′ ″	° ′ ″	° ′ ″	° ′ ″	° ′ ″
½	100	20 8 11	20 58 24	21 48 5	22 37 12	23 25 43	24 13 40
¾	150	13 44 11	14 20 7	14 55 54	15 31 28	16 6 49	16 41 58
1·0	200	10 23 20	10 51 0	11 18 36	11 46 6	12 13 30	12 40 49
¼	250	8 20 38	8 43 4	9 5 25	9 27 45	9 50 1	10 12 14
½	300	6 58 6	7 16 54	7 35 41	7 54 26	8 13 9	8 31 51
¾	350	5 58 50	6 15 2	6 31 12	6 47 21	7 3 28	7 19 35
2·0	400	5 14 15	5 28 27	5 42 39	56 49	6 10 58	6 25 8
¼	450	4 39 30	4 52 8	5 4 47	5 17 25	5 30 1	5 42 39
½	500	4 11 39	4 23 4	4 34 26	4 45 49	4 57 12	5 8 34
¾	550	3 38 51	3 59 13	4 9 35	4 19 57	4 30 18	4 40 39
3·0	600	3 29 49	3 39 20	3 48 51	3 58 21	4 7 51	4 17 21
¼	650	3 13 43	3 22 30	3 31 17	3 40 4	3 48 51	3 57 37
½	700	2 59 55	3 8 4	3 16 14	24 23	3 32 32	3 40 42
¾	750	2 47 56	2 55 33	3 3 10	3 10 47	3 18 24	3 26 1
4·0	800	2 37 27	2 44 36	2 51 45	2 58 54	3 6 2	3 13 10
¼	850	2 28 12	2 34 56	2 41 40	2 48 23	2 55 7	3 1 50
½	900	2 19 59	2 26 20	2 32 41	2 39 2	2 45 23	2 51 45
¾	950	2 12 37	2 18 38	2 24 40	2 30 41	2 36 42	2 42 43
5·0	1000	2 6 0	2 11 43	2 17 26	2 23 9	2 28 53	2 34 36
¼	1050	2 0 0	2 5 27	2 10 54	2 16 21	2 21 48	2 27 15
½	1100	1 54 33	1 59 45	2 4 57	2 10 10	2 15 21	2 20 33
¾	1150	1 49 33	1 54 33	1 59 32	2 4 30	2 9 28	2 14 27

NOTE.—When observed Masthead Angles are not more than 5° they should be taken both on and off the arc, and the mean will give angle free of index error.

Table of Masthead Angles.—(CONTINUED.)

Distance in		HEIGHT OF MASTHEAD.					
Cables.	Yards.	110 Feet.	115 Feet.	120 Feet.	125 Feet.	130 Feet.	135 Feet.
		° ′ ″	° ′ ″	° ′ ″	° ′ ″	° ′ ″	° ′ ″
6·0	1200	1 45 1	1 49 47	1 54 33	1 59 19	2 4 5	2 8 51
¼	1250	1 40 49	1 45 23	1 49 58	1 54 33	1 59 8	2 3 43
½	1300	1 36 56	1 41 20	1 45 45	1 50 9	1 54 33	1 58 57
¾	1350	1 33 21	1 37 35	1 41 50	1 46 4	1 50 18	1 54 33
7·0	1400	1 30 1	1 34 6	1 38 12	1 42 17	1 46 22	1 50 28
¼	1450	1 26 55	1 30 52	1 34 49	1 38 46	1 42 43	1 46 40
½	1500	1 24 1	1 27 50	1 31 39	1 35 28	1 39 17	1 43 6
¾	1550	1 21 19	1 25 0	1 28 42	1 32 23	1 36 5	1 39 46
8·0	1600	1 18 46	1 22 21	1 25 56	1 29 30	1 33 5	1 36 40
¼	1650	1 16 23	1 19 51	1 23 19	1 26 48	1 30 16	1 33 44
½	1700	1 14 8	1 17 30	1 20 53	1 24 15	1 27 37	1 30 59
¾	1750	1 12 1	1 15 17	1 18 34	1 21 50	1 25 6	1 28 23
9·0	1800	1 10 1	1 13 12	1 16 23	1 19 34	1 22 45	1 25 56
¼	1850	1 8 7	1 11 13	1 14 19	1 17 25	1 20 30	1 23 36
½	1900	1 6 20	1 9 21	1 12 22	1 15 23	1 18 23	1 21 24
¾	1950	1 4 38	1 7 34	1 10 31	1 13 27	1 16 23	1 19 19
1 Mile	2000	1 3 1	1 5 53	1 8 45	1 11 36	1 14 28	1 17 20
¼	2500	0 50 25	0 52 42	0 55 0	0 57 17	0 59 35	1 1 52
½	3000	0 42 1	0 43 56	0 45 50	0 47 45	0 49 39	0 51 34
¾	3500	0 36 1	0 37 39	0 39 17	0 40 56	0 42 34	0 44 12
2·0	4000	0 31 31	0 32 57	0 34 23	0 35 48	0 37 14	0 38 40
¼	4500	0 28 1	0 29 17	0 30 33	0 31 50	0 33 6	0 34 23
½	5000	0 25 12	0 26 21	0 27 30	0 28 39	0 29 48	0 30 56
¾	5500	0 22 55	0 23 58	0 25 0	0 26 3	0 27 5	0 28 8
3·0	6000	0 21 0	0 21 58	0 22 55	0 23 52	0 24 50	0 25 47

NOTE.—When observed Masthead Angles are not more than 5° they should be taken both on and off the arc, and the mean will give angle free of index error.

Table of Masthead Angles.—(CONTINUED.)

Distance in		HEIGHT OF MASTHEAD.					
Cables.	Yards.	140 Feet.	145 Feet.	150 Feet.	155 Feet.	160 Feet.	165 Feet.
		° ′ ″	° ′ ″	° ′ ″	° ′ ″	° ′ ″	° ′ ″
½	100	25 1 0	25 47 46	26 33 54	27 19 26	28 4 21	28 48 39
¾	150	17 16 55	17 51 37	18 26 6	19 0 22	19 34 24	20 8 10
1·0	200	13 8 2	13 35 10	14 2 11	14 29 5	14 55 53	15 22 35
¼	250	10 34 25	10 56 32	11 18 36	11 40 36	12 2 33	12 24 27
½	300	8 50 30	9 9 9	9 27 44	9 46 18	10 4 50	10 23 20
¾	350	7 35 40	7 51 46	8 7 49	8 23 51	8 39 51	8 55 50
2·0	400	6 39 16	6 53 24	7 7 30	7 21 36	7 35 40	7 49 44
¼	450	5 55 14	6 7 50	6 20 25	6 33 0	6 45 33	6 58 6
½	500	5 19 56	5 31 17	5 42 39	5 53 59	6 5 19	6 16 38
¾	550	4 50 59	5 1 20	5 11 40	5 22 0	5 32 19	5 42 39
3·0	600	4 26 51	4 36 21	4 45 49	4 55 18	5 4 47	5 14 15
¼	650	4 6 23	4 15 10	4 23 55	4 32 41	4 41 26	4 50 12
½	700	3 48 51	3 57 0	4 5 8	4 13 17	4 21 25	4 29 33
¾	750	3 33 38	3 41 14	3 48 51	3 56 27	4 4 3	4 11 39
4·0	800	3 20 19	3 27 27	3 34 35	3 41 42	3 48 51	3 55 58
¼	850	3 8 33	3 15 16	3 21 59	3 28 42	3 35 25	3 42 8
½	900	2 58 6	3 4 27	3 10 47	3 17 8	3 23 29	3 29 50
¾	950	2 48 44	2 54 45	3 0 46	3 6 47	3 12 48	3 18 48
5·0	1000	2 40 19	2 46 2	2 51 45	2 57 28	3 3 10	3 8 53
¼	1050	2 32 41	2 38 8	2 43 35	2 49 1	2 54 28	2 59 55
½	1100	2 25 45	2 30 57	2 36 9	2 41 21	2 46 33	2 51 45
¾	1150	2 19 26	2 24 24	2 29 22	2 34 21	2 39 19	2 44 17

NOTE.—When observed Masthead Angles are not more than 5° they should be taken both on and off the arc, and the mean will give angle free of index error.

Table of Masthead Angles.—(CONTINUED.)

Distance in		HEIGHT OF MASTHEAD.					
Cables.	Yards.	140 Feet.	145 Feet.	150 Feet.	155 Feet.	160 Feet.	165 Feet.
		° ′ ″	° ′ ″	° ′ ″	° ′ ″	° ′ ″	° ′ ″
6.0	1200	2 13 37	2 18 24	2 23 9	2 27 55	2 32 41	2 37 27
¼	1250	2 8 17	2 12 52	2 17 26	2 22 1	2 26 35	2 31 10
½	1300	2 3 21	2 7 46	2 12 9	2 16 33	2 20 57	2 25 21
¾	1350	1 58 47	2 3 2	2 7 16	2 11 30	2 15 45	2 19 59
7.0	1400	1 54 33	1 58 39	2 2 43	2 6 49	2 10 54	2 14 59
¼	1450	1 50 36	1 54 33	1 58 30	2 2 27	2 6 23	2 10 20
½	1500	1 46 55	1 50 44	1 54 33	1 58 22	2 2 11	2 6 0
¾	1550	1 43 28	1 47 10	1 50 51	1 54 33	1 58 14	2 1 56
8.0	1600	1 40 12	1 43 49	1 47 24	1 50 58	1 54 33	1 58 8
¼	1650	1 37 12	1 40 40	1 44 8	1 47 37	1 51 5	1 54 33
½	1700	1 34 21	1 37 43	1 41 5	1 44 27	1 47 49	1 51 11
¾	1750	1 31 39	1 34 56	1 38 12	1·41 28	1 44 44	1 48 0
9.0	1800	1 29 6	1 32 17	1 35 28	1 38 39	1 41 50	1 45 1
¼	1850	1 26 42	1 29 48	1 32 53	1 35 59	1 39 5	1 42 10
½	1900	1 24 25	1 27 26	1 30 27	1 33 28	1 36 28	1 39 29
¾	1950	1 22 15	1 25 12	1 28 8	1 31 4	1 34 0	1 36 56
1 Mile	2000	1 20 12	1 23 4	1 25 56	1 28 47	1 31 39	1 34 31
¼	2500	1 4 10	1 6 27	1 8 45	1 11 2	1 13 20	1 15 37
½	3000	0 53 28	0 55 23	0 57 17	0 59 12	1 1 6	1 3 1
¾	3500	0 45 50	0 47 28	0 49 7	0 50 45	0 52 23	0 54 1
2.0	4000	0 40 6	0 41 32	0 42 58	0 44 24	0 45 50	0 47 16
¼	4500	0 35 39	0 36 55	0 38 12	0 39 28	0 40 45	0 42 1
½	5000	0 32 5	0 33 14	0 34 23	0 35 31	0 36 40	0 37 49
¾	5500	0 29 10	0 30 13	0 31 15	0 32 18	0 33 20	0 34 23
3.0	6000	0 26 55	0 27 42	0 28 39	0 29 36	0 30 33	0 31 31

NOTE.—When observed Masthead Angles are not more than 5° they should be taken both on and off the arc, and the mean will give angle free of index error.

Table of Masthead Angles.—(CONTINUED.)

Distance in		HEIGHT OF MASTHEAD.					
Cables.	Yards.	170 Feet.	175 Feet.	180 Feet.	185 Feet.	190 Feet.	195 Feet.
		° ′ ″	° ′ ″	° ′ ″	° ′ ″	° ′ ″	° ′ ″
½	100	29 32 19	30 15 21	30 57 50	31 39 30	32 20 51	33 1 25
¾	150	20 41 44	21 15 1	21 48 6	22 20 54	22 53 26	23 25 43
1·0	200	15 49 9	16 15 36	16 41 57	17 8 11	17 34 17	18 0 15
¼	250	12 46 16	13 8 2	13 29 45	13 51 23	14 12 57	14 34 27
½	300	10 41 47	11 0 12	11 18 36	11 36 56	11 55 15	12 13 30
¾	350	9 11 48	9 27 44	9 43 40	9 59 34	10 15 25	10 31 15
2·0	400	8 3 47	8 17 49	8 31 51	8 45 51	8 59 50	9 13 47
¼	450	7 10 38	7 23 9	7 35 41	7 48 11	8 0 40	8 13 9
½	500	6 27 57	6 39 16	6 50 34	7 1 51	7 13 9	7 24 25
¾	550	5 53 57	6 3 15	6 13 33	6 23 51	6 34 8	6 44 24
3·0	600	5 23 43	5 33 10	5 42 38	5 52 5	6 1 32	6 10 58
¼	650	4 58 57	5 7 42	5 16 26	5 25 11	5 33 54	5 42 39
½	700	4 37 41	4 45 49	4 53 57	5 2 4	5 10 11	5 18 18
¾	750	4 19 15	4 26 50	4 34 26	4 42 2	4 49 37	4 57 12
4·0	800	4 3 6	4 10 13	4 17 21	4 24 28	4 31 35	4 38 42
¼	850	3 48 51	3 55 33	4 2 16	4 8 58	4 15 40	4 22 22
½	900	3 36 10	3 42 30	3 48 51	3 55 11	4 1 31	4 7 51
¾	950	3 24 49	3 30 49	3 36 50	3 42 51	3 48 51	3 54 51
5·0	1000	3 14 36	3 20 19	3 26 1	3 31 44	3 37 26	3 43 8
¼	1050	3 5 21	3 10 47	3 16 14	3 21 40	3 27 6	3 32 32
½	1100	2 56 56	3 2 8	3 7 20	3 12 31	3 17 43	3 22 54
¾	1150	2 49 17	2 54 14	2 59 12	3 4 10	3 9 8	3 14 6

NOTE.—When observed Masthead Angles are not more than 5° they should be taken both on and off the arc, and the mean will give angle free of index error.

Table of Masthead Angles.—(CONTINUED.)

Distance in		HEIGHT OF MASTHEAD.					
Cables.	Yards.	170 Feet.	175 Feet.	180 Feet.	185 Feet.	190 Feet.	195 Feet.
		° ′ ″	° ′ ″	° ′ ″	° ′ ″	° ′ ″	° ′ ″
6·0	1200	2 42 13	2 46 59	2 51 45	2 56 30	3 1 16	3 6
¼	1250	2 35 44	2 40 19	2 44 53	2 49 28	2 54 2	2 58 36
½	1300	2 29 45	2 34 9	2 38 33	2 42 57	2 47 21	2 51 45
¾	1350	2 24 13	2 28 27	2 32 41	2 36 56	2 41 9	2 45 24
7·0	1400	2 19 4	2 23 9	2 27 15	2 31 20	2 35 25	2 39 28
¼	1450	2 14 17	2 18 14	2 22 10	2 26 7	2 30 4	2 34 0
½	1500	2 9 48	2 13 37	2 17 26	2 21 15	2 25 4	2 28 53
¾	1550	2 5 37	2 9 19	2 13 1	2 16 42	2 20 23	2 24 5
8·0	1600	2 1 42	2 5 17	2 8 52	2 12 26	2 16 1	2 19 35
¼	1650	1 58 1	2 1 29	2 4 57	2 8 25	2 11 53	2 15 22
½	1700	1 54 33	1 57 55	2 1 17	2 4 39	2 8 1	2 11 23
¾	1750	1 51 16	1 54 33	1 57 49	2 1 5	2 4 21	2 7 38
9·0	1800	1 48 12	1 51 22	1 54 33	1 57 44	2 0 54	2 4 5
¼	1850	1 45 16	1 48 22	1 51 27	1 54 33	1 57 39	2 0 44
½	1900	1 42 30	1 45 31	1 48 31	1 51 32	1 54 33	1 57 34
¾	1950	1 39 52	1 42 38	1 45 45	1 48 41	1 51 37	1 54 33
1 Mile	2000	1 37 22	1 40 14	1 43 6	1 45 58	1 48 50	1 51 41
¼	2500	1 17 55	1 20 12	1 22 30	1 24 47	1 27 4	1 29 22
½	3000	1 4 56	1 6 50	1 8 45	1 10 39	1 12 34	1 14 28
¾	3500	0 55 39	0 57 17	0 58 56	1 0 34	1 2 12	1 3 50
2·0	4000	0 48 42	0 50 8	0 51 34	0 53 0	0 54 26	0 55 52
¼	4500	0 43 17	0 44 34	0 45 50	0 47 7	0 48 23	0 49 39
½	5000	0 38 58	0 40 6	0 41 15	0 42 24	0 43 33	0 44 41
¾	5500	0 35 25	0 36 28	0 37 30	0 38 33	0 39 35	0 40 38
3·0	6000	0 32 28	0 33 25	0 34 23	0 35 20	0 36 17	0 37 15

NOTE.—When observed Masthead Angles are not more than 5° they should be taken both on and off the arc, and the mean will give angle free of index error.

Horizon Method of ascertaining from a Fort, or Ship, the Distance of an Enemy's Ship.

WHEN the height of the enemy's masthead above her water line is known, the Table of Masthead Angles will give the distance; but in the case of Monitors without masts, gun-boats, mortar-boats, &c., this method is not available.

If there are two forts sufficiently far apart to allow of the distance between them, forming a base of sufficient length, and telegraphic communication is established between the forts, then Professor Sieman's Table, on which a chart of the approach, sub-divided into small squares is placed, will show with great accuracy, the exact position of the enemy's ship; it will be at the intersection of two rulers, one of which is controlled by a telescope pointing at the enemy's ship at one fort, and the other ruler at the same fort is by electricity constrained to follow exactly, and be parallel to the direction of the telescope at the other fort, also kept pointing at the enemy. This Table may be usefully employed for torpedo purposes as well as for gunnery.

In the absence of such a table, the bearing of the enemy and the changes in it may be telegraphed from one fort to the other and pencilled on a chart.

If there is no second fort or suitable position for a second observer, and the enemy is mastless, the distance from the enemy may be found from the dip angle, observed with a theodolite, or a method styled by its inventor (Admiral A. P. RYDER), the "horizon" method,[*] may be found useful (if there is no theodolite), by the aid of a sextant or pocket sextant, provided the horizon is in sight above the ship. This method was more especially designed for use on board ship, but it is equally available on shore.

Every fort should have its own "horizon" table calculated and hung up, the difference being due to the varying height.

---

[*] This method was first suggested and the table prepared in June, 1841, by Lieut. A. P. Ryder. It has been adopted in foreign Navies and found useful. One copy is specially supplied to each of our Men-of-War.

Table IV. in Admiral Ryder's work has been calculated especially for this method, in which the angle subtended between the horizon and the enemy's water-line is ascertained by an observer aloft, or if on shore, from a fort (the higher the better); and is corrected by the addition of the dip angle due to the height.

Table IV. cannot be abbreviated, but up to a certain point it is nearly identical with Table III. in the same work, and therefore, with the following table, which has been abbreviated from Table III. by Admiral Ryder, for the Pocket Book. Although Table III. ends by diverging widely from Table IV., nevertheless, for Naval Gunnery purposes, viz., range *at sea*, the error is not very material—its limit is as follows :—

When the observer is at a height above the water of		The error introduced by using Table III. in Admiral Ryder's work, or Table I. in the Pocket Book instead of Table IV. in Admiral Ryder's work, will not exceed 100 yards, until the distance exceeds	
	50 feet		2000 yards
	100 ,,		3000 ,,
	150 ,,		4000 ,,
	200 ,,		4700 ,,

For greater distances or lower heights, or if an error of 100 yards is too great, as may be the case for "fuze cutting," recourse must be had to Admiral Ryder's work.

This abbreviated table is sufficiently compact to allow of its being copied into the watch-bill.

Example. (1). Let the angle subtended between the horizon and the enemy's water-line, ascertained by an observer at the *Sultan's* crosstrees, or the fort in Drake's Island, (say a height of 205 feet) be 2°.0. Required the distance :—

The dip of the horizon due to a height of 205 feet, see Table. II, is 15′, 11″ 2°.0 + 15′. 11 = 2° 15′, neglecting the seconds.

The distance due to an angle of 2° 15′, if the height had been 20 feet, would have been, from Table I. column 2, 169 yards. The correction for each foot of additional height, column 3, Table I., is 8·5 yards.

Distance = 169 + (205 − 20) 8·5 = 1741 yards.

This is a sufficiently close approximation to the distance given in Table IV. in Admiral Ryder's work, viz., 1743 yards showing that Table I. in the Pocket Book can be safely used with the "horizon" method, within the prescribed limits.

HORIZON TABLE.

Example. (2). Let the angle subtended between the horizon and the enemy's water-line, ascertained by an observer in *Devastation's* top, or the upper battery in the Breakwater fort, (say 95 feet above the water), be 9′.

The dip of the horizon due to a height of 95 feet, see Table II., is 10′. 19″. 9′ + 10′. 19″ = 19′, neglecting seconds.

The distance due to an angle of 19′, if the height had been 20 feet, would have been, from Table I. column 2, 1204 yards.

Table I. of Masthead Angles.—Height 20 Feet.

Masthead Angle.	Distance in Yards.	Correction for 1 foot in Height.	Masthead Angle.	Distance in Yards.	Correction for 1 foot in Height.	Masthead Angle.	Distance in Yards.	Correction for 1 foot in Height.	Masthead Angle.	Distance in Yards.	Correction for 1 foot in Height.
° ′	yds.	yds.	° ′	yds.	yds.	° ′	yds.	yds.	° ′	yds.	yds.
	76	3·8	1 52	204	10·2	1 4	358	17·9	0 38	602	30·2
	85	4·2	1 50	208	10·4	1 2	369	18·5	0 37	618	31·0
	95	4·8	1 48	212	10·6	1 0	382	19·1	0 36	636	31·8
	99	5·0	1 46	216	10·8	0 59	388	19·4	0 35	654	32·2
3 40	104	5·2	1 44	220	11·0	0 58	395	19·7	0 34	673	33·7
3 30	109	5·4	1 42	224	11·2	0 57	402	20·1	0 33	694	34·7
3 20	114	5·7	1 40	229	11·4	0 56	409	20·5	0 32	715	55·8
3 10	120	6·0	1 38	234	11·7	0 55	416	20·8	0 31	739	37·0
3 0	127	6·4	1 36	238	11·9	0 54	424	21·2	0 30	763	38·2
2 55	131	6·5	1 34	243	12·2	0 53	432	21·6	0 29	790	39·5
2 50	135	6·7	1 32	249	12·5	0 52	440	22·0	0 28	818	40·9
2 45	139	6·9	1 30	254	12·7	0 51	449	22·5	0 27	848	42·4
2 40	143	7·1	1 28	260	13·0	0 50	458	22·9	0 26	881	44·1
2 35	148	7·4	1 26	266	13·3	0 49	467	23·4	0 25	916	45·8
2 30	152	7·6	1 24	273	13·6	0 48	477	23·9	0 24	954	47·7
2 25	158	7·9	1 22	279	14·0	0 47	487	24·4	0 23	997	49·8
2 20	163	8·2	1 20	286	14·3	0 46	498	25·0	0 22	1041	52·1
15	169	8·5	1 18	293	14·7	0 45	509	25·5	0 21	1090	54·6
10	176	8·8	1 16	301	15·1	0 44	520	26·0	0 20	1145	57·3
5	183	9·2	1 14	309	15·5	0 43	532	26·6	0 19	1204	60·3
	191	9·5	1 12	318	15·9	0 42	545	27·3	0 18	1272	63·6
	194	9·7	1 10	327	16·4	0 41	558	28·0	0 17	1347	67·4
	197	9·9	1 8	337	16·8	0 40	572	28·9	0 16	1431	71·6
1 54	201	10·0	1 6	347	17·4	0 39	587	29·4	0 15	1526	76·4

Table II.—A Dip Table.

Height of eye above water in feet.	Amount of dip.	Height of eye above water in feet.	Amount of dip.	Height of eye above water in feet.	Amount of dip.
20	4·40	85	9·45	150	13·0
25	5·11	90	10·3	155	13·13
30	5·43	95	10·19	160	13·27
35	6·11	100	10·35	165	13·39
40	6·40	105	10·51	170	13·51
45	7·5	110	11·7	175	14·3
50	7·30	115	11·22	180	14·15
55	7·51	120	11·38	185	14·26
60	8·12	125	11·52	190	14·38
65	8·32	130	12·6	195	14·49
70	8·52	135	12·19	200	15·1
75	9·10	140	12·33	205	15·11
80	9·28	145	12·46	210	15·22

Note.—This differs from the Dip Table, page 78 in that it has not been corrected for refraction.

The correction for each foot of additional height, column 3 Table 1., is 60·3 yards; distance = 1204 + (95 − 20) 60·3 = 5726 yards.

On ascertaining the distance for the same height and angle from Table IV. in Admiral Ryder's work, it will be found to be 7229 yards, showing an error of 1503 yards, introduced by using Table I. with the "horizon" method beyond the prescribed limits.

This example is given to enforce the necessity of confining the use of Table I. for the "horizon" method, to the limits laid down above.

Of course if there is any index correction it must be applied, but it will be found more convenient and less embarrassing to the observer aloft to use a sextant that has no index error.

A. P. R.

FLEET SAILING.

Definitions of Technical Terms.

A COLUMN means any number of ships in a distinct group, whether in line ahead, abreast, or otherwise.

The LEADING COLUMN is the headmost Column in any formation.

The STARBOARD WING COLUMN is the Column on the extreme right of any formation.

The PORT WING COLUMN is the Column on the extreme left of any formation.

The REAR COLUMN is the sternmost Column in any formation.

The LEADER OF A COLUMN is the headmost ship.

The STARBOARD WING SHIP of a Column is the ship on its extreme right.

The PORT WING SHIP of a Column is the ship on its extreme left.

The REAR SHIP of a Column is the sternmost ship.

The ship NEXT AHEAD of another, is that immediately before her, or before her beam.

The ship NEXT ASTERN of another, is that immediately behind her, or abaft her beam.

The STARBOARD COLUMNS of a formation are the alternate Columns, commencing from the right.

The PORT COLUMNS of a formation are the alternate Columns, commencing with the second column from the right.

The foregoing terms when used in the Evolutionary Signals refer solely to the position of the ships at the moment of making the signal, and no ship or column is ever alluded to by a term expressive of position, unless such ship or column actually holds that position at the time.

The terms used to designate certain portions of the fleet, such as 1st, 2nd, 3rd Divisions; 1st, 2nd, 3rd, 4th, 5th, and 6th Sub-divisions, &c.; are fixed terms, and always refer to the portions of the fleet to which they have been originally attached, whatever be their positions at the moment.

A Column is said to be in LINE AHEAD when the ships are in one line ahead of each other.

A Column is said to be in LINE ABREAST when the ships are ranged in one line abeam of each other.

A Column is said to be in QUARTER LINE when the ships are ranged in one line abaft each others' beam, but not right astern. Generally this line is formed four points abaft the beam of the leader.

A Column is said to be in TWO QUARTER LINES when the ships are ranged on each quarter of a single ship.

A Column is said to be in TWO BOW LINES when the ships are ranged on each bow of a single ship.

The DIRECTION of a Column is the bearing on which the ships are formed taken from their leaders.

The formation or disposition of a fleet is termed its ORDER.

The arrangement of the fleet in divisions and sub-divisions is termed its ORGANIZATION.

The term COMMANDER OF COLUMN indicates the Senior Officer in that Column.

Trying Rate of Sailing.

To find the distance from a ship? To the log of the height of the masthead, add the log cot, of the observed angle.

To find the distance of a ship to windward or to leeward? To the log of the distance add log sin. of angle between a line at right angles to the wind and bearing of ship.

COLOMB'S FLASHING SIGNALS.

FLASHING SIGNALS are made by the motion of any single object. In most instances the object is made to appear and disappear; and in others it is made to change its position, so that one position shall represent the appearance and the other the disappearance of the object. The symbols are determined by successive appearances and disappearances at regulated intervals, constantly recurring after a fixed pause.

The appearances of the object are termed FLASHES, and are of two lengths, termed "short" and "long" flashes, which are used in combination, to express the signs required, and are usually written thus : ——— to express the SHORT flash ; ———————— to express the LONG flash ; the interval of obscuration, or of the disappearance of the object, being left blank.

The *long* flash should be at least three times as long as the *short* one.

The disappearances of the object are termed INTERVALS, and are of three lengths. That between the flashes composing a figure is equal in duration to a *short* flash; that between two figures is equal to a *long* flash; and that between any two repetitions of a signal is equal to one-third of the whole length of the signals, or from about seven to ten seconds.

At *night* those signals are made by the obscuration and exposure of a single light; in the *daytime*, by the different apparatus which may be employed.

At short distances no special apparatus is necessary, the simple waving of the arm with a hat, flag, handkerchief, &c., being sufficient.

The system is equally applicable in fogs; long and short sounds on a fog horn, bugle, or steam whistle, representing the *long* and *short flashes*.

In all cases the signals or combinations have the same signification, so that an observer, having learnt the use of one apparatus, can read and make signals with any other description of apparatus without further instruction.

The following Tables exhibit the Signs necessary for use with this Code:—

Numerals.

1. —		6. ———	
2. — —		7. — ———	
3. — — —		8. ——— — —	
4. — — — —		9. — — ———	
5. — — — — —		0. ——— — —	

Auxiliary Signs.

Compass	——— — ———
Pendants	——— . — — ———
Numeral	——— — — —
Special and Repeat	— ——— — —
Horary	— — ——— —
Interrogative	— — — ———
Negative	——— — — —
List of Navy	— ——— —
Alphabetical	——— ———

Answer ——— — ——— — ——— — *a continuation of* long *and* short *flashes.*

Spelling — ——— ——— — —

Preparative — — — — — *a continuation of* short *flashes.*

Stop ——— ——— ——— *a continuation of* long *flashes.*

FLASHING SIGNALS.

Alphabetical Table.

THE following Alphabet, &c., can be used under circumstances when it is not convenient or possible to refer to the Signal Book, and forms in itself a perfect *telegraphic* system; necessarily rather slow in its application, but having the advantage of requiring very little previous knowledge and practice to work with correctness.

		A 5 — — — —		
B 6 —	C 7 — —	D 8 — —	E 9 — — —	F 10 — — — —
G 11 — —	H 12 — — —	I 13 — — — —	J 14 — — — — —	K 15 — — — — — —
L 16 —	M 17 — —	N 18 — — —	O 19 — — — —	P 20 — — — — —
Q 21 — — —	R 22 — — — —	S 23 — — — — —	T 24 — — — — — —	U 25 — — — — — — —
V 26 — — — —	W 27 — — — — —	X 28 — — — — — —	Y 29 — — — — — — —	Z 30 — — — — — — — —

All other lights in the vicinity of a Flashing Signal at night should be concealed. Signals should not be answered until thoroughly comprehended.

THE MORSE ALPHABET.

For use in communicating with Telegraph Stations not provided with a Code.

The system of signalling by flashes may be used for communicating messages by the Morse Alphabet, now in very general use for telegraphic purposes.

```
A · —              N — ·
B — · · ·          O — — —
C — · — ·          P · — — ·
D — · ·            Q — — · —
E ·                R · — ·
F · · — ·          S · · ·
G — — ·            T —
H · · · ·          U · · —
I · ·              V · · · —
J · — — —          W · — —
K — · —            X — · · —
L · — · ·          Y — · — —
M — —              Z — — · ·
```

The Preparative and Erasure, Stop, General Answer, and Repeat and Interrogative, are the same in both codes.

The signals should be made in precisely the same manner as when using the numerical code, a pause being made between each letter as between each figure, and treating each word as a number.

Figures should not be used; but numbers should be sent in words.

When a word is wrongly sent, the erasure signal must be made and acknowledged, and the word sent again.

FLASHING SIGNALS WITH FLAGS.

From the fact that flags are not fully exposed to view unless kept in motion, and as the plan of exposure and concealment cannot be employed in using them, a different arrangement is adopted to make flashing signals.

Fig. 1.   Fig. 2.

The signalman may work from left to right, or from right to left, as shown in Figs. 1 and 2, according to convenience, and the direction of the wind. To make a *short* flash—the flag is waved from a to b, and back to the normal position a. To make a *long* flash—the flag is waved from a to c, and after a distinct pause, brought back to the normal position, a.

The numerals 1 to 5 are, therefore, denoted by one to five waves of the flag from a to b, recovering to a. The numeral 6, by a wave from a to c, recovering to a. The numeral 7, by a wave from a to b, back to a, and then to c, recovering to the normal position, a. The numeral 8 is denoted by a wave from a to c, back to a, and then to b, recovering to the normal position, a. The numeral 9 is denoted

THE INTERNATIONAL CODE OF SIGNALS.

This Code has been adopted by all the following Maritime Powers for their Imperial as well as for their Mercantile Navies, viz:—

Great Britain.	Denmark.	Russia.	Austria.	Portugal.
France.	Holland.	Greece.	Germany.	Brazil.
America.	Sweden.	Italy.	Spain.	

SIGNAL STATIONS.

SIGNAL Stations have been established, with the approval of Her Majesty's Government, at some of the most important points on the coasts of the United Kingdom, and it is in contemplation to establish others where experience may prove them to be necessary or desirable.

The Signal Stations already established on the coast of the United Kingdom are as follows:—

Aldborough.	Dover.	Penzance.
Bridlington.	Dungeness.	The Scilly Islands.
Flamboro' Head.	Yarmouth, I. of W.	Roche's Point,
Grimsby.	St. Catherine's Point,	Queenstown.
Yarmouth, Norfolk.	(I. of W.)	Holyhead.
Broadstairs.	Prawle Point, near	Caldy Island, (Tenby.)
Deal.	the Start.	Cardiff.

Signal Stations have also been established at—

Straits of Sunda.	Ascension.	Oxo, Christiansand.
Straits of Messina.	Cape Point, C.G.H.	Elsinore.
Gibraltar.	Heligoland.	Palais, Belle Isle.
St. Helena.	Skagen, N. of Jutland	Sulina, Danube.

At these Signal Stations the International Code is the only Code recognized, and Vessels of any nation which make their names known by means of the International Code in passing these Stations will be reported in the "Shipping Gazette."

SIGNAL FLAGS FOR BRITISH MERCHANT SERVICE.

Answering 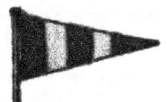 Pendant.

When used as the "Code Signal" this Pendant is to be hoisted under the "Ensign. For "Answering Pendant" where best seen.

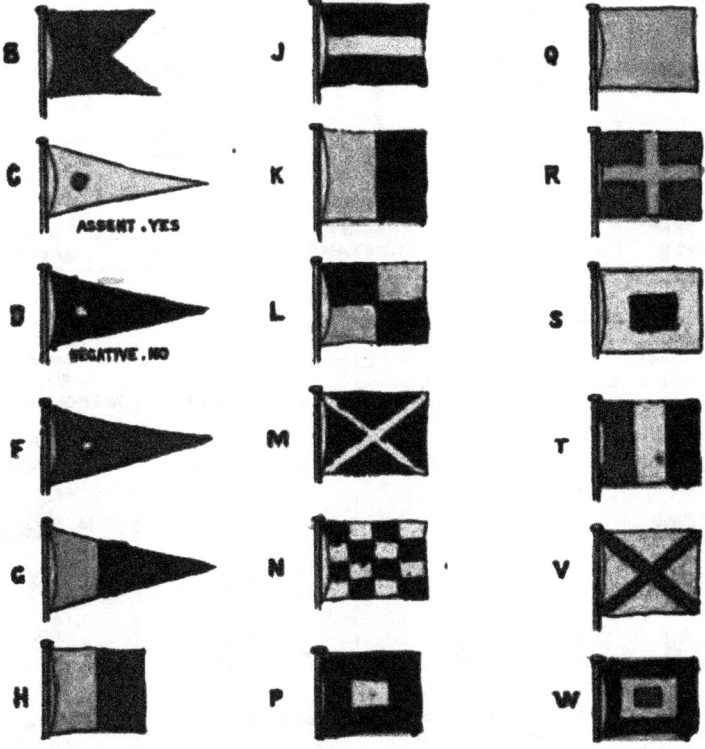

32 CODE OF SIGNALS. Sec. I.

The following examples will serve to illustrate how the form of a Hoist will usually denote the nature of the Signal made.

Two Flags.			Three Flags.	
Burgee uppermost, 'Attention' Signals.	**Pendant uppermost, 'Compass' Signals.**	**Square Flag uppermost, 'Danger' Signals.**	**General Signals.**	
B / D	F / T	N / C	K / C / P	P / M / V
WHAT SHIP IS THAT?	W. BY S.	IN DISTRESS, WANT ASSISTANCE.	ENGINE BROKEN DOWN.	CALL FOR ORDERS OFF

Four Flags.			
Burgee uppermost, Geographical Signals.	**Pendants C, D, F, uppermost, National Vocabulary.**	**Pendant G uppermost, Names of Men-of-War.**	**Square Flag uppermost, Names of Merchant Ships.**
B / D / T / M	D / V / C / H	G / S / M / W	J / M / C / Q
FALMOUTH.	WHAT SHIPS HAVE YOU SPOKEN?	"MARLBOROUGH" SCREW (3) GUNS.	"MARCO POLO" No. 6025

SEC. I.   AREA OF SAILS.   33

Table for Computing the Area of Sails.

LIST OF SAILS.	NO. OF YARDS IN EACH.	AREA IN SQUARE FEET.	TOTAL AREA.
Square Sails.			
Courses — { Fore / Main			
Topsail — { Fore / Main / Mizen			
Top-Gallant Sails { Fore / Main / Mizen			
Royals — { Fore / Main / Mizen			
	TOTAL AREA OF SQUARE SAILS.		
Fore and Aft Sails.			
Flying Jib			
Jib			
Fore Topmast Staysail ..			
Fore Trysail			
Main Trysail			
Spanker			
Studding Sails.			
	TOTAL AREA WITH ALL SAIL.		

Deduction for Reefs.

For single reefs. deduct   sq. feet.
  ,,   double   ,,   ,,
  ,,   treble   ,,   ,,
  ,,   close   ,,   ,,

This Table is inserted in Section I. as it is constantly required.

D

Section 2.

THE STANDARD COMPASS.

THE STANDARD COMPASS.

Instructions for the use of the Admiralty Standard Compass.

This instrument is so constructed as to answer the purpose of a Steering compass and an Azimuth compass.

The chief points in its construction are the following :—

1. The *bowl* is of stout copper, with the view to calm the vibrations of the needle, and the intersecting point of the axis of its gimbals is made to coincide with the point of suspension of the card, and also with the centre of the azimuth circle.

2. In observing amplitudes and azimuths, the bearings are read from the card without reference to the azimuth circle, the card being graduated to twenty minutes of arc. The *azimuth circle* is accurately graduated to minutes of arc, and may be used on shore for surveying purposes. When so used, and when accurate magnetic bearings are required, the zero of the circle may be adjusted to the magnetic north, shown by the card, then clamped, and any number of magnetic bearings may then be obtained round the circle. Or by adjusting the zero of the circle to any given object, and clamping the compass to its stand, the angles of objects round the horizon may be observed and read off to the nearest minute.

3. The *magnetic needles* employed are compound bars, formed of laminæ of that kind of steel (clock-spring) which has been ascertained by numerous experiments to be capable of receiving the greatest magnetic power. Each compass has two cards, A and J; the former is used at all times, except in stormy weather; with much motion in the ship, the heavy or J card is substituted.

Each card is fitted with four needles fixed vertically and equidistant on a light framework of brass screwed to the card; the pair of central needles are 7·3 inches long, and the pair of external ones 5·3 inches, the whole weight of the A card being 1525 grains.

The extremities of the needles are 15° and 45° from the extremities of the diameter of the card, which is parallel to them.

Two of the pivots for card A are pointed with "native alloy," a material which is harder than steel, and which does not corrode by exposure to the atmosphere; and the ruby and agate caps are worked to a form to suit these points.

Two spare pivots pointed with hardened steel are likewise supplied, and these are gilded by the electrical process.

The two ruby headed pivots are made exclusively for the heavier card J. *and no other kind of pivot is to be used with this card.*

4. The *impressions of the cards* are taken after the paper has been cemented to the mica plate forming the basis; distortions from shrinking are thus prevented, and a more perfect centering attained.

The zero points of the cards and the magnetic axes of the needles are adjusted to each other with great care at the Compass Observatory for Her Majesty's navy at Debtford. The various adjustments for centering, and the elimination of errors due to the displacement of sight vanes and prism, are always made at this observatory, the compass being afterwards supplied to H.M. ships free from error.

1. When used as an Azimuth Compass.

This compass may be immediately converted to this purpose by removing the glass or steering cover, and fixing the azimuth circle with its glass on its upper margin.

In observing amplitudes and azimuths at sea, the bearings are read from the card, without reference to the external graduated circle.

2. When used either as an Azimuth or Steering Compass.

Be very careful to preserve the *pivot point* from injury when screwing the pivot into the bowl; and place the card gently upon it : and never move the compass without first having lifted the card (by means of the side screw) against the centre-pin. This pin is so contrived as to leave the card entirely free when in its level or proper position, and it cannot touch any part of the collar, except under some violent motions of the ship, for which the gimbals do not compensate.

THE STANDARD COMPASS.

The pivots, caps, and margins of the cards should be examined occasionally to see that their free working is not impeded by dust or fibres from the paper. When card A is not sufficiently steady, the heavy card J is to be used with one of its own ruby headed pivots.

Whenever the card works sluggishly, or injury by accident or long wear be suspected, a new cap or pivot should be screwed in or (as the case may be) card J with its appropriate pivot may be used. The central cap-screw, or nut on the face of the A card, must be taken off before the ruby cap is attempted to be unscrewed.

The cards should be adjusted for dip by the balancing slides, when necessary.

When the bowl does not work freely on its gimbals, the axes and their bushes should be examined and slightly rubbed with plumbago.

Caution.

Experience having proved that both the pivots and caps of the A card, as also the ruby headed pivot of the J card, have received injury by the concussion arising both from the firing and exercise of heavy guns in the vicinity of the compass, it is especially enjoined that the card should on these occasions (unless it is very inconvenient to do so) be lifted off the pivot by the apparatus at the side of the bowl.

Should it be necessary to remove the prism for the purpose of cleaning it from dust or moisture, the cover is arranged to slide upwards on guides, and can easily be removed and replaced by hand, taking care in replacing the prism that the lens or rounded part is placed below. All the screws attached to the prism plate have received special adjustment, and must not, therefore, be altered.

The Standard Compass should be placed as far as possible from the extremity of any elongated mass of iron, *especially if vertical*, such as the spindle of the capstan, iron stanchions, the iron quarter-davits, iron funnels, etc., and not less than five feet above the deck.

No stand of arms or other iron, *subject to occasional removal*, should be placed within at least fourteen feet of this Standard Compass, whether on the same deck, or immediately below it.

When the ship is ready for sea, and with davits and other iron work secured in the positions in which it is intended they shall remain at sea, then the deviation of the Standard Compass from the magnetic meridian should be ascertained by one of the following means :—

Process by Bearing of a distant object.

The requisite warps being prepared, the ship is to be gradually swung round, so as to bring her head successively upon each of the 32 points of the Compass; and as her head approaches each of these points, so gently to check her motion as to prevent any continued swing of the card. When the ship is quite steady, and her head exactly on any one point, the Direct Bearing of some distant but well-defined object is to be observed with the Standard Compass, and registered.

The ship's head is then to be gently warped round in the same manner to the next point, and when duly stopped and steadied there, the bearing of the same object is to be again set, and again recorded; and so on, point after point, till the exact bearing of the one object has been ascertained with the ship's head on every separate point of the compass.

The object selected for that purpose should be at such a distance from the ship, that the diameter of the space through which she revolves shall make no sensible difference in its real bearing. If she be at anchor in a tideway; a steeple or a detached house, or a large single tree, at not less than from 6 to 8 miles distance, will well answer the purpose; but if in a basin, a less distance will suffice.

The next step is to determine the *real* or *correct* magnetic bearing of the selected object from the ship; or, in other words the compass bearing which it would have from on board, if the compass were not disturbed by the attraction of the iron in the ship. This may generally be effected by taking the mean of all the observed bearings, if observed on equi distant points; or of four or more bearings, if observed also on equi-distant points; but a surer result will be obtained by carrying the Standard Compass to some place on the adjacent wharf or shore, from whence the ship (that is to say, the part of the ship where the compass stood) and the distant object, of which the bearings had been observed, shall both be exactly in one line with the observer's eye. The bearing of the object from that point will

THE STANDARD COMPASS.

evidently be the same as its *correct* magnetic bearing from the ship by that compass. The difference between this *correct* magnetic bearing of the object, and the bearing observed with the Standard Compass on board, when the ship's head is on any particular point, will show the error on that point which was caused by the ship's iron, —or in other words, the *Deviation* of the Standard Compass, according to the direction in which the ship's head was placed.

Process by Azimuths and Amplitudes of the Sun.

Compute the Apparent Time at Ship :—Knowing this, and with the aid of Burdwood's Tables,* the sun's true bearing can be obtained by inspection. The difference between the sun's *true* bearing and its *compass* bearing, (as observed with the ship's head on any particular point of the compass) is the amount of the ship's deviation on that particular point, combined with the variation of the compass due to the geographical position. By deducting the latter from the difference of bearings observed, the deviation is obtained.

By this process, during the course of a voyage, especially near sunset or sunrise, and with a tranquil sea, a deviation table can be formed by steaming, or tacking and sailing round a circle, in the course of half-an-hour. The ship should be kept as upright as possible, to avoid heeling error.

Should the Standard Compass be corrected by magnets, it can never be considered as entirely compensated; and the deviation must be expected to change on change of latitude and other causes.

It will thus be seen that the mariner can have no absolutely safe guide, except in the system of actual and unceasing observations, which has been enjoined in the foregoing pages.

Process by Reciprocal Bearings.

Should there be no suitable object visible from the ship, and at the requisite distance as stated above, the deviations must be ascertained by the process of Reciprocal Bearings. A careful observer must go ashore with a second compass, and place its tripod in some open spot, where it may be distinctly seen from the Standard Compass on

---

* The Sun's True Bearings, or Azimuth Tables, 1873, computed for intervals of time, between the parallels of lat. 30° and 60° inclusive, by the late Staff Commander John Burdwood, R.N., Naval Assistant in the Hydrographic Department, Admiralty.

board. Then, by means of preconcerted signals, the mutual bearings of those two compasses from each other are to be observed at the moment when the ship's head is quietly steady on each of the 32 points successively, as before directed.

To ensure the success of this operation, the compass on shore should not be more distant from the ship than is consistent with the most distinct visibility with the naked eye of both compasses from each other. The observations should be made as strictly simultaneously as possible. And to guard against any mistake, such as might be occasioned by a signal being misinterpreted, the time at which each bearing is taken should be noted, both on shore and on board, by compared watches.

Before this process is complete, the Standard Compass should be carried on shore, in order to be compared with the compass which had been employed there, by means of the bearing of some distant object; and the difference, if any, is to be recorded. This comparison of the two compasses may as well be made previously to the observations; and in all cases when compasses are compared, the caps, pivots, &c., should be first carefully examined.

Experience has shown that the deviations in iron vessels (also wood-built vessels with iron beams) are affected by the heeling of the ship; the magnitude of the error so resulting being proportional to the amount of heel, and the *maximum* disturbance when heeling, being found when the ship's head is *North* or *South*, by the observing compass, the disturbance vanishing when the ship's head is *East* or *West* by the observing compass.

It becomes therefore an early duty of those in charge of iron ships to ascertain by repeated observations at sea the degrees of error arising from the various conditions of heeling both to starboard and to port.

In steam vessels fitted with telescopic funnels, and when from necessity, the Standard Compass is placed near to the steam machinery the deviation is materially altered by the elongated or compressed state of the funnel. For instance, with her head east, H.M.S. *Blenheim* has 9° deviation with the funnel *up*, and 15¾° when *down*. In H.M.S. *Vulcan* (of iron) there is a contrary result. In swinging vessels, therefore, so fitted, and when the funnel is within 30 or 35 feet of the compass, it is necessary that on the eight principal points of the compass at least, the deviations should be ascertained in both conditions of the funnel.

The mariner must remember that the corrections found for the Standard Compass belong to that compass alone, and to that compass only while it is in its proper place; and that those corrections will furnish no guide whatever, to the effects of the ship's iron on a Compass placed in any other part of the ship. It is essential, therefore, that the ship's course should not only be *invariably* directed by the Standard Compass, but that all the courses and bearings inserted in the log should be those shown by that compass alone; the Binnacle Compass being regarded solely as a guide to the helmsman. When the ship has been placed on her proper course by the Standard Compass, the helmsman will notice the point shown by the Binnacle Compass as being that to which *he* has to attend; and a comparison of the two Compasses should be frequently repeated by the officer of the watch or the quartermaster, and should always be made when any alteration occurs in the direction of the course, or when the ship heels over. When the ship is by the wind, the apparent course by the Standard Compass is that which must be inserted in the log.

During either of the processes for ascertaining the deviations of the Standard Compass, when the ship's head pauses at each point, it is desirable to note, in a separate Table, the corresponding directions of her head by the Binnacle Compasses.

DEVIATION TABLE, H.M.S.

Ship's Head by *Standard Compass*.	Deviation at	Deviation at	Deviation at	Deviation at
NORTH.				
N. by E.				
N.N.E.				
N.E. by N.				
N.E.				
N.E. by E.				
E.N.E.				
E. by N.				
EAST.				
E. by S.				
E.S.E.				
S.E by E.				
S.E.				
S.E. by S.				
S.S.E.				
S. by E.				
SOUTH.				
S. by W.				
S.S.W.				
S.W. by S				
S.W.				
S.W. by W.				
W.S.W.				
W. by S.				
WEST.				
W. by N.				
W.N.W.				
N.W. by W.				
N.W.				
N.W. by N.				
N.N.W.				
N. by W.				

Sec. II.   THE STANDARD COMPASS.   45

STEERING TABLE, H.M.S.

Magnetic Course proposed be made.	Course therefore to be steered by Standard Compass, in order to make the proposed Correct Magnetic Course.
NORTH. N. by E. N.N.E. N.E. by N. N.E. N.E. by E. E.N.E. E. by N. EAST. E. by S. E.S.E. S.E. by E. S.E. S.E. by S. S.S.E. S. by E. SOUTH. S. by W. S.S.W. S.W. by S. S.W. S.W. by W. W.S.W. W. by S. WEST. W. by N. W.N.W. N.W. by W. N.W. N.W. by N. N.N.W. N. by W. North.	

Archibald Smith's (Straight Line) Course Table.

Standard Compass Courses		Correct Magnetic Courses
NORTH.		NORTH.
N. by E.		N. by E.
N.N.E.		N.N.E.
N.E. by N.		N.E. by N.
N.E.		N.E.
N.E. by E.		N.E. by E.
E.N.E.		E.N.E.
E. by N.		E. by N.
EAST.		EAST.
E. by S.		E. by S.
E.S.E.		E.S.E.
S.E. by E.		S.E. by E.
S.E.		S.E.
S.E. by S.		S.E. by S.
S.S.E.		S.S.E.
S. by E.		S. by E.
SOUTH.		SOUTH.

STANDARD COMPASS.

HOW TO FILL UP THE TABLE.—This Table may be filled up in two ways; viz.—Either by drawing the lines from all the points, half-points and quarter-points in the *Correct Magnetic Course* Column, as in the Example : or from all the points, &c., in the *Standard Compass Course* Column.

The first will be the most useful to the Officers of the Watches in Tactical Manœuvres, and has been adopted in the Example. A correct method of drawing these lines is to Project the Deviation Curve on Ryder's Graphic Form (which is supplied with all Chart boxes) and then draw the Straight Lines on this Table, connecting those degrees in each column, lines from which intersect one another at the Curve on the Graphic Form.

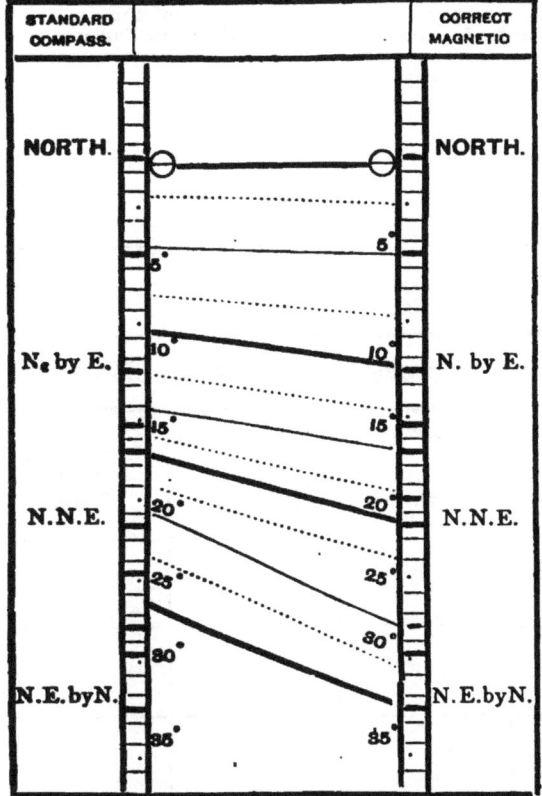

This table should be filled up from the Deviation Table, page 45, as soon as the ship has been swung, by drawing the lines from all the points, half-points, and quarter-points in the Correct Magnetic Course column, as in the example. By many it will be preferred to that at page 44, especially for deck use in squadrons performing steam evolutions. It shows, at a glance, the Standard Compass Course to be steered corresponding to the Correct Magnetic Course required, and *vice versa*.

EXPLANATION.

HOW TO USE THE TABLE.

First Case. If the fleet is steering a given course by signal, (which always means a correct magnetic course), and a signal is made to alter course a given number of points to starboard or port, run the finger up or down on the correct Magnetic Course Column the required number of points, and then across on the straight line to the Standard Compass Course Column ; and the Standard Compass Course to be steered will be found.

Second Case. If the fleet is not steering a correct magnetic course given by signal, but the ships are keeping station, then run the finger across from the Standard Compass Course, which is being steered on the straight line to the opposite column, and proceed as in Case 1.

The following are convenient Forms for Registering the Observations for Determining the Deviation of the Standard Compass.

Form for Registering the Process by Direct Bearing.

Ships' Head by the Standard Compass.	Bearing of.......... by the Standard Compass.	Deviation of the Standard Compass.

Form for Registering the Process by Reciprocal Bearings.

Time.	Ships' Head by the Standard Compass.	Simultaneous Bearings.		Deviation of the Standard Compass.
		From Standard Compass on board.	From the Shore Compass.	

Table for converting Points of the Compass, and their Fractional Parts, into Degrees, &c.

POINTS.	No.	DEGREES, &c. °	′	″	No.	POINTS.
NORTH.	0	0	0	0	0	SOUTH.
	⅛	1	24	22	⅛	
	¼	2	48	45	¼	
	⅜	4	13	7	⅜	
N. ½ E.....N. ½ W....	½	5	37	30	½	S. ½ W.........S. ½ E.
	⅝	7	1	52	⅝	
	¾	8	26	15	¾	
	⅞	9	50	37	⅞	
N. by E...N. by W....	1	11	15	0	1	S. by W.......S. by E.
	⅛	12	39	22	⅛	
	¼	14	3	45	¼	
	⅜	15	28	7	⅜	
N.byE.½E..N.byW.½W.	½	16	52	30	½	S.byW.½W...S.byE.½E.
	⅝	18	16	52	⅝	
	¾	19	41	15	¾	
	⅞	21	5	37	⅞	
N.N.E.....N.N.W....	2	22	30	0	2	S.S.W...........S.S.E.
	⅛	23	54	22	⅛	
	¼	25	18	45	¼	
	⅜	26	43	7	⅜	
N.N.E.½E. N.N.W.½W.	½	28	7	30	½	S.S.W.½W....S.S.E.½E.
	⅝	29	31	52	⅝	
	¾	30	56	15	¾	
	⅞	32	20	37	⅞	
N.E.byN....N.W.byN.	3	33	45	0	3	S.W. by S...S.E. by S.
	⅛	35	9	22	⅛	
	¼	36	33	45	¼	
	⅜	37	58	7	⅜	
N.E. ½N....N.W. ½N.	½	39	22	30	½	S.W. ½S......S.E. ½S.
	⅝	40	46	52	⅝	
	¾	42	11	15	¾	
	⅞	43	35	37	⅞	

Converting Points of the Compass into Degrees. *(Continued)*.

Points.	No.	Degrees, &c.			No.	Points.
		°	′	″		
N.E........N.W......	4	45	0	0	4	S.W............S.E.
	⅛	46	24	22	-⅛	
	¼	47	48	45	-¼	
	⅜	49	13	7	-⅜	
N.E. ½ E....N.W. ½ W.	½	50	37	30	-½	S.W. ½ W..:. S.E. ½ E.
	⅝	52	1	52	-⅝	
	¾	53	26	15	-¾	
	⅞	54	50	37	-⅞	
N.E by E...N.W.byW.	5	56	15	0	5	S.W. byW....S.E.byE.
	⅛	57	39	22	-⅛	
	¼	59	3	45	-¼	
	⅜	60	28	7	-⅜	
N.E. by E. ½ E...... ⎫	½	61	52	30	-½	⎧ S.W. by W. ½ W......
N.W. by W. ½ W..... ⎭	⅝	63	16	52	-⅝	⎩ S.E. by E. ½ E.
	¾	64	41	15	-¾	
	⅞	66	5	37	-⅞	
E.N.E.....W.N.W....	6	67	30	0	6	W.S.W........ E.S.E.
	⅛	68	54	22	-⅛	
	¼	70	18	45	-¼	
	⅜	71	43	7	-⅜	
E.byN.½N.. W.byN.½N.	½	73	7	30	-½	W.by S.½S... E.byS.½S.
	⅝	74	31	52	-⅝	
	¾	75	56	15	-¾	
	⅞	77	20	37	-⅞	
E. by N..... W. by N.	7	78	45	0	7	W. by S...... E. by S.
	⅛	80	9	22	-⅛	
	¼	81	33	45	-¼	
	⅜	82	58	7	-⅜	
E. ½ N.... W. ½ N....	½	84	22	30	-½	W. ½ S...... ..E. ½ S.
	⅝	85	46	52	-⅝	
	¾	87	11	15	-¾	
	⅞	88	35	37	-⅞	
East......West	8	90	0	0	8	West..........East.

To Correct the Compass for Variation.

Course by Compass given.
If Variation East, allow to right.
If Variation West, allow to left.
Will give true course.

True Course given.
If Variation East, allow to left.
If Variation West, allow to right.
Will give magnetic course.

To Correct Compass Courses.

Easterly Variation is + to all points between N. and E. ......S. and W.
Westerly Variation is − from all points between N. and E. ....S. and W.
Easterly Variation is − from all points between N. and W. ....S. and E.
Westerly Variation is + to all points between N. and W ......S. and E.

To Convert True Course into Compass Course.

Easterly Variation is − from all points between N. and E....S. and W.
Westerly Variation is + to all points between N. and E.......S. and W.
Easterly Variation is + to all points between N. and W. ....S. and E.
Westerly Variation is − from all points between N. and W....S. and E.

To Correct the Course at once for Variation and Deviation.

If they are both of the same name, *i.e.*, both E. or both W., add one to the other, and apply the sum according to their joint name. Example :—Compass course N.N.E. ; given, the deviation 8° 10′ E., and the variation 20° 10′ E., the sum thereof 28° 20′ E., being applied to the right hand of N.N.E., gives the true course, N. 50° 50′ E. But if one be E. and the other W., take their difference, give it the name of the greater, and apply it according to that name. Example :—Suppose the deviation be still 8° 10′ E., but with the variation of 20° 10′ W., their difference 12° 0′ West, applied to the left hand, will give N. 10° 30′ E. for the true course.

Mariner's Compass of Various Nations.

ENGLISH	FRENCH.	ITALIAN.	SPANISH	GERMAN	DUTCH.	SWEDISH, NORWEGIAN DANISH.
NORTH	Nord	Tramontana	Norte	Nord	Noord	Nord
N. b E.	N. q. N.E.	T. quarto G.	N. cuarto N.E.	N. zu O.	N. ten O.	N. til O.
N.N.E.	N.N.E.	G.T.	N.N.E.	N.N.O.	N.N.O.	N.N.O.
N.E. b N.	N.E. quart N.	G. quarto T.	N.E. cuarto N.	N.O. zu N.	N.O. ten N.	N.O. til N
N.E.	N.E.	Greco	N.E.	N.O.	N.O.	N.O.
N.E. b E.	N.E. quart E.	G. quarto L.	N.E. cuarto E.	N.O. zu O.	N.O. ten O.	N.O. til O
E.N.E.	E.N.E.	G.L.	E.N.E.	O.N.O.	O.N.O.	O.N.O.
E. b N.	E. quart N.E.	L. quarto G.	E. cuarto N.E.	O. zu N.	O. ten N.	O. til N.
EAST	Est	Levante	Este	Ost	Oost	Ost
E. b S.	E. quart S.E.	L. quarto S.	E. cuarto S.E.	O. zu S.	O. ten Z.	O. til S.
E.S.E.	E.S.E.	S.L.	E.S.E.	O.S.O.	O.Z.O.	O.S.O.
S.E. b E.	S E. quart S.	S. quarto L.	S.E. cuarto S.	S.O. zu O.	Z.O. ten O.	S.O. til O.
S.E.	S.E.	Scirocco	S.E.	S.O.	Z.O.	S.O.
S E. b S.	S.E. quart S.	S. quarto O.	S.E. cuarto S.	S.O. zu S.	Z.O. ten Z.	S.O. til S.
S.S.E.	S.S.E.	O.S.	S.S.E.	S.S.O.	Z.Z.O.	S.S.O.
S. b E.	S. quart S.E.	O. quarto S.	S. cuarto S.E.	S. zu O.	Z ten O.	S. til O.
SOUTH	Sud.	Ostro	Sur	Sud	Zuid	Syd
S. b W.	S. quart S.O.	O. quarto L.	S. cuarto S.O.	S. zu W.	Z. ten W.	S til V.
S.S.W.	S.S.O.	O.L.	S.S.O.	S.S.W.	Z.Z.W.	S.S.V.
S.W. b S.	S.O. quart S.	L. quarto O.	S.O. cuarto S.	S.W. zu S.	Z.W. ten Z.	S.V. til S
S.W.	S.O.	Libeccio	S.O.	S.W.	Z.W.	S.V.
S.W. b W.	S.O. quart O.	L. quarto P.	S.O. cuarto O.	S.W. zu W.	Z.W. ten W.	S.V. til V.
W.S.W.	O.S.O.	P L.	O.S.O.	W.S.W.	W.Z.W.	V.S.V.
W. b S.	O. quart S.O.	P. quarto L.	O. cuarto S.O.	W. zu S.	W. ten Z.	V. til S.
WEST	Ouest	Ponente	Oeste	West	West	Vest
W b N.	O. quart N.O.	P. quarto M.	O. cuarto N.O.	W. zu N.	W. ten N.	V. til N.
W.N W.	O.N.O.	P.M.	O.N.O.	W.N.W.	W.N.W.	V.N.V.
N.W. b W	N.O. quart O.	M. quarto P.	N.O. cuarto O	N.W. zu W.	N.W. ten W.	N.V. til V.
N.W.	N.O.	Maestro	N.O.	N.W.	N.W.	N.V.
N.W. b N	N.O. quart N.	M. quarto T.	N.O. cuarto N.	N.W. zu N.	N.W. ten N.	N.V. til N
N.N.W.	N.N.O.	M.T.	N.N.O.	N.N.W.	N.N.W.	N.N.V.
N. b W.	N. quart N.O.	T. quarto M.	N. cuarto N.O.	N. zu W.	N. ten W.	N til V.

In the French Compass the abbreviation is q as N. q N.E. or N¼ N.E.
  ,,   Italian   ,,   ,,   q as T. q G. or T. ¼ G.
  ,,   Spanish   ,,   c as N. c N.E. or N. ¼ N.E
  ,,   German   ,,   z as N. z O.
  ,,   Dutch   ,,   t as N. t O.
  ,,   Swedish   ⎫
  ,,   Norwegian ⎬   ,,   t as N. t O.   equivalent to the English by.
  ,,   Danish   ⎭

Mariner's Compass of Various Nations.

ENGLISH.	CHINESE, SHANGHAI, AND NING-PO.	JAPANESE.	GREEK.	HINDOOSTANEE.	TURKISH.
North	Póh	Kita	O. Borras	Guy	Yeldiz
N.E.	Tóng Póh	Higachi Kita	O. Mesēs	Guy weejow Arkrop	Porias
East	Tóng	Higachi	O. Apēliōtēs	Mutly	Gun-doghûsû
S.E.	Tóng-nen	Higachi minami	O. Euros	Sooly dow Arkrop	Ketsehichleme
South	Nèn	Minami	O. Notos	Sooly	Kible
S.W.	Si-nèn	Michi minami	O. Lips? Libas?	Sooly weejow Arkrop	Lodos
West	Si	Nichi	O. Zephuros	Cably	Bâti
N.W.	Si-póh	Nichi Kita	O. Skirōn	Guy dow Arkrop.	Kavayel

The Russians generally use the English or Dutch Compass.

Section 3.

RULE OF THE ROAD AT SEA.

EXTRACTS FROM MERCHANT SHIPPING ACTS.

THE RULE OF THE ROAD AT SEA.

In the preface to a small pamphlet on this subject, published by Captain A. de Horsey, R. N., the following passage occurs :—

"The cause of collision in 99 cases out of 100, I believe to be either ignorance of the regulations or neglect of them. Under the heads of ignorance and neglect may be included the frequent practice of altering course prematurely without having ascertained *which way* it should be altered, and also of deviation from your course *when it is your duty to keep it.*"

Mr. Gray, the Secretary to the Board of Trade, in some remarks on the regulations for preventing collisions at sea, says :—

"The very great majority of collisions happen through bad look-out and neglect to show lights. No rule of the road can meet these cases."

"Many collisions are caused through the fixed belief of some sailors that it is right, under the present rule of the road, to *port* under all circumstances. This is an entire misapprehension of the rule. The rule is not to blame for these collisions."

"Many collisions are caused through neglect, misapplication, and utter ignorance of the rule of the road. No rules, however perfect, can meet these cases."

"The rules are good; but some seamen have failed to make themselves acquainted with them, or have not acted on them when they ought, or as they ought. The legislature cannot make careless people careful, nervous people strong, ignorant people wise, dull people bright, or sleepy people wakeful. Let them enact rules for ever, collisions will continue to happen, through ignorance, bad look-out, and carelessness, just in the same way that ships will continue to be wrecked and stranded from the same causes, and from neglect of the lead, and other omissions."

General Rule for Steam Ships Meeting, and Particular Rule for Steam Ships Crossing.

The *general* rule of the road for steamers is precisely the same as the general rule of the pavement for foot passengers in London, and in all our large towns, viz., that in all ordinary cases two steam ships, like two pedestrians, meeting face to face, or "end on or nearly end on," so as to involve risk of collision, shall port, that is to say, shall keep to the right, so that each may pass on the port (left) side of the other. Nothing can be more simple than this; but the man who will persist in crossing right over the pavement, if, when proceeding along the left-hand side, he sees another man coming along to his own right on the other side, cannot justify his proceeding by the rule. He will obviously get in the way of the other.

The *particular* rule of the road for steamers is, that if they are crossing, then the steamer that has another steamer on her own right-hand side shall get out of the way.

Steam-ships *crossing* so as to involve risk of collision, always show to each other a different coloured light—green to red, and red to green; unless, therefore, a steamer sees another steamer's green light on her own port side, or another steamer's red light on her own starboard side, there is no danger so far as steamers crossing are concerned.

There are six cases in which it is your duty to alter course to avoid risk of collision—

1. In a steamer, meeting a steamer end on or nearly end on.

2. In a steamer, nearing a sailing vessel.

3. In a steamer, approaching another on your starboard side.

   NOTE.—This case should be carefully considered as it is the one requiring the most caution and judgment.

4. If under sail on the port tack, nearing a vessel under sail—on the starboard tack.

5. If under sail going free, nearing a vessel under sail—close hauled.

6. If under sail going free, and nearing another vessel to leeward—going also free.

In the *first case only* is it right to *port* the helm without further consideration. In the other five cases, the course should not be altered, until, either by bearings taken with an interval between them, or by bringing the vessel on with some part of the rigging, and watching whether she draws aft or forward, it is ascertained that the vessels are converging on one point—and *which is the best way to alter it*, to avoid collision.

Seamen are to be found who port at every light seen ahead, or nearly ahead; but if they port when they should not, for example, with a green light, say two points on their starboard bow, and say they do it because the *light* is nearly ahead or nearly end on with them—that is no fault of the rule, and has no reference to the rule, for the rule does not apply in a case where there is no risk of collision; and there is no risk of collision, as has already been admitted, if a green light is seen ahead or anywhere on the starboard side.

One of the most fruitful causes of collision is, that the ship that has by the rules to alter course, does not do so promptly and sufficiently to show to the other ship clearly, and evidently, that she knows her duty and is performing it. When this is not done, the other ship is often led to adopt some wrong course to avoid collision, and thus bring it to pass. If under steam, a slight yaw with the helm will serve to show the direction you intend to take: if under sail and about to tack, let fly the jib-sheet; if to bear up, shiver the mizen topsail or brail up the spanker.

So long as you keep a Green Light opposed to a Green Light, or a Red Light opposed to a Red Light, no Collision can happen between passing Ships.

The reckless use of Port Helm leads to Collision.

Aid to Memory, in four verses, by Thomas Gray.

1. Two Steam Ships Meeting.

> When both side-lights you see a-head,
> Port your helm, and show your RED.

2. Two Steam Ships Passing.

> GREEN to GREEN, or RED to RED—
> Perfect safety—Go a-head!

3. Two Steam Ships Crossing.

NOTE.—This is the position of greatest danger; there is nothing for it but good look-out, caution, and judgment.

> If to your starboard RED appear,
> It is your duty to keep clear;
> To act as judgment says is proper;—
> To Port—or Starboard—Back—or Stop her!

> But, when upon your Port is seen
> A steamer's Starboard light of GREEN,
> There's not so much for you to do,
> For GREEN to Port keeps clear of you.

4. All ships must keep a good look-out, and Steam Ships must stop and go-astern if necessary.

> Both in safety and in doubt
> Always keep a good look out;
> In danger, with no room to turn,
> Ease her!— Stop her!—Go astern!

Extracts from the Board of Trade Regulations for Preventing Collisions at Sea.

In the following Rules every steam ship which is under sail and not under steam, is to be considered a sailing ship; and every steam ship which is under steam, whether under sail or not, is to be considered a ship under steam.

STEERING AND SAILING RULES.

Two Sailing Ships Meeting.

Article 11. If two sailing ships are meeting end-on, or nearly end-on, * so as to involve risk of collision, the helms of both shall be put to port, so that each may pass on the port side of the other.

Two Sailing Ships Crossing.

Article 12. When two sailing ships are crossing so as to involve risk of collision, then, if they have the wind on different sides, the ship with the wind on the port side shall keep out of the way of the ship with the wind on the starboard side; except in the case in which the ship with the wind on the port side is close hauled and the other ship free, in which case the latter ship shall keep out of the way; but if they have the wind on the same side, or if one of them has the wind aft, the ship which is to windward shall keep out of the way of the ship which is to leeward.

Two Ships under Steam, Meeting.

Article 13. If two ships under steam are meeting end-on, or nearly end-on,* so as to involve risk of collision, the helms of both shall be put to port, so that each may pass on the port side.

---

* The two articles, numbered 11 and 13 respectively, only apply to cases where ships are meeting end-on, or nearly end-on, *in such a manner as to involve risk of collision.* They consequently do not apply to two ships which must, if both keep on their respective courses, pass clear of each other.

Two Ships under Steam Crossing.

Article 14. If two ships under steam are crossing so as to involve risk of collision, *the ship which has the other on her own starboard side shall keep out of the way of the other*.

Sailing Ship and Ship under Steam.

Article 15. If two ships, one of which is a sailing ship, and the other a steam ship, are proceeding in such directions as to involve risk of collision, the steam ship shall keep out of the way of the sailing ship.

LIGHTS.

The mast-head light ordered to be carried by steamers when underweigh, is to throw the light from right ahead to two points abaft the beam on each side, and should be visible in a clear atmosphere five miles.

The Starboard—green, and Port—red, side lights ordered to be carried by all vessels when underweigh, are to throw the light from right ahead to two points abaft the beam, and should be visible two miles.

Lights for Steam Tugs.

Article 4. Steam ships, when towing other ships, shall carry two bright white masthead lights vertically, in addition to their side lights, so as to distinguish them from other steam ships. Each of these masthead lights shall be of the same construction and character as the masthead lights which other steam ships are required to carry.

NOTE.—The ship being towed carries side lights only.

The only cases in which the two Articles apply, are when each of the two ships is end-on, or nearly end-on to the other, in other words, to cases in which, *by day*, each ship sees the masts of the other in a line, or nearly in a line, with her own; and *by night*, to cases in which each ship is in such a position as to see both the side lights of the other.

The said two articles do not apply *by day* to cases in which a ship sees another *a-head* crossing her own course; or *by night* to cases where the red light of one ship is opposed to the red light of the other; or where the green light of one ship is opposed to the green light of the other; or where a red light without a green light, or a green light without a red light, is seen ahead; or where both green and red lights are seen anywhere but ahead.

Sec. III — RULE OF THE ROAD AT SEA.

Article 8.—Sailing pilot vessels shall not carry the lights required for other sailing vessels, but shall carry a white light at the masthead visible all round the horizon, and shall also exhibit a flare up light every fifteen minutes.

Fishing vessels and open boats when at anchor, or attached to their nets and stationary, shall exhibit a bright white light. Fishing vessels and open boats shall, however, not be prevented from using a flare-up in addition if considered expedient.

French or English boats fishing with drift nets shall carry on one of their masts, two lights; one over the other, three feet apart.

Rules concerning Fog Signals.

Article 10.—Whenever there is a fog, whether by day or night, the fog signals described below shall be carried and used, and shall be sounded at least every five minutes, viz. :—

(a.) Steam ships under weigh shall use a steam whistle placed before the funnel, not less than eight feet from the deck.

(b.) * Sailing ships under weigh shall use a fog horn.

(c.) Steam ships and sailing ships when not under weigh shall use a bell.

Extract from Article 16.—" Every steam ship shall, when in a fog, go at a moderate speed."

Extract from the Catechism or Heads of Examination for the use of Examiners in Seamanship, issued by the Board of Trade.

49. What should be the result if you ported to a green light ahead?

I should probably run right across the path of the vessel carrying the green light.

---

* In the Naval Fog Signals it is directed that :—" Single sounds of the fog horn at about 3-second intervals will denote being on the STARBOARD Tack ; and double sounds at 3-second intervals will denote being on the PORT Tack.

50. If a steamer, A, sees the *three* lights of another steamer B ahead or nearly ahead, are the two steamers meeting, passing, or crossing?

Meeting end on, or nearly end on.

51. Do the regulations expressly require the helm of a ship to be put to port in any case; and if so, when?

Yes; in the case of two steamers or two sailing vessels meeting end on, or nearly end on.

52. Do they expressly require the helm of a ship to be put to port in any other case; and if so, in what other?

No. The use of the port helm is not in any other case expressly required by the regulations.

[The Examiner should then explain that the only case in which port-helm is mentioned in the regulations is in Articles 11 and 13 for two ships meeting end on, or nearly end on.]

53. If you port to a green light ahead, or anywhere on your starboard bow, and if you get into collision by doing so, do you consider that the regulations are in fault?

No; because the regulations do not expressly require me to port in such a case, and because by porting I know that I should probably and almost certainly run across the other vessel's path, or run into her.

54. If a steamer A sees another steamer's red light B on her own starboard side, are the steamers meeting, passing, or crossing; and how do you know?

Crossing; because the red light of one is opposed to the green light of the other; and whenever a green light is opposed to a red light, or a red light to a green light, the ships carrying the lights are crossing ships.

55. Is A to stand on; and if not, why not?

A has the other vessel B on her starboard side. A knows she is crossing the course of B because she sees the red light of B on her (A's) own starboard side. A also knows she must get out of the way of B, because Article 14 expressly requires that the steamer that has the other on her own starboard side shall keep out of the way of the other.

56. Is A to starboard or to port in such a case?

A must do what is right, so as to get herself out of the way of B, she must starboard if necessary, or port if necessary; and she must stop and reverse if necessary.

57. If A (referred to in question 54) gets into collision by porting, will it be because she is acting on any rule?

No; the rule does not require her to port. If she ports, and gets into collision by porting, it is not the fault of any rule.

58. If a steamer A sees the green light of another steamer B on her own (A's own) port bow, are the two steamers meeting, passing, or crossing; and how do you know?

Crossing, because the green light of one ship is shown to the red light of the other.

59. What is A to do, and why?

By the rule contained in Article 18 of the Regulations, A is required to keep her course, subject only to the qualification that due regard must also be had to any special circumstances which may exist in any particular case rendering a departure from that rule necessary in order to avoid immediate danger. The crossing ship B on A's port side must get out of the way of A, because A is on B's starboard side.

60. A, a steamer going east, sees the green light of another steamer, B, a point on her (A's) port bow. Is there any regulation requiring A to port in such a case, and if so, where is it to be found?

There is not any.

61. Are steam ships to get out of the way of sailing ships?

If a steamer and a sailing ship are proceeding in such a direction

F

as to involve risk of collision, the steamer is to get out of the way of the sailing ship.

62. What is to be done by A, whether a steamer or sailing ship, if overtaking B?

A is to keep out of the way of B.

63. When by the rules, one of two ships is required to keep out of the way of the other, what is the other to do?

To keep her course.

64. Is there any qualification or exception to this?

Yes. Due regard must be had to all dangers of navigation, and to any special circumstances which may exist in any particular case to avoid immediate danger.

Official Notice.—Merchant Shipping Act, 1873.
Collisions.

The Board of Trade give notice, that on and after the 1st of November, 1873, the following Sections (16 and 17) of the Merchant Shipping Act, 1873, come into operation.

16. "In every case of collision between two vessels, it shall be the duty of the Master or person in charge of each vessel, if and so far as he can do so without danger to his own Vessel, Crew, and Passengers (if any), to stay by the other vessel until he has ascertained that she has no need of further assistance, and to render to the other Vessel, her Master, Crew and Passengers (if any), such assistance as may be practicable and as may be necessary in order to save them from any danger caused by the collision; and also to give to the Master or person in charge of the other Vessel the name of his own Vessel, and of her port of registry, or of the port or place to which she belongs, and also the names of the ports and places from which and to which she is bound.

"If he fails so to do, and no reasonable cause for such failure is shown, the collision shall, in the absence of proof to the contrary, be deemed to have been caused by his wrongful act, neglect, or default.

"Every Master or person in charge of a British Vessel who fails without reasonable cause to render such assistance, or give such information as aforesaid, shall be deemed guilty of a misdemeanour; and if he is a certificated Officer, an inquiry into his conduct may be held, and his certificate may be cancelled or suspended.

17. "If in any case of collision it is proved to the Court before which the case is tried that any of the regulations for preventing collisions contained in or made under the Merchant Shipping Acts, 1854 to 1873, have been infringed, the ship by which such regulation has been infringed shall be deemed to be in fault, unless it is shown to the satisfaction of the Court that the circumstances of the case made departure from the regulations necessary."

THOMAS GRAY, one of the Assistant Secretaries.
Marine Department, Board of Trade, September, 1873.

Signals for Pilots.

The Board of Trade give notice, that on and after the 1st of November, 1873, if a vessel requires the services of a Pilot, the Signals to be used and displayed shall, in accordance with the 19th Section of the Merchant Shipping Act, 1873, be the following, viz :—

"**In the Daytime.**—The following Signals, numbered 1 and 2, when used or displayed together or separately, shall be deemed to be Signals for a Pilot in the daytime, viz :—

> "1. To be hoisted at the fore, the Jack or other national colour usually worn by Merchant ships, having round it a white border one-fifth of the breadth of the flag; or
>
> "2. The International Code Pilotage Signal indicated by P.T.

"**At Night** :—The following Signals, numbered 1 and 2, when used or displayed together or separately, shall be deemed to be Signals for a Pilot at night, viz :—

> "1. The Pyrotechnic Light, commonly known as a Blue Light, every 15 minutes; or
>
> "2. A bright White Light, flashed or shown at short or frequent intervals, just above the bulwarks, for about a minute at a time."

And "any Master of a vessel who uses or displays, or causes or permits any person under his authority to use or display, any of the said Signals for any other purpose than that of summoning a Pilot, or uses or causes or permits any person under his authority to use, any other Signal for a Pilot, shall incur a Penalty not exceeding Twenty Pounds."

C. CECIL TREVOR, one of the Assistant Secretaries.

Harbour Department, Board of Trade, September, 1873.

Signals of Distress.

The Board of Trade give notice, that on and after the 1st of November, 1873, the following Signals shall, in accordance with the 18th section of the Merchant Shipping Act, 1873, be deemed to be Signals of Distress.

"**In the Daytime** :—The following Signals, numbered 1, 2, and 3, when used or displayed together or separately, shall be deemed to be Signals of Distress in the Daytime :—

"1. A gun fired at intervals of about a minute.

"2. The International Code Signal of Distress indicated by N. C.

"3. The Distant Signal, consisting of a square flag, having either above or below it a ball, or anything resembling a ball.

"**At Night** :—The following Signals, numbered 1, 2, and 3, when used or displayed together or separately, shall be deemed to be Signals of Distress at Night :—

"1. A gun fired at intervals of about a minute.

"2. Flames on the ship (as from a burning tar barrel, oil barrel, &c.)

"3. Rockets or shells of any colour or description, fired one at a time, at short intervals."

And "any Master of a vessel who uses or displays, or causes or permits any person under his authority to use or display any of the said signals, except in the case of a vessel being in distress, shall be liable to pay compensation for any labour undertaken, risk incurred, or loss sustained, in consequence of such signal having been supposed to be a signal of distress ; and such compensation may, without prejudice to any other remedy, be recovered in the same manner in which salvage is recoverable."

THOMAS GRAY,
Assistant Secretary of the Marine Department of the Board of Trade.

August, 1873.

Section 4.

WIND, WEATHER, BAROMETER, THERMOMETER.

CURRENTS; ICE; PASSAGE TABLES.

The Beaufort Notation, to Indicate the Force of the Wind.

0—Denotes calm.
1—Light air: just sufficient to give steerage way.

Beaufort		Sail carried	Speed
2—Light breeze	⎫ With which a well-conditioned Man-of-war under all sail and clear full, would go in smooth water. ⎬		1 to 2 knots
3—Gentle breeze			3 to 4 knots
4—Moderate breeze	⎭		5 to 6 knots
5—Fresh breeze			Royals, &c.
6—Strong breeze			Single reefs and top-gallant sails
7—Moderate gale	⎫ In which the same ship could just carry close-hauled ⎬		Double reef, jib, &c.
8—Fresh gale			Triple reefs, courses, &c.
9—Strong gale	⎭		Close reefs and courses
10—Whole gale	With which she could only bear		Close-reefed main top-sail, and reefed fore-sail
11—Storm	With which she would be reduced to		Storm staysail
12—Hurricane	To which she could show		No canvas

Rate		Pressure	Description of Wind.
Miles per hour	Feet per minute	Force in lbs. per square foot	
1	88	·005	Hardly perceptible
2	176	·020	Just perceptible
3	264	·044	
4	352	·079	Gentle breeze
5	440	·123	
10	880	·492	Pleasant breeze
15	1320	1·107	
20	1760	1·970	Brisk gale
25	2200	3·067	
30	2640	4·429	High wind
35	3080	6·027	
40	3520	7·870	Very high wind
45	3960	9·900	
50	4400	12·304	Storm
60	5280	17·733	Great storm
70	6160	24·153	
80	7040	31·490	Hurricane
100	8800	49·200	

INFORMATION ON THE AVERAGE LIMITS OF THE REGIONS OF TRADE WINDS AND MONSOONS; WITH THE LOCALITIES OF CYCLONIC STORMS AND RAINY SEASONS IN THE ATLANTIC, PACIFIC, AND INDIAN OCEANS, THROUGHOUT THE YEAR.

ATLANTIC OCEAN.
AVERAGE LIMITS OF THE TRADE WINDS.

Months.		Jan., Feb., March.	April, May, June.	July, Aug., Sept.	Oct., Nov., Dec.
N.E. Trade	Northern Boundary	25° N.	27° N.	28° N.	24° N.
	Southern Boundary*	2° N.	4° N.	11° N.	6° N.
Breadth of Variable Belt		120 Miles	180 Miles	500 Miles	200 Miles
S.E. Trade	Northern Boundary†	The Equator	1° N.	3° N.	3° N.
	Southern Boundary‡	A line drawn from the Cape of Good Hope to the Isles of Trinidad and Martin Vaz.			
RAINY SEASONS		Guayana & North Brazil — Africa south of the Equator.‡	Guayana, Brazil, and Africa, north of the Equator in May and June. Carribean sea in June	West Indies and Africa, north of the Equator. — Brazil in July and Aug.	Guayana in Dec., Africa south of the Equator in Nov. & Dec.
CYCLONES.				West Indies.	West Indies in Oct., and occasionally in Nov.

* Between Cape Verde and Sierra Leone, extending 50 to 70 leagues from the shore, winds and currrents change with seasons. From June to September squally S.W. winds, and a N.E. or Northerly current. From October to May, when N.E. and Northerly winds prevail, a South Easterly current is experienced.

† On the Brazilian coast, Southward of Bahia, the winds blow from N.N.E. to East, between October and March; and between April and September from S.W. and S.E.

‡ South and South-west winds prevail on the West Coast of Africa all the year round. A line drawn from Walfisch Bay in about 23° S., to Cape Palmas will form the general boundary between these winds and the S.E. trade. December to February is the Harmattan season on African coast, Northward of the Equator.

1

WIND CHART FO
Compiled by Staff Command
Arrows indicate direction of winds. — Tracks of storm centres, shewn ti

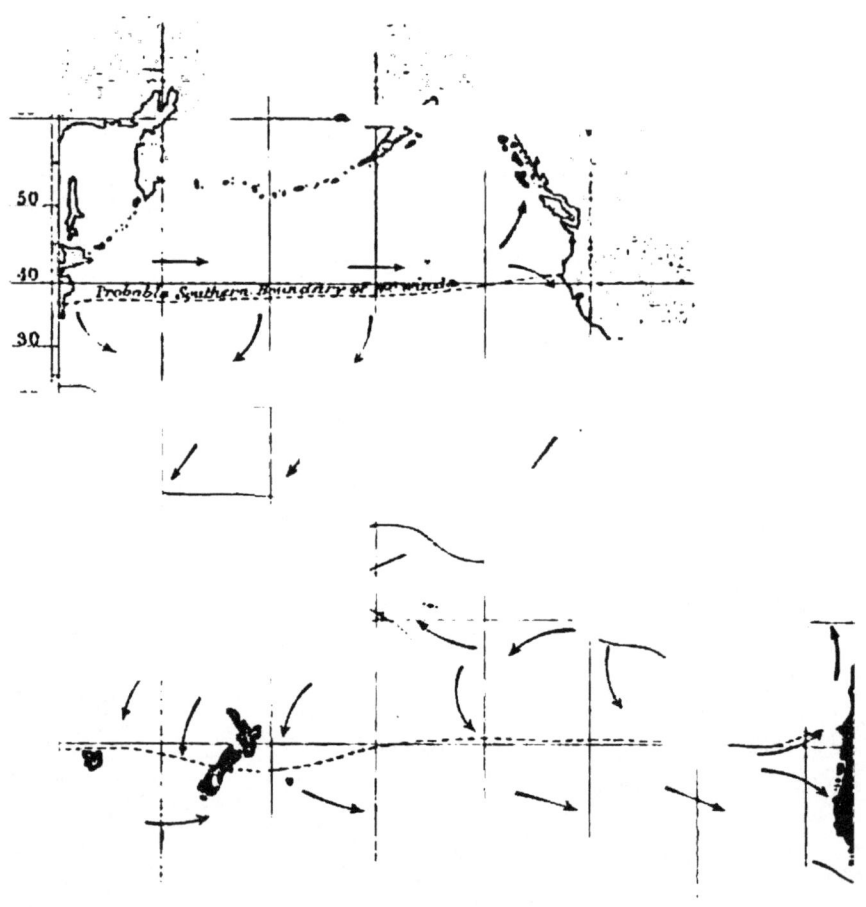

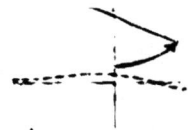

SEC. IV. WIND. 75

PACIFIC OCEAN.

Average limits of the Trade Winds Eastward of the Meridian of 160° W.

Months.		Jan., Feb., March.	April, May, June.	July, Aug., Sept.	Oct., Nov., Dec.
N.E. Trade	Northern Boundary	28° N.	29° N.	31° N.	28° N.
	Southern Boundary	8° N.	10° N.	12° N.	11° N.
Breadth of Variable Belt		*250 Miles	450 Miles	†180 Miles	300 Miles
S.E. Trade	Northern Boundary	4° N.	4° N.	9° N.	6° N.
	Southern Boundary	A line drawn from Juan Fernandez through Easter Ids. and the Marquesas	A line drawn from St. Felix Island to Tahiti	A line drawn from St. Felix Island to Tahiti	A line drawn from Valparaiso to Tahiti
RAINY SEASONS.		Sandwich Islands and Coast of Ecuador — South Pacific Ids.	Mexico and Central America in June	Mexico and Central America	Sandwich Ids. and Coast of Ecuador in December — South Pacific Islands in Nov. & Dec.
CYCLONES.				Mexico in Aug. & Sept.	Mexico in October

* This is near the American Coast. In Long. 160° W., but little Variables will be experienced.

† Eastward of the Meridian of 133° W., the region of the Variables falls within a triangle formed by the American Coast between Mazatlan and the Equator, and lines drawn from these points to Lat. 10° N., and Long. 130° W.

PACIFIC OCEAN.

Winds Westward of the Meridian of 160° W.

The N.E. Trade in this part of the Pacific is much affected by the extensive Archipelagos of the Marshall, Caroline, and Gilbert Islands; while to the Westward of the Mariana Islands, the S.W. Monsoon of the China Sea extends in April, May, and June to the Meridian of 136° E., and in July, August, and September to the Meridian of 146° E. The general limits of the N.E. Trade may be said to lie between the parallels of 22° N. and 28° N., and of 4° N. to 10° N., according to the season of the year. Southward of the N.E. Trade, from October to April, the winds blow fresh from the Northward and Westward, with violent gales from the S.W., and heavy rain.

In a similar manner the S.E. Trade of this part of the Pacific is affected by the numerous Islands lying between the Marquesas and the Australian coast. The S.E. Trade can be said to blow fairly across the South Pacific only in July, August, and September; and is then found between the parallels of 8° N. and 22° S. During the remaining 9 months of the year, the direction of the Trade in the neighbourhood of the Islands becomes more N.E. than S.E.

To the Southward of the Islands and towards the Australian coast, the S.E. Trade will again be found extending, between December and March, as far South as the parallel of 28°.

Between the months of November and March, this part of the S. Pacific may be said to be invaded by the N.W. or middle Monsoon of the Indian Ocean, blowing with less certainty than in that ocean, but constantly experienced as far East as the Marquesas and as far South as the 20° parallel; this wind blows with diminished force, and less prevalence as it approaches those Islands. This is the bad and wet season of Polynesia; and even where the S.E. Trades prevail, they are often found to be light and uncertain. Hurricanes have been experienced in January, February, and March. *

---

* For more detailed information on the winds, weather, and currents on the Australian station, reference must be made to the wind and current charts for the Pacific, Atlantic, and Indian Oceans, published by the Hydrographic Office.

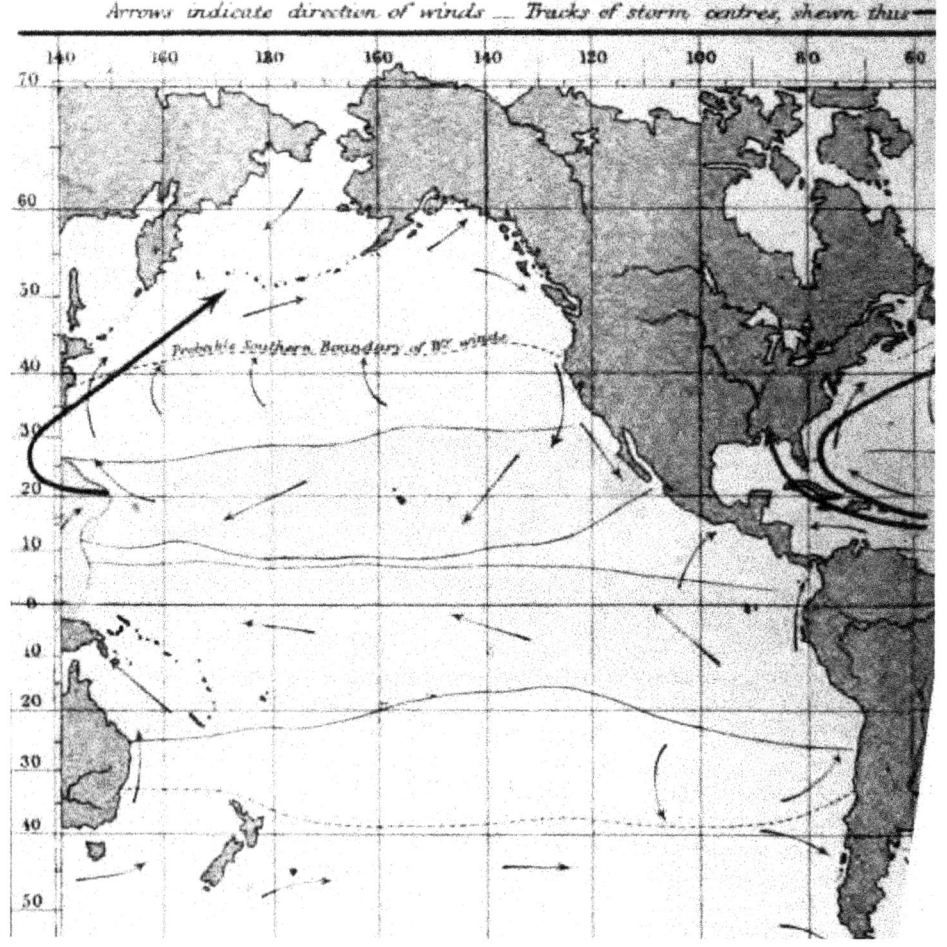

SEC. IV. WIND. 77

Winds in the Indian Ocean.

ARABIAN SEA.	BAY OF BENGAL.	CHINA SEA.
November to March. N.E. MONSOON. *Moderate and Fine.*	November to March. N.E. MONSOON. *Moderate and Fine.*	October to April. N.E. MONSOON. *Blows fresh in Nov. Dec. and Jan.*
May to September. S.W. MONSOON. Blowing fiercely, with bad weather in June and July, moderating in August. Cyclones in April and May.	May to September. S.W. MONSOON. Blowing fresh with bad weather in June and July, moderating in August. October and December.	May to September. S.W. MONSOON. Moderate with rain, strongest in June, July and August. Typhoons from July to Nov.

East Coast of Africa and the Mozambique Channel.

December to March. April to November.
Northerly Winds. Southerly Winds.

Between the Equator and the Parallel of 10° S.

NOVEMBER TO MARCH—N.W., or Middle Monsoon. *A light wind, with squalls, rains, and frequent calms.*
APRIL TO SEPTEMBER. — S. E. TRADE.

Between the Parallels of 10° and 25° or 30° S.*
Constant S.E. Trade.—Cyclones from December to April.

* According to the season of the year.

HURRICANES, CYCLONES, TYPHOONS.

These storms are progressive revolving gales of unusual violence, and may be generally described as great whirlwinds turning round and rolling forward at the same time. The average rate of the progressive movement of the centre or focus of the West Indian hurricanes is about 300 miles a day; of those on the Malabar Coast, Bay of Bengal, and China sea, 200 miles a day, and often less; whilst the Cyclones of the South Indian Ocean vary in their rate of progress from 200 to 50 miles in the twenty-four hours.

Within the tropics the effects of these storms are often felt 100 miles on either side of the central track, this limit expanding in the extra-tropical regions.

The Seasons in which these Storms prevail are as follows:

Hurricanes.—West Indies and American Coast. Coasts of Mexico and Lower California in North Pacific Ocean—July to October, and occasionally November. In South Pacific Ocean, between Australian Coast and Low Archipelago—December to March.

Cyclones.—Malabar Coast and Bay of Bengal—April, May, October, November. South-Indian Ocean—December to April.

Typhoons.—China Sea—July to November. Coasts of Japan—August, September, October.

The peculiar characteristic of the revolving action of these storms is, that in each hemisphere of the world the gyration invariably takes place in one direction, and that direction contrary to the apparent course of the sun; so that in north latitudes these storms revolve from right to left, and in south latitudes from left to right. The knowledge of this law is the more especially important, as it not only supplies the seaman with direct means of distinguishing them from common gales, but it reveals to him the actual position of the centre or vortex with respect to the place of his vessel, and therefore points out with certainty the way to escape from them.

Sec. IV.   WIND.   79

In all cases within the tropics, they commence to the eastward.
For some days they travel along a path not exactly west, but inclining
a point or two towards the pole of that hemisphere which they are
crossing; and, as they advance, they seem to be more inclined to
curve away from the equator. When they reach the 25th degree of
latitude, they generally curve still more until they move to the N.E.,
in the northern hemisphere, and to the S.E. in the southern hemisphere. Occasionally they are found to cross the line of the shore,
and to sweep over the land that opposes their progress, as appears
to be generally the case in the East Indies and China Sea; but
but by far the greater number seem to be repelled by the land, so as
to be deflected back to the N.E. in the Northern hemisphere and to
the S.E. on the other side of the equator. The Atlantic and Japan
storms, for instance, almost always wheel round to the northward,
and follow the sea-board of North America or Japan.

Another remarkable feature of these storms, is their increasing
violence in the neighbourhood of their centre or vortex; and as
this is approached (unless on the direct line of its own progressive
motion), the more rapid become the changes of the wind, which at
length, instead of veering point by point, as on entering the storm-
now flies round at once to the opposite point.

With that threatening aspect of the sky which generally foretells
bad weather, every seaman is acquainted; in addition to this a continued and troubled agitation of the sea often precedes these revolving
storms, and always shows that they are at no great distance.* The
best and surest of all warnings will, however, be found in the barometer; which, within the tropics varies so slightly under ordinary
circumstances that any fall greater than ·35 inches is a sure sign of
an atmospheric disturbance; but in more temperate latitudes where
the barometer varies considerably with no apparent atmospheric
change, the indications are less certain.† If these combined prognos-
tics should occur within the limits of *those regions* which have been

---

* The Chinese boatmen of Hongkong frequently predict the approach of a
Typhoon 24 hours before its commencement: they are seldom in error.

† The Aneroid is of great value at such a time, especially at night, for it can be
registered with great facility, and being portable, may be watched constantly,

pointed out, let the seaman immediately consider the possibility at least of his being about to encounter a storm of the revolving type.

The first care should be to discover the position of the storm with respect to the vessel. This is readily determined, by noting the direction of the wind at the commencement of the storm, as the bearing of the centre generally lies eight points of the compass from the direction of the wind. The rule is therefore: face the wind and take the eighth point to the RIGHT, that will be about the bearing to the centre of the storm if in north latitude; or if in south latitude, the eighth point to the LEFT. For example: suppose the vessel to be in 14° N. latitude, the wind from the North, and the barometer and sky indicating a coming gale,—then, look at the compass, take the eighth point to the right of North, and East will be the bearing of the vortex. Or, if in 14° S. latitude, with the wind South, take eight points to the left, and East will be the direction of the centre.

Knowing the bearing of the centre, it is next required to discover on which side of the storm's path the vessel lies. In the Northern Hemisphere if the ship is on the right-hand side of the path (looking in the direction towards which the storm is advancing) the wind veers N., N.E., E., S.E., or with the hands of a watch, whilst if she is on the left-hand side of the path, the wind veers N., N.W., W., S.W., or against the hands of a watch. And in like manner in the Southern Hemisphere, if the ship is on the right-hand side of the path (looking in the direction towards which the storm is advancing), the wind veers S., S.W., W., N.W., or with the hands of a watch, whilst if she is on the left-hand side of the path, the wind veers S., S.E., E., N.E., or against the hands of a watch.

In using the above rules it is essential that the ship's position be as nearly as possible stationary; for if a ship advances somewhat faster than, and in the same direction as, an approaching storm, having the wind also in the same direction (ship running), the wind may shift in a direction exactly opposite to that which would have

---

when the marine barometer may not be accessible; moreover, its variations occur simultaneously with their causes, showing minute changes, unobservable in the best constructed marine barometers, owing to the pumping of the quicksilver, when the motion of the ship is violent.

been observed had the vessel been stationary, and an erroneous conclusion as to her position in the storm would be drawn. The advisibility of heaving-to on the first approach of a hurricane, cannot therefore be too strongly urged.

Although a cyclone may be made use of by the experienced seaman when making a passage, and a good run frequently ensured; and, where sea room will admit, a vessel may be able to run out of the direct influence of the coming storm; yet, as a general rule, it is advisable to HEAVE THE VESSEL TO ON THE TACK ON WHICH SHE WOULD COME UP AS THE WIND SHIFTS; therefore, when the ship is in the right-hand semicircle, *i.e.*, on the right-hand side of the storm's path, she should be hove-to on the *starboard* tack; and if in the left-hand semicircle, on the *port* tack.

If, with the ship hove-to, the wind continues steady in the same direction, increasing in violence with a rapidly falling barometer, it may be presumed that the ship lies in the direct path of the storm's centre, and it will be necessary to run before the wind.

If the vessel is unable, from want of sea room, to run, her position becomes one of great danger, and every precaution should be taken to prepare for the passage of the centre over the ship; for the steady wind will be succeeded by a short interval of calm, to be followed by a sudden and most violent burst of wind from the opposite point of the compass to that first experienced. After the calm the barometer will commence slowly to rise.

Caution.—In the practical application of the theory of revolving storms, it must be borne in mind that, although the region and season of the year would render the seaman extremely cautious, yet every strong wind or gale met with, especially in extra tropical regions, must not be treated as cyclonic. For example: the master of a vessel running up the Bay of Bengal in the month of May, with a S.W. gale, heavy rain, and a falling barometer, must not at once assume that he is in a cyclone, the centre of which bears N.W., and therefore that a North or N.E. course will carry his vessel clear of danger.

Similarly, in the neighbourhood of Mauritius, in the months of January or February, with the wind from S.E., steady in the direction, but increasing, accompanied with squalls, rain, and falling barometer—to assume that the vessel is in a hurricane, the centre of which bears N.E., and to run to the N.W., would be dangerous.*

The barometer must, in these cases, be closely watched, and a decided fall of at least ·3 inches from the normal average should be experienced before the conclusion is arrived at, that the vessel is within the influence of the cyclone proper.

In either of the above cases the safe proceeding is to *heave to*, and carefully watch for the shifting of the wind, and the changes of the barometer; then, if the veering of the former, and the marked *fall* of the latter prove the gale to be rotating, or cyclonic, the position of the centre and probable direction of its path, could be determined with some degree of precision.

The following Tables, taken from Professor Dove's "Law of Storms," gives in detail how a ship should be handled under any circumstances in these storms.

---

* Such a proceeding, in fact, proved fatal to several vessels which, under these *circumstances*, weighed from Reunion and stood to the N.W. in February, 1860. See "*Notes on Cyclones in the Southern Ocean*," by C. MELDRUM, M.A., F.R.A.S.

DOVE'S TABLES.

Northern Hemisphere.

Direction of Wind at beginning of Storm.	Bearing of Centre from Ship.	Shift of Wind.			Corresponding Course to be held.		Shift of Wind.			Course to be adopted.
1 N.W.	N.E.	N.W. towards		W.	S.E.		N.W. towards		N.	
2 N.W. byN.	N.E.byE.	N.W.byN.	,,	W.	S.E.byS.		N.W.byN.	,,	N.	
3 N.N.W.	E.N.E.	N.N.W.	,,	W.	S.S.E.		N.N.W	,,	N.	
4 N.byW.	E.byN.	N.byW.	,,	W.	S.byE.		N.byW.	,,	N.	
5 N.	E.	N.	,,	W.	S.		N.	,,	E.	
6 N.byE.	E.byS.	N.byE.	,,	N.	S.byW.		N.byE.	,,	E.	
7 N.N.E.	E.S.E.	N.N.E.	,,	N.	S.S.W.		N.N.E.	,,	E.	
8 N.E.byN.	S.E.byE	N.E.byN.	,,	N.	S.W. by S.	Or else heave the Ship to, on the port tack.	N.E.byN.	,,	E.	Ship to be hove to, on the starboard tack.
9 N.E.	S.E.	N.E.	,,	N.	S.W.		N.E.	,,	E.	
10 N.E.byE.	S.E.byS.	N.E.byE.	,,	N.	S.W.byW		N.E.byE.	,,	E.	
11 E.N.E.	S.S.E.	E.N.E.	,,	N.	W.S.W.		E.N.E.	,,	E.	
12 E.byN.	S.byE.	E.byN.	,,	N.	W.byS.		E.byN.	,,	E.	
13 E.	S.	E.	,,	N.	W.		E.	,,	S.	
14 E.byS.	S.byW.	E.byS.	,,	E.	W.byN.		E.byS.	,,	S.	
15 E.S.E.	S.S.W.	E.S.E.	,,	E.	W.N.W.		E.S.E.	,,	S.	
16 S.E.byE.	S.W byS.	S.E.byE.	,,	E.	N.W. byW.		S.E.byE.	,,	S.	
17 S.E.	S.W.	S.E	,,	E.	N.W.		S.E.	,,	S.	
18 S.E.byS.	S.W.byW.	S.E.byS.	,,	E.	N.W. byN.		S.E.byS.	,,	S.	
19 S.S.E.	W.S.W.	S.S.E.	,,	E.	N.N.W.		S.S.E.	,,	S.	
20 S.byE.	W.byS.	S.byE.	,,	E.	N.byW.		S.byE.	,,	S.	
21 S.	W.	S.	,,	E.	N.		S.	,,	W.	
22 S.byW.	W.byN.	S.byW.	,,	S.	N.byE.		S.byW.	,,	W.	
23 S.S.W.	W.N.W.	S.S.W.	,,	S.	N.N.E.		S.S.W.	,,	W.	
24 S.W.byS.	N.W. byW.	S.W.byS.	,,	S.	N.E.byN.		S.W.byS.	,,	W.	
25 S.W.	N.W.	S.W.	,,	S.	N.E.		S.W.	,,	W.	

DOVE'S TABLES.

Southern Hemisphere.

	Direction of Wind at beginning of Storm.	Bearing of Centre from Ship.	Shift of Wind.		Corresponding Course to be held.		Shift of Wind.		Course to be adopted.
1	S.	E.	S. towards W.		N.		S. towards E.		
2	S.byE.	E.byN.	S.byE.	,,	S.	N.byW.	S.byE.	,,	E.
3	S.S.E.	E.N.E	S.S.E.	,,	S.	N.N.W.	S.S.E.	,,	E.
4	S.E.byS.	N.E.byE.	S.E.byS.	,,	S.	N.W.byN.	S.E.byS.	,,	E.
5	S.E.	N.E.	S.E.	,,	S.	N.W.	S.E.	,,	E.
6	S.E.byE.	N.E.byN.	S.E.byE.	,,	S.	N.W.byW.	S.E.byE.	,,	E.
7	E.S.E.	N.N.E.	E.S.E.	,,	S.	W.N.W.	E.S.E.	,,	E.
8	E.byS.	N.byE.	E.byS.	,,	S.	W.byN.	E.byS.	,,	E.
9	E.	N.	E.	,,		W.	E.	,,	N.
10	E.byN.	N.byW.	E.byN.	,,	E.	W.byS.	E.byN.	,,	N.
11	E.N.E.	N.N.W.	E.N.E.	,,	E.	W.S.W.	E.N.E.	,,	N.
12	N.E.byE.	N.W.byN.	N.E.byE.	,,	E.	S.W.byW.	N.E.byE.	,,	N.
13	N.E.	N.W.	N.E.	,,	E.	S.W.	N.E.	,,	N.
14	N.E.byN.	N.W.byW.	N.E.byN.	,,	E.	S.W.byS.	N.E.byN.	,,	N.
15	N.N.E.	W.N.W.	N.N.E.	,,	E.	S.S.W.	N.N.E.	,,	N.
16	N.byE.	W.byN.	N.byE.	,,	E.	S.byW.	N.byE.	,,	N.
17	N.	W.	N.	,,		S.	N.	,,	W.
18	N.byW.	W.byS.	N.byW.	,,	N.	S.byE.	N.byW.	,,	W.
19	N.N.W.	W.S.W.	N.N.W.	,,	N.	S.S.E.	N.N.W.	,,	W.
20	N.W.byN.	S.W.byW.	N.W.byN.	,,	N.	S.E.byS.	N.W.byN.	,,	W.
21	N.W.	S.W.	N.W.	,,	N.	S.E.	N.W.	,,	W.

{ Ship to, on the starboard tack. Or else heave the Ship to, on the port tack. Ship to be hove to. }

WEATHER.
Beaufort Notation.

b	Blue sky.
c	Clouds, (detached.)
d	Drizzling rain.
f	Foggy.
g	Gloomy.
h	Hail.
l	Lightning.
m	Misty, (hazy.)
o	Overcast.
p	Passing showers.
q	Squally
r	Rain.
s	Snow
t	Thunder.
u	Ugly, (threatening.)
v	Visibility, { (objects at a distance.) (unusually visible.) }
w	Wet, (dew.)

A bar (—) under any letter augments its signification; thus: <u>f</u> very foggy, <u>r</u> heavy rain, <u>r</u> heavy and continuing rain.

GENERAL REMARKS.
Colour of Sky.

A few of the more marked signs of weather—useful alike to seaman, farmer, and gardener, are the following:—

Whether clear or cloudy—a rosy sky at sunset presages fine weather; a sickly, greenish hue—wind and rain; tawny, or coppery clouds—wind; a dark (or Indian) red—rain; a red sky in the morning—bad weather, or much wind (perhaps also rain); a grey sky in the morning—fine weather; a high dawn—wind; a low dawn—fair weather.

Soft looking or delicate clouds foretell fine weather, with moderate or light breezes ; hard edged oily-looking clouds—wind. A dark, gloomy blue sky is windy ; but a light bright blue sky indicates fine weather. Generally, the softer clouds look, the less wind (but perhaps more rain) may be expected; and the harder, more "greasy," rolled, tufted, or ragged,—the stronger the coming wind will prove. Also, a bright yellow sky at sunset, presages wind ; a pale yellow— wet ; orange or coppered colour—wind and rain ; and thus by the prevalence of red, yellow, green, grey, or other tints, the coming weather may be foretold very nearly—indeed, if aided by instruments, almost exactly.

CLOUDS.

The scale generally adapted in this country for denoting the amount of cloud, is 0 to 10 :—0, indicating a clear sky ; 5, a sky half covered ; and 10, the sky wholly obscured.

Light, delicate, quiet tints or colours, with soft indefinite forms of clouds, indicate and accompany fine weather ; but gaudy or unusual hues, with hard definitely outlined clouds, foretell rain, and probably strong wind.

Small inky-looking clouds foretell rain ; light scud clouds driving across heavy masses show wind and rain ; but if alone, may indicate wind only—proportionate to their motion.

High *upper* clouds crossing the sun, moon, or stars, in a direction different from that of the lower clouds, or the wind then felt below, foretell a change of wind toward *their* direction.

After fine clear weather, the first signs in a sky, of a coming change, are usually light streaks, curls, wisps, or mottled patches of white distant cloud, which increase and are followed by an overcasting of murky vapour that grows into cloudiness. This appearance, more or less oily or watery, as wind or rain will prevail, is an infallible sign.

Usually the higher and more distant such clouds seem to be, the more gradual, but general, the coming change of weather will prove.

Misty clouds forming, or hanging on heights, show wind and rain coming—if they remain, increase or descend. If they rise or disperse, the weather will improve, or become fine.

Description of Clouds.

Cirrus cloud consists of streaks, wisps and fibres, vulgarly called "mare's tails," which may increase in any or all directions. Of all clouds it has the least density, the greatest elevation, and the greatest variety of extent and direction, or figure.

It remains for a short time when formed in the lower parts of the atmosphere and near other clouds, and longest when alone in the sky, and at a great height. When streaks of cirrus run quite across the sky in the direction in which a light wind happens to blow, the wind will probably soon blow hard but remain steady. When the fine threads of the cirrus appear blown or brushed backward at one end, as if by a wind prevailing in these elevated regions, the wind on the surface will ultimately veer round to that point.

Cumulus, a cloud in dense convex heaps in rounded forms definitely terminated above; the lower surface remains roughly horizontal.

When of moderate height and size, of well-defined curved outline, and appear only during the heat of the day, they indicate a continuance of fair weather. But when they increase with great rapidity, sink down to the lower parts of the atmosphere, and do not disappear towards evening, *rain* may be expected.

Stratus is a continuous extended sheet of cloud, increasing from below upwards. It is the lowest sort of cloud. It generally forms about sunset, grows denser during the night, and disappears about sunrise.

Cirro-Cumulus is composed of well-defined, small, rounded masses, lying near each other, and quite separated by intervals of sky. It is commonly known as a *"mackerel sky,"* it occurs frequently in summer, and is attendant on warm and dry weather.

Cirro-Stratus. This cloud partakes partly of the characteristics of the Cirrus and Stratus. In distinguishing it, attention must be paid, not so much to the form, which is very variable, but to the structure, which is dense in the middle and thin towards the edges. It is a precursor of storms; and from its greater or less abundance and permanence, it gives some indication of the time when the storm may be expected.

Cumulo-Stratus. This cloud is formed by the Cirro-Stratus blending with the Cumulus, either among its piled up heaps, or spreading underneath its base as a horizontal layer of vapour.

Cumulo-Cirro-Stratus or Nimbus. This is the *rain cloud*. At a considerable height a sheet of Cirro-Stratus cloud is spread out, under which Cumulus clouds drift from windward; these rapidly increasing unite at all points, forming one continuous mass, from which rain falls.

When a rain cloud is seen approaching at a distance, Cirri appear to shoot out from its top in all directions, and it has been observed that the more copious the rainfall, the greater is the number of Cirri thrown out from the cloud.

"A rainbow in the morning—
Sailors, take warning;
A rainbow at night
Is the sailor's delight."

Morning rainbows are always seen in the west, and indicate the advance of rain cloud from that quarter when it is clear in the east; and the fall of rain at the time of day when the temperature should be rising, is regarded as a prognostic of a change to wet, stormy weather.

On the contrary, the conditions under which a rainbow can appear in the evening are: the passing of the rain cloud to the east, and a clearing-up in the west at the time of day when the temperature has begun to fall, thus indicating a change from wet to dry weather.

"The evening grey and the morning red,
Put on your hat, or you 'll wet your head."

This does not refer to a *high* red dawn, which may be regarded as a prognostic of settled weather. But if the clouds be red and lowering later in the morning, it may be accepted as a sign of rain.

SQUALLS.

Generally, squalls are preceded, or accompanied, or followed by clouds: but the very dangerous "white squall" (of the West Indies and other regions) is indicated by a rushing sound, and by white wave crests.

A squall cloud that can be seen through or under is not likely to bring, or be accompanied by so much wind as a dark continued cloud extending beyond the horizon.

The rapid or slow rise of a squall cloud—its more or less disturbed look,—that is, whether its body is much agitated, and changing form continually, with broken clouds, or scud, flying about—or whether the mass of cloud is shapeless or nearly quiet, though floating onwards across the sky—foretells more or less wind accordingly.

In some of the saws about wind and weather, there is so much truth, that, though trite and simple, their insertion here can do no harm.

Adverting to the barometer :—

When rise begins, after low,
Squalls expect and clear blow.

Or :—

First rise, after low,
Indicates a stronger blow.

Also :—

Long foretold, long last :
Short notice, soon past.

To which may be added :—In Squalls—

When rain comes before wind,
Halyards, sheets, and braces mind.

And :—

When wind comes before rain,
Soon you may make sail again.

Also, generally speaking :—

> When the glass falls low,
> Prepare for a blow ;
> When it rises high,
> Let all your kites fly.

THE BAROMETER.

The barometer, feeling the pressure of the air, shows at once when that pressure is changing. If the pressure at one place on the earth's surface be greater than at another, the air has a tendency to move from the place where the pressure is greater towards that where it is less, and thus *wind* is caused.

A change of weather comes almost always with a change of wind, and the extent of this change of weather depends on the fact of the new wind being warmer or colder, damper or drier, than that which has been blowing. Any conclusions drawn from its movements must be checked by observations of temperature, moisture of the air, present direction and force of wind, and state of the sky, before any correct opinion can be formed as to what may be expected. In general, whenever the level of the mercury continues steady, settled weather may be expected ; but when it is unsteady a change must be looked for, and perhaps a gale.

A sudden rise of the barometer is very nearly as bad a sign as a sudden fall, because it shows that atmospherical equilibrium is unsteady. In an ordinary gale the wind often blows hardest when the barometer is just beginning to rise, directly after having been very low.

Besides these rules for the instruments, there is a rule about the way in which the wind changes, which is very important. It is well-known to every sailor, and is contained in the following couplet :—

> When the wind shifts against the sun,
> Trust it not, for back it will run.

The wind usually shifts *with the sun*, *i.e.*, from left to right, in the *Northern Hemisphere*. A change in this direction is called *veering*.

Thus an East wind shifts to West through South-East, South, South-West; and a West wind shifts to East through North-West, North, and North-East.

If the wind shifts the opposite way, viz., from West to South-West, South, and South-East, the change is called *backing*, and it seldom occurs unless when the weather is unsettled.

However, slight changes of wind do not follow this rule exactly; for instance, the wind often shifts from South-West to South and back again.

In the *Southern Hemisphere* the motion *with the sun* is, of course, from right to left, and therefore the above rules will necessarily be reversed.

Admiral FITZROY proposed the following words for barometer scales:—

RISE.	FALL.
For	For
North	South
N.W.—N.—E.	S.E.—S.—W.
Dry	Wet
or	or
less	more
Wind	Wind.
Except	Except
Wet from	Wet from
North.	North.

Buys-Ballot's Law.

THE law connecting barometric pressure with the direction of wind which has been proposed by Professor BUYS-BALLOT may be stated thus:—If there be a difference between the barometric readings at any two stations, the wind will blow at *right angles to the* line joining these two places, and the observer standing with his back to the wind will have the place where the reading is lowest on the left-hand side in the Northern Hemisphere, but on the right-hand side in the Southern Hemisphere.

Reading the Barometer.

THE lower edge of the vernier should be brought exactly on a level with the top of the mercurial column. Great care should be taken to acquire the habit of reading with the eye exactly on a level with the top of the mercury, that is, with the line of sight at right angles to the scale. A piece of white paper placed behind the slit in the tube, so as to reflect the light, assists in setting the vernier accurately.

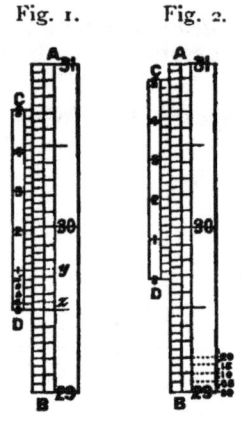

Fig. 1.   Fig. 2.

The mode of reading off may be learned from a study of the following diagrams, in which A B represents part of the scale, and C D the vernier, the lower edge D denoting the top of the mercurial column. The scale is readily understood; B is 29·000 inches; the second line 29·100, and so on. The first thing is to note the scale line just below D, and the next is to find out the line of the vernier which is in one and the same direction with a line of the scale. In figure (1), the lower edge of the vernier, D, is represented in exact coincidence with scale line 29·5; the barometer therefore reads 29·500 inches. Studying it attentively in this position it will be perceived that the vernier line a is ·002 inch below the next line of the scale.

If, therefore, the vernier be moved so as to place a in a line with z, the edge D would read 29·502. In like manner it is seen that b is ·004 inch away from the line next above it *on the scale* ; c, ·006 inch apart from that next above it ; d ·008 from that next above it ; and 1, on the vernier, is 010 below y. Hence, if 1 be moved into line with y, D. would read 29·510· Thus the numbers 1, 2, 3, 4, 5, on the vernier, indicate hundredths, and the intermediate lines the even *thousandths of* an inch. Referring now to figure (2), the scale line just below D is 29·650. Looking carefully up the vernier, the third line above the figure 3 is seen to lie evenly with a line on the scale. The number 3 indicates ·030, the third subdivision ·006: and thus we get—

Reading on scale	... ...	29·650
Reading on vernier	... ...	·030
		·006
Actual reading	... ...	29·686 inches.

Sometimes two pairs of lines will appear to be coincident; in which case the intermediate thousandth of an inch should be set down as the reading. Thus, suppose the reading appears to be 29·684 or 29·686, the mean 29·685 should be adopted.

No reading from a barometer which is not hanging truly vertically should ever be recorded.

THE ANEROID.

In the aneroid, atmospherical pressure is measured by its effect in altering the shape of a small, hermetically sealed, metallic box, from which almost all the air has been withdrawn, and which is kept from collapsing by a spring.

When atmospherical pressure rises above the amount which was recorded when the instrument was made, the top is forced inwards, and *vice versa*; when pressure falls below that amount the top is forced outwards by the spring.

These motions are transferred by a system of levers and springs to a hand, which moves on a dial like that of a wheel barometer.

This instrument is very sensitive, showing minute changes that are concealed by the pumping of the quicksilver even in the best constructed marine barometers, when the motion of the ship is violent. The Aneroid can be registered with great facility, and, being portable, may be watched constantly when the marine barometer may not be accessible; nevertheless, it should, when opportunity offers, be compared with *a good* mercurial barometer.

The average range of the barometer in the higher latitudes (60°—50°) is about 1·5 inches, but on extraordinary occasions, ranges of 2·75 and 3 inches have been recorded.

In the intertropical regions the range varies from 0·4 to 0·2 inches, and in the neighbourhood of the Equator it seldom exceeds 0·15 inches, this small change being due in great part to a regular diurnal variation. The average movement of the barometer within the tropics, being thus confined within small limits, any interruption to the law may be deemed a warning of the approach of bad weather.

The fall of the barometer in hurricanes ranges from 1·0 to 2·0, and even 2·5 inches, the rapidity of the fall and the depression of the mercury increases as the centre of the storm approaches.

The following Tables of Mean Barometric Pressures in the North and South Atlantic Oceans are reduced to 32° Fahrenheit, and must be corrected by the annexed table.

Temp. of air.	Correction.
35°	+0·018
40	0·031
50	0·058
60	0·085
70	0·111
80	0·138
90	0·164

For example: in the Atlantic Ocean on the Equator, with the barometric table shewing 29·90, and the temperature 80° Fahrenheit, the navigator's barometer would be 30·038.*

---

* For the lines of equal barometric pressure for the world, see the Wind Charts published by the Hydrographic Office, 1872.

NORTH ATLANTIC OCEAN.
Mean Barometric Pressure reduced to 32° Fahrenheit.

	Latitude.	Jan. In.	Feb. In.	March. In.	April. In.	May. In.	June. In.
For Eastern half or to 40th Meridian.	50° to 45°	30·15	30·12	29·90	29·96	29·94	30·00
	45 - 40	30·16	30·11	30·10	30·01	30·02	30·05
	40 - 35	30·18	30·07	30·12	30·06	30·07	30·20
	35 - 30	30·22	30·14	30·10	30·16	30·12	30·20
	30 - 25	30·17	30·23	30·13	30·17	30·22	30·23
	25 - 20	30·10	30·11	30·10	30·10	30·14	30·16
	20 - 15	30·01	30·06	30·04	30·00	30·06	30·03
	15 - 10	29·95	29·97	29·96	29·97	29·99	29·96
	10 - 5	29·90	29·94	29·91	29·92	29·94	29·94
	5 - 0	29·88	29·91	29·89	29·90	29·92	29·93

	Latitude.	July. In.	August. In.	Sept. In.	Oct. In.	Nov. In.	Dec. In.
For Eastern half or to 40th Meridian.	50° to 45°	30·03	30·03	30·00	29·93	30·07	30·06
	45 - 40	33·17	30·10	30·04	30·01	29·98	30·03
	40 - 35	30·19	30·16	30·11	30·09	30·02	30·06
	35 - 30	30·24	30·19	30·15	30·18	30·03	30·18
	30 - 25	30·19	30·16	30·10	30·14	30·06	30·18
	25 - 20	30·09	30·07	30·32	30·07	30·01	30·06
	20 - 15	30·00	30·00	29·99	29·99	29·98	30·01
	15 - 10	29·97	29·93	29·93	29·95	29·96	29·96
	10 - 5	29·98	29·97	29·95	29·93	29·94	29·91
	5 - 0	29·98	29·97	29·98	29·95	29·92	29·91

In the Northern Hemisphere the effect of the veering of the wind on the barometer is according to the following law :—

With East, S.E. South Winds...Barometer falls.
,,   S.W.   ,,   ceases to fall, begins to rise.
,,   West, N.W. North ,,   ,,   rises.
       N.E.   ,,   ceases to rise, begins to fall.

SOUTH ATLANTIC OCEAN.

Mean Barometric Pressure, reduced to 32° Fahrenheit.

Latitude	Jan. In.	Feb. In.	March In.	April In.	May In.	June In.
0° to 5°	29·89	29·91	29·90	29·92	29·94	29·94
5 - 10	29·95	29·94	29·94	29·94	30·04	30·05
10 - 15	29·97	29·98	29·96	29·99	30·04	30·05
15 - 20	30·02	30·01	30·01	30·03	30·09	30·09
20 - 25	30·06	30·05	30·07	30·05	30·06	30·14
25 - 30	30·07	30·05	30·06	30·03	30·14	30·09
30 - 35	30·05	30·05	30.04	30·03	30·10	30·04
35 - 40	29·98	30·04	30·02	29·98	29·90	29·90
40 - 45	29·92	29·95	29·99	29·95	29·88	29·89
45 - 50	29·71	29·78	29·77	29·76	29·72	29·65
50 - 55	29·41	29·44	29·49	29·44	29·41	29·48
55 - 60	29·25	29·23	29·25	29·20	29·26	29·28

Latitude	July In.	August In.	Sept. In.	Oct. In.	Nov. In.	Dec. In.
0° to 5°	29·99	30·00	30·01	29·96	29·94	29·93
5 - 10	30·02	30·03	30·03	30·02	29·99	29·96
10 - 15	30·05	30·06	30·05	30·07	30·04	30·00
15 - 20	30·09	30·13	30·11	30·10	30·05	30·05
20 - 25	30·11	30·16	30·17	30·18	30·08	30·08
25 - 30	30·13	30·18	30·13	30·16	30·08	30·08
30 - 35	30·12	30·10	30·10	30·08	30·09	30·00
35 - 40	30·04	29·94	29·94	30·05	30·05	29·97
40 - 45	29·95	29·93	29·96	30·02	29·94	29·95
45 - 50	29·82	29·83	29·87	29·77	29·70	29·67
50 - 55	29·53	29·56	29·57	29·48	29·31	29·43
55 - 60	29·25	29·28	29·29	29·10	29·11	29·21

In the Southern Hemisphere, the effect of the veering of the wind on the Barometer is according to the following law :—

With East, N.E. North Winds...Barometer falls.
,,   N.W.   ,,   ,,   ceases to fall, begins to rise.
,,   West, S.W. South ,,   ,,   rises.
       S.E.   ,,   ceases to rise, begins to fall.

SEC. IV. BAROMETER. 97

BAROMETER.

Table for converting Millimetres into English Inches.

Mill.	Inches.	Mill.	Inches.	Mill.	Inches.	Mill.	Inches.
712	28·032	731	28·780	750	29·528	769	30·276
3	·071	2	·819	1	·568	770	·316
4	·111	3	·859	2	·607	1	·355
5	·150	4	·898	3	·646	2	·394
6	·190	5	·938	4	·686	3	·434
7	·229	6	28·977	5	·725	4	·473
8	·268	7	29·016	6	·764	5	·512
9	·308	8	·056	7	·804	6	·552
720	28·347	9	·095	8	·843	7	·591
1	·386	740	29·134	9	·882	8	·631
2	·426	1	·174	760	·922	9	·670
3	·465	2	·213	1	29·961	780	30·709
4	·505	3	·253	2	30·001	1	·749
5	·544	4	·292	3	·040	2	·788
6	·583	5	·331	4	·079	3	·827
7	·623	6	·371	5	·119	4	·867
8	·662	7	·410	6	·158	5	·906
9	·701	8	·449	7	·197	6	·945
730	28·741	749	29·489	768	30·237	787	30·985

H

BAROMETER.

Table for converting English Inches into Millimeters.

Ins.	·00	·01	·02	·03	·04	·05	·06	·07	·08	·09	Ins.
	mm.	mm.	mm.	mm.	mm.	mm.	mm.	mm.	mm.	mm	
28·0	711·2	711·4	711·7	712·0	712·2	712·5	712·7	713·0	713·2	713·5	28·0
·1	13·7	14·0	14·2	14·5	14·7	15·0	15·3	15·5	15·8	16·0	·1
·2	16·3	16·5	16·8	17·0	17·3	17·5	17·8	18·0	18·3	18·6	·2
·3	18·8	19·1	19·3	19·6	19·8	20·1	20·3	20·6	20·8	21·1	·3
·4	21·3	21·6	21·9	22·1	22·4	22·6	22·9	23·1	23·4	23·6	·4
·5	23·9	24·1	24·4	24·7	24·9	25·2	25·4	25·7	25·9	26·2	·5
·6	26·4	26·7	26·9	27·2	27·4	27·7	28·0	28·2	28·5	28·7	·6
·7	29·0	29·2	29·5	29·7	30·0	30·2	30·5	30·7	31·0	31·3	·7
·8	31·5	31·8	32·0	32·3	32·5	32·8	33·0	33·3	33·5	33·8	·8
·9	34·0	34·3	34·6	34·8	35·1	35·3	35·6	35·8	36·1	36·3	·9
29·0	736·6	736·8	737·1	737·4	737·6	737·9	738·1	738·4	738·6	738·9	29·0
·1	39·1	39·4	39·6	39·9	40·1	40·4	40·7	40·9	41·2	41·4	·1
·2	41·7	41·9	42·2	42·4	42·7	42·9	43·2	43·4	43·7	44·0	·2
·3	44·2	44·5	44·7	45·0	45·2	45·5	45·7	46·0	46·2	46·5	·3
·4	46·7	47·0	47·3	47·5	47·8	48·0	48·3	48·5	48·8	49·0	·4
·5	49·3	49·5	49·8	50·1	50·3	50·6	50·8	51·1	51·3	51·6	·5
·6	51·8	52·1	52·3	52·6	52·8	53·1	53·4	53·6	53·9	54·1	·6
·7	54·4	54·6	54·9	55·1	55·4	55·6	55·9	56·1	56·4	56·7	·7
·8	56·9	57·2	57·4	57·7	57·9	58·2	58·4	58·7	58·9	59·2	·8
·9	59·4	59·7	60·0	60·2	60·5	60·7	61·0	61·2	61·5	61·7	·9
30·0	762·0	762·2	762·5	762·8	763·0	763·3	763·5	763·8	764·0	764·3	30·0
·1	64·5	64·8	65·0	65·3	65·5	65·8	66·1	66·3	66·6	66·8	·1
·2	67·1	67·3	67·6	67·8	68·1	68·3	68·6	68·8	69·1	69·4	·2
·3	69·6	69·9	70·1	70·4	70·6	70·9	71·1	71·4	71·6	71·9	·3
·4	72·1	72·4	72·7	72·9	73·2	73·4	73·7	73·9	74·2	74·4	·4
·5	74·7	74·9	75·2	75·5	75·7	76·0	76·2	76·5	76·7	77·0	·5
·6	77·2	77·5	77·7	78·0	78·2	78·5	78·8	79·0	79·3	79·5	·6
·7	79·8	80·0	80·3	80·5	80·8	81·0	81·3	81·5	81·8	82·1	·7
·8	82·3	82·6	82·8	83·1	83·3	83·6	83·8	84·1	84·3	84·6	·8
30·9	784·8	785·1	785·4	785·6	785·9	786·1	786·4	786·6	786·9	787·1	9
	·00	·01	·02	·03	·04	·05	·06	·07	·08	·09	

THE THERMOMETER.

As the barometer shews weight or pressure of the air, so the thermometer shows heat and cold, or temperature.

The result of numerous observations shows, that in the NORTHERN HEMISPHERE :—

The thermometer rises with E., S.E., and S. winds; with a S.W. wind it ceases to rise and begins to fall; it falls with W., N.W., and N. winds; and with a N.E. wind it ceases to fall and begins to rise.

And in the SOUTHERN HEMISPHERE :—

The thermometer, rises with E., N.E., and N. winds; with a N.W. wind it ceases to rise and begins to fall; it falls with W., S.W., and S. winds; and with a S.E., wind it ceases to fall and begins to rise.

Besides the use of the thermometer, in conjunction with the barometer, in foretelling the changes of weather, the seaman, by its aid, may frequently derive information when passing from one Ocean current to another; it also may give warning of the vicinity of Ice.

Fahrenheit's thermometer is generally used in this country, and Reaumur's and the Centigrade abroad.

The different scales are easily convertible by the following rules :—*

Centigrade to *Fahrenheit,* Multiply by 9, and divide by 5.

$$\frac{9}{5} C. + 32° = F., \text{ or } \overset{\text{Cent.}}{100°} \times 9 = 900 \div 5 = 180 + 32 = 212° \text{ Faht.}$$

Réaumur to *Fahrenheit,* multiply by 9, and divide by 4.

$$\frac{9}{4} R. + 32° = F., \text{ or } \overset{\text{Reaum.}}{80°} \times 9 = 720 \div 4 = 180 + 32 = 212° \text{ Faht.}$$

---

* When F is between zero and 32°, the values of R and C are negative, and express the required number of degrees below zero on Reaumur's and the Centigrade scale.

Fahrenheit to *Centigrade*, multiply by 5, and divide by 9.

$$(F.-32°)\tfrac{5}{9}=C., \text{ or } \overset{\text{Faht.}}{212°}-32=180\times5=900\div9=100° \text{ Cent.}$$

Fahrenheit to *Réaumur*, multiply by 4, and divide by 9.

$$(F.-32°)\tfrac{4}{9}=R., \text{ or } \overset{\text{Faht.}}{212°}-32=180\times4=720\div9=80° \text{ Réaum.}$$

Centigrade to *Réaumur*, multiply by 4, and divide by 5.

$$\tfrac{4}{5}C.=R., \text{ or } \overset{\text{Cent.}}{100°}\times4=400\div5=80° \text{ Réaum.}$$

Réaumur to *Céntigrade*, multiply by 5, and divide by 4.

$$\tfrac{5}{4}R.=C., \text{ or } \overset{\text{Reaum.}}{80°}\times5=400\div4=100° \text{ Cent.}$$

Comparison of the Fahrenheit, Centigrade, and Reaumur's Thermometers.

Fahren-heit.	Centi-grade.	Reau-mur.	Fahren-heit.	Centi-grade.	Reau-mur.	Fahren-heit.	Centi-grade.	Reau-mur.
1	−17·2	−13·8	35	1·7	1·3	69	20·6	16·4
2	16·7	13·3	36	2·2	1·8	70	21·1	16·9
3	16·1	12·9	37	2·8	2·2	71	21·7	17·3
4	15·6	12·4	38	3·3	2·7	72	22·2	17·8
5	15·0	12·0	39	3·9	3·1	73	22·8	18·2
6	14·4	11·6	40	4·4	3·6	74	23·3	18·7
7	13·9	11·1	41	+5·0	+4·0	75	23·9	19·1
8	13·3	10·7	42	5·6	4·4	76	24·4	19·6
9	12·8	10·2	43	6·1	4·9	77	25·0	20·0
10	12·2	9·8	44	6·7	5·3	78	25·6	20·4
11	−11·7	−9·3	45	7·2	5·8	79	26·1	20·9
12	11·1	8·9	46	7·8	6·2	80	26·7	21·3
13	10·6	8·4	47	8·3	6·7	81	27·2	21·8
14	10·0	8·0	48	8·9	7·1	82	27·8	22·2
15	9·4	7·6	49	9·4	7·6	83	28·3	22·7
16	8·9	7·1	50	+10·0	+8·0	84	28·9	23·1
17	8·3	6·7	51	10·6	8·4	85	29·4	23·6
18	7·8	6·2	52	11·1	8·9	86	30·0	24·0
19	7·2	5·8	53	11·7	9·3	87	30·6	24·4
20	6·7	5·3	54	12·2	9·8	88	31·1	24·9
21	−6·1	−4·9	55	12·8	10·2	89	31·7	25·3
22	5·6	4·4	56	13·3	10·7	90	32·2	25·8
23	5·0	4·0	57	13·9	11·1	91	32·8	26·2
24	4·4	3·6	58	14·4	11·6	92	33·3	26·7
25	3·9	3·1	59	15·0	12·0	93	33·9	27·1
26	3·3	2·7	60	15·6	12·4	94	34·4	27·6
27	2·8	2·2	61	16·1	12·9	95	35·0	28·0
28	2·2	1·8	62	16·7	13·3	96	35·6	28·4
29	1·7	1·3	63	17·2	13·8	97	36·1	28·9
30	1·1	0·9	64	17·8	14·2	98	36·7	29·3
31	−0·6	−0·4	65	18·3	14·7	99	37·2	29·8
32	0·0	0·0	66	18·9	15·1	100	37·8	30·2
33	+0·6	+0·4	67	19·4	15·6			
34	1·1	0·9	68	20·0	16·0			

CURRENTS.

The Currents of the ocean are properly distinguished by the different and significant names—STREAM and DRIFT.

The Drift Current is merely the effect of the wind on the surface of the water; as, for example, in the region of the Trade winds, where the whole surface of the sea (generally speaking) is converted into a slow current moving to leeward. A drift current is, therefore, shallow and slow, and can run in no other direction than to leeward.

The Stream Current has been described as an accumulation of the parts of the drift into a collective mass by the intervention of some obstacle; the mass then running off by means of its own gravity, and taking the direction imposed on it by the obstacle, becomes a stream of current, and in many cases a powerful stream, pursuing its way like a vast river through the ocean.

These "Oceanic rivers" may vary in breadth like the Gulf Stream, from 50 to 250 miles; and are sufficiently deep to be turned aside by banks which do not rise within 60 or 80 fathoms of the surface—as the Agulhas current is deflected by the Agulhas bank off the Cape of Good Hope—and run with such rapidity as to be uninfluenced materially by the wind except near their borders.

Changes of temperature of the surface water are frequently abrupt. These changes are especially marked on the edges of stream currents, as, for example, on the north and west edges of the Gulf Stream, where it is met by the current from Davis Strait, and the two streams run nearly side by side; also off the Cape of Good Hope, at the junction of the Agulhas Current and the cold water flowing into low latitudes from the Antarctic Sea; and again off the East coast of South America, South of Rio de la Plata, where the Brazil current is met by the cold stream from Cape Horn.

Caution.—It is especially to be observed that the seamen must be prepared for unsettled weather and a cross and often turbulent sea, when passing through alternating bands of warm and cold water.

The Currents of the Atlantic Ocean.

THE ARCTIC OR LABRADOR...............	Cold.
THE GULF STREAM	Warm.
THE N.E. TRADE DRIFT.....................	
THE GUINEA CURRENT	Warm.
THE EQUATORIAL CURRENT	Cold at the beginning.
THE EQUATORIAL COUNTER CURRENT ...	Warm.
THE BRAZIL CURRENT......................	Warm.
THE CAPE HORN CURRENT................	Cold.

A general set to the northward from the Antarctic regions.

The Gulf Stream :—This remarkable "Ocean river" issues with great velocity, and in great volume from a sea of unusually high temperature, and deposits a vast body of warm water in the central part of the North Atlantic Ocean; a warmer region of the atmosphere is thus formed in the midst of a colder one, across two-thirds of the Atlantic, between the parallels of 35° and 45°.

The water of the Gulf Stream is a deep indigo blue in colour, and its junction with ordinary sea water distinctly marked. In moderate weather the edges are marked also by ripplings, and, in the higher latitudes, frequently by evaporation. At the surface—especially so off the Coasts of the United States—it is divided into several bands of higher and lower temperature, of which the axis of the stream, or line of greatest velocity is the hottest; the temperature of the water falling rapidly in shore, and more slowly outside this line; the same features are observed in a modified form at considerable depths.

Between the axis of the stream and the coast, the fall of temperature is so sudden that the line of separation has received the distinctive name of the "cold wall;" at the surface a difference of 30° has been observed within a cables' length. The presence of warm water is, however, not always an indication that a vessel is under the influence of the Easterly current, as the Gulf stream, like a swollen river, frequently overflows its banks, and a counter current of the returning warm water prevails; hence the westerly and south-westerly currents so often experienced to the southward and eastward of the Gulf Stream.

The warm water limits are very variable, as are also in a lesser degree those of the stream, for N.E. and S.E. gales force the latter to the banks skirting the American shores, whilst N.W. and Westerly gales drive the stream from the shores, so that the anomaly of the stream running in cold water and its absence in the warm water has been observed.

The velocity of the Gulf Stream varies with the seasons, running strongest in July, August, and September. On entering Florida Strait, from the Gulf of Mexico, the general rate is from 2½ to 4 miles an hour; in the narrows of the Straits 5 miles an hour has been observed in August; beyond this, to the parallel of 35°, the rate is about 3½ miles, gradually decreasing as the stream expands to the N.E.; although, even near the meridian of 45° W. and on the 43rd parallel, the exceptional rate of 4 miles an hour in August has been recorded. Eastward of 35° W. it becomes a North-Easterly surface drift, felt even on the coast of Norway.

On issuing from the Gulf of Mexico the stream has a maximum temperature of 85°, or 5° to 6° above the Ocean temperature due to that latitude; and off the banks of Newfoundland is higher by 20° or 30° than the adjoining ocean. The following table gives the average temperature in the axis of the stream or warmest band of water.

	Winter.	Spring.	Summer.	Autumn.
FLORIDA STRAIT	77°	78°	83°	82°
OFF CHARLESTON	75	77	82	81
OFF CAPE HATTERAS	72	73	80	76
S.E. OF NANTUCKET SHOALS	67	68	80	72
SOUTH OF NOVA SCOTIA	62	67	78	69

The weather in the Gulf Stream is warm, squally, and unsettled, and with S.W. or West gales the air is sultry. Beyond its north and west limits the weather is extremely cold. The sea in bad weather is heavy and irregular.

The evidence of the old navigators is against taking advantage of the favorable set of the Gulf Stream on the voyage from the West Indies to Europe; they deemed this route ill-judged, as not compensating for the wear and tear of the rough weather experienced; and adopted in preference one South of the 33rd parallel, as the great storms were considered to happen to the North of 32° and 33°.

The Guinea Current is a stream current, running to the Eastward, along that part of the African coast comprised chiefly between Cape Roxo and the Bight of Biafra; extending Southward to the 3rd and 2nd parallels of North latitude. Its Western limit can be traced at all seasons of the year as far as the 23rd Meridian of West Longitude; but in the summer and autumn months especially, an Easterly current, extending beyond this meridian to that of 53° W. is frequently experienced;—this is probably an expansion of the Guinea Current proper, and due to influences not yet investigated.

The greatest velocity of the Guinea Current is stated to be off Cape Palmas, where at a few miles from the shore it has been found to run more than 3 knots an hour. The space separating this current from the Equatorial is generally limited, thus presenting the remarkable feature of too well marked streams running in exactly opposite directions side by side.

The Guinea current proper may be considered a warmer stream than the Equatorial at all seasons, but careful observations on this head are yet wanting.

Between Cape Verde and Sierra Leone, extending 50 to 70 leagues from the shore, winds and currents change with seasons. From June to September squally S.W. winds, and a N.E. or Northerly current. From October to May, when N.E. and Northerly winds prevail, a South Easterly current is experienced.

Between Sierra Leone and Cape Palmas, the current is influenced by the wind; setting N.W. with winds South of S.W.; and setting S.E., with winds west of S.W. In November generally a N.W. current; from December to May, a S.E. current.

In the Harmattan season (December to February) the Guinea Current near the land is checked, and in shore a Westerly set is found. Heavy Tornadas have a similar effect.

The Equatorial Current in its course between the Continents of Africa and America, may be considered chiefly as a "drift" current formed of water brought from a cooler region by the S.E. trade wind. It may be said to commence in the neighbourhood of Anno Bom, or just South of the Equator between the 2nd and 8th degrees of East Longitude, although from this locality a continuity of the Northerly drift along the coast of South Africa, as well as from the River Congo, may be traced.* The surface temperature in its Eastern part is several degrees colder for a great part of the year than the adjacent Guinea Current, affording evidence of receiving waters from a remote and colder parallel.

The Equatorial Current appears to attain its greatest volume and velocity during the season of the Northern summer. From the African Coast to about the 15th degree of West longitude, the maximum strength has been observed in June and July; Westward of that meridian at successive later periods, or between July and October. Being, however, mainly a "drift" current it is probably subject to irregularities in strength, depending on the winds.

The Northern boundary, or rather the well marked line of separation between it and the Guinea Current has been well traced in the space extending from the meridian of Greenwich to 23° W., and is found to vary little at the several seasons of the year. For example, in the 20th West meridian the "line of separation" in October and November is in 5° N.; in March and April in 2½ N., in the 5th degree of West Longitude, the "line of separation" appears to be generally constant in 2° N. Approaching the African coast, Anno Bom island is considered to be at all seasons in the Equatorial current; Princes Island in the Guinea current; and St. Thomas, situated nearly midway between the two, as within the influence of one or the other current, according to the seasons.

---

* The stream of the river Congo is occasionally felt 300 miles from the entrance, and the discoloration of the ocean perceptible. It runs with almost undeviating regularity to the N.W. and N N.W., decreasing in strength as it extends seaward; its rate is about 2 miles an hour until absorbed in the Equatorial current.

From Anno Bom Island to about the 15th meridian of West longitude the following are the average surface temperatures of the Equatorial current:—

December to March, 78° to 82°.   March to July, 82° to 72°
July to October, 72° to 75°.   October to December, 75° to 78°

Westward of the 15th meridian, the surface temperatures at the several seasons lessen materially in their range, and the Equatorial current gradually loses its earlier features of being a cold water stream at one particular season of the year.

The Equatorial Counter Current.—Between the months of July and November the North-westerly drifts from the Equatorial Current appear to be suspended, and an Easterly Current prevails (probably an expansion of the Guinea Current). Between 53° and 40° W. it attains a rate of 60 miles a day; Eastward of 40° it decreases in strength, and between 30° and 20° W., it runs from 30 to 15 miles a day.

North Coast of Brazil and Coast of Guiana.—A great part of the Equatorial Current runs along this part of South America on its way to the West Indies. The rate of this current may vary from one-half to 4 miles an hour, according to the distance from middle of the stream and season of the year; it probably attains its greatest strength at 100 or 120 miles from the coast. This stream is about 200 miles wide, the inner edge varying from 15 to 45 miles from the coast. Within a few miles of the coast tidal streams are perceptible.

During the prevalence of N.E. winds a current runs E.S.E. along and near the north coast of Brazil; this fact is well known to the masters of the coasters.

River Amazon.—The waters of the Amazon attain their highest elevation in May, after a gradual rising of six months duration; and then gradually fall six months.

The stream from the Amazon (the ebb) at first sets E.N.E., then inclines to North and N.W., as it unites with the Equatorial Current; increasing the velocity of the latter. The surface of the ocean is discoloured, by the waters of the Amazon, for a considerable distance to the Northward and Westward of the mouth of that river.

Off Cape North, the stream of flood into the Amazon runs with double the strength of the ebb between January and April; in May, the flood and ebb streams are equal; from July to October the ebb has the greater strength; from November to January the two streams are again equal.

The Brazil Current is a branch of the Equatorial, setting along the coast of the South American Continent to about the parallel of 42° S., where it is met by the cold Cape Horn current setting Northward.

For about 150 miles from the Brazilian coast, between the parallel of 10° S. and Cape Frio, the current is influenced by the wind. Between October and January it generally sets to the S.W., at the rate of 25 to 30 miles a day; between March and September it sets to the Northward, running strong in July, when it occasionally attains a velocity of 48 miles a day.

Rio de la Plata.—The Currents of the Rio de la Plata are governed by the winds, the water rising with Southerly and falling with Northerly winds. Off the entrance they generally set to the N.N.W. before and with Southerly winds, and to the S.S.E. before and with Northerly, at rates varying from one to three miles an hour.

An East and E.N.E. current of one mile an hour, a supposed outfall from the Rio de la Plata, has been experienced, extending nearly to the 40th meridian.

The Currents of the Indian Ocean.

The currents of this ocean are generally with the Monsoons. There is, however, a constant stream setting to the Westward past the North end of Madagascar, striking the African coast about Cape Delgado, where it divides; one part trending to the Northward, in the neighbourhood of Zanzibar, and then (in the N.E. Monsoon,) running to the Eastward to the Southward of the Equator, but in the S.W. Monsoon, setting to the North-east towards the island of Sokótra. The other branch flows Southward through the Mozambique Channel *but is constant only* in the N.E. Monsoon.

1

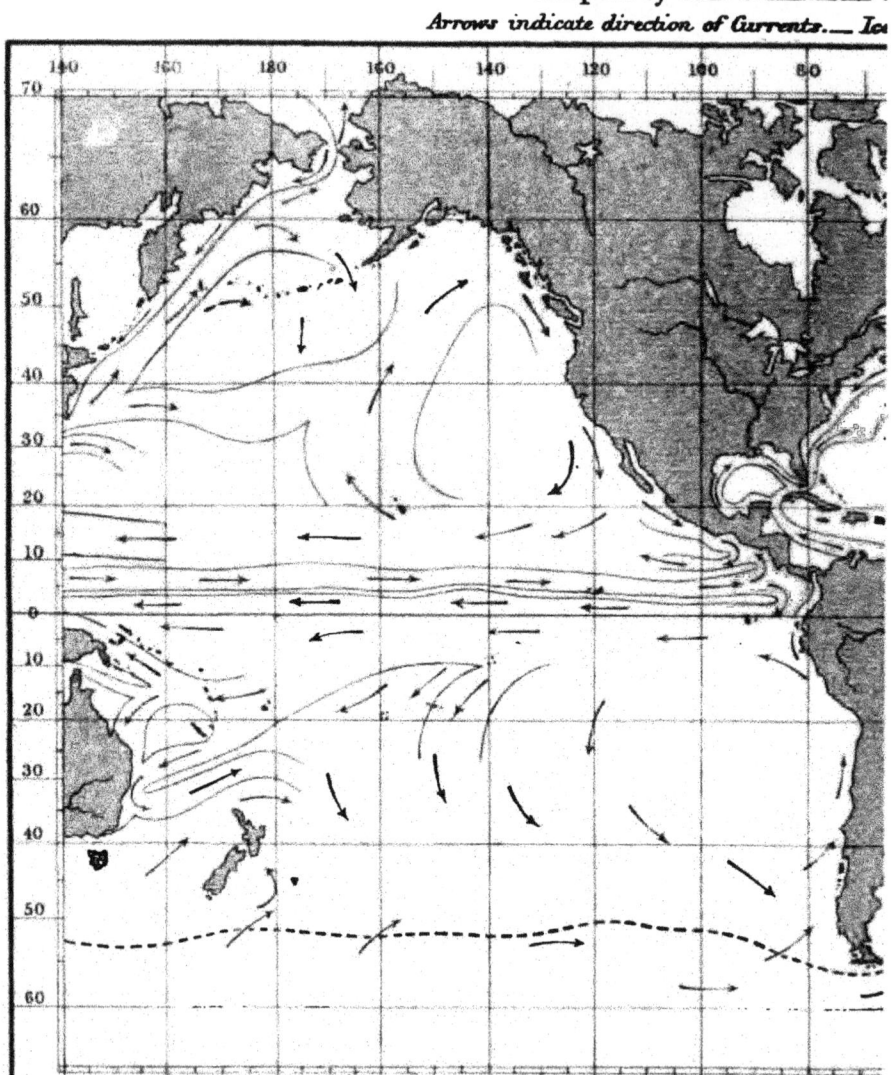

Drawn for the "Sailor's Pocket Book", July, 1874.

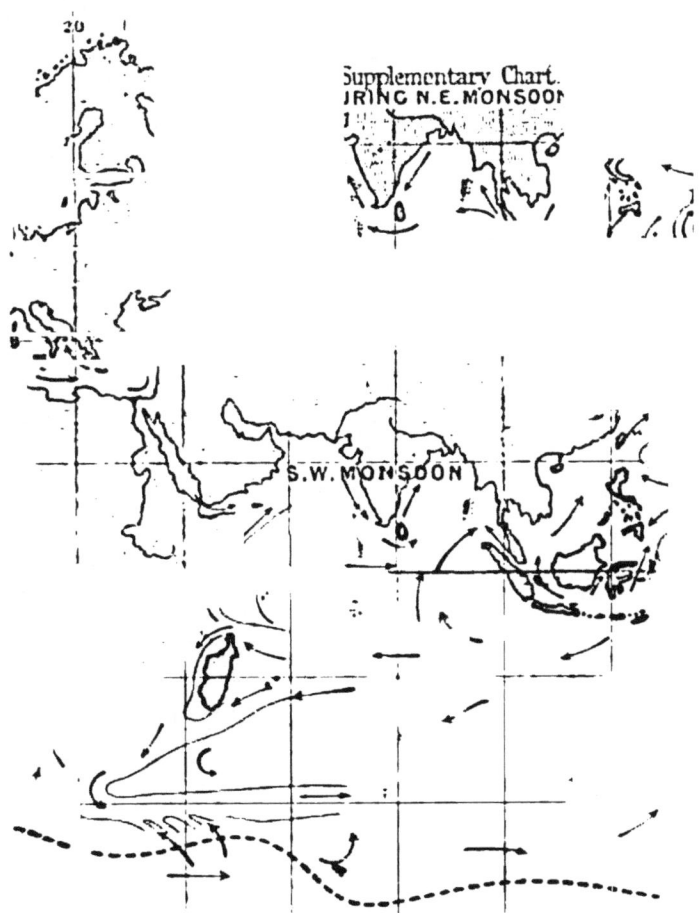

CURRENTS.

The Currents in the Arabian Sea in South-west Monsoon are regular in direction, their velocity depending much on the force of wind and local circumstances. The general course of the current in the middle of the sea is about East, inclining to S.E. as it nears the western coast of India; its velocity varies from half a mile to 2 miles per hour.

On the Eastern coast of Africa the current sets along the coast to the N.N.E. at a velocity of 2 to 4 miles per hour, passes through the channel between Sokótra and the North-east point of Africa at a rate of 1½ to 2 miles per hour, pursuing a course Northerly and Easterly, until it impinges on the Arabian coast about Rás Kosaïr, whence it takes a North-easterly course along that coast to Rás-al-Hadd, at a velocity of half to 1½ miles per hour.

To the south of Sokótra, at a distance of about 150 miles, is a great whirl of current, caused possibly by the interposition of the island; or, it may be, that shoal water exists at that spot; this eddy commences about the parallel of Rás Hafún, strikes off to the Eastward, as far as the 55th meridian, then turns to the Southward, to the 5th parallel, whence it again curves up to the North-eastward, forming a complete whirl. At the Northern limit the velocity is very great, being 4 to 5 miles per hour, while at its Southern extreme it is only ¾ to one mile per hour. A heavy confused sea is created by this whirl. Care should be taken to avoid the strongest portion of the current in making the coast of Africa from the Eastward, by keeping well to the Southward.

Little is known of the currents at this season close to the Northward of Sokótra, but there is said to exist a whirl similar to, but of less magnitude than that South of the island. Horsburgh remarks that the currents on the North side run with the prevailing breezes to the North-Eastward, but, when the wind moderates, an almost equally strong current runs in the opposite direction. This most probably applies to within a few miles of the land.

Throughout the S.W. Monsoon, or from June to September inclusive, the water runs out of the Red Sea, while from November to May the contrary is the case. During the S.W. Monsoon, the current on the Arabian side of the gulf of 'Aden runs to the

Eastward, as far as Rás Rehmat, or Hisn Ghorab, whence it strikes off to the South-Eastward to Rás 'Asír or Cape Guardafui, the North-east point of Africa; it then turns to the Westward, close along the African coast, as far as the 47th meridian, whence it again curves to the N.W. From the Straits of Bab-el-Mandeb the current sets along the African coast to the 47th meridian, where it meets the Westerly set from Cape Guardafui, and turns with it to the North and N.W. The velocity of the current throughout the gulf varies from half a mile to 2 miles per hour.

Currents in North-East Monsoon.—The current in the Arabian Sea generally sets to the South-Westward, its velocity depending on the force of the wind. When the wind is light there is little or no current. There is, however, a general set of from 1 to 2 miles an hour, to the South-west, along the coast of Arabia.

The Currents off the South Coast of Ceylon from the month of November, or during the North-east Monsoon, set S.S.W. out of the Bay of Bengal for five months and a half without variation—running with a velocity of from 1 to 3½, and sometimes of 5 knots an hour—thus, a steamer, steering N.N.E., has not only to oppose a 3-knot current, but also has the wind dead against her; some steamers make barely 2 miles an hour when bound to the Northward.

From May to September, or during the South-west Monsoon, the current runs N.E. and N.N.E., at from 1 to 3 knots an hour. During these months, in the event of a vessel making Dondra Head, when bound to Galle, which is not uncommon, the master should never attempt to beat to Galle, but recross the line.

In the neighbourhood of the Basses reefs a remarkable circumstance occurs with regard to the current during the South-west Monsoon. After running 3 knots to the N.E. for ten days, suddenly it slackens, and runs 2 to 3½ knots S.W., or from the Bay of Bengal, lasting sometimes only one day, at other times for a week; these changes happen at all times of the moon, and appear to obey no recognized law.

Agulhas Current.—An enormous body of warm water running to the S.W. from the Mozambique Channel and Indian Ocean, *and known as* the Agulhas Current, skirts the coast of South Africa, approaching very near the shore between the 30th and 28th meridians of East longitude. The stream has here an average width of 50 miles, with an occasional velocity of 4 miles an hour. As it progresses to the S.W., and arrives at the Eastern side of the Agulhas bank, the main body is deflected to the Southward, and a large part recurves to the Eastward, thus flowing into the Indian Ocean—but with diminished strength and temperature—on a high parallel of latitude (40° S.), where its influence is felt as far as the Islands of St. Paul's and Amsterdam. The cold Antarctic current is frequently found crossing this part of the Agulhas Current.

A small portion passes round and over the Southern part of the Agulhas bank, and branching off to the N.W., is joined by the Easterly or "connecting" current of the South Atlantic Ocean, collectively forming a wide stream running to the N.W., at the rate of 1 or 1½ miles an hour, which is perceptible as far as the parallel of 25° South.

The remarkable recurving of the main body of the current is due to the action of a polar or cold water current flowing from the S.W.; the junction, or intermingling of the hot and cold waters of the two streams notably taking place off the Agulhas bank, giving rise to the troubled and confused sea, the irregular and uncertain set of the currents, and by their effect on the atmosphere to those severe and fitful gales so well known to seamen rounding the Cape of Good Hope.

In the summer season (January—March) the warm water of the Agulhas current appears to be pushed a considerable distance to the Westward (as far as the longitude of about 10° East), the stream at this time attains its maximum strength and volume. In the winter season the current diminishes in force and extent, and the cold water from the higher latitudes repels and permeates it easily.

Temperature of the Agulhas Current in :—	Spring.	Summer.	Autumn.	Winter.
Off Natal	71°	76°	75°	70°
Off Algoa Bay	69	74	73	61
S.E. of Agulhas Bank ...	65	71	69	65
Meridian of Cape Agulhas	63	68	67	61

Bands of warm and cold water are found at all seasons, extending South as far as the 42nd and 44th parallels of latitude ; but the space included between the parallels of 36° and 38° S. and the meridians of 22° and 26° E. is that where the violence of the conflicting currents of cold and warm water, and cold and warm air appears to be centred.

There is generally a constant set to the Northward along the Western shore of Australia, and a general set to the Northward from the Antartic regions.

The Currents of the Pacific.

The Aleutian: a cold current setting to the Southward through those Islands from Behring Sea.

Japan Stream (Kuro Siwo).* The north-east trade drift current of the Pacific, which flows to the Westward between the parallels of 9° N. and 20° N., on reaching the Eastern shores of the Philippine islands is deflected to the Northward, forming in latitude 21° N. between the meridian of 125° E. and the East coast of Formosa, the commencement of the great oceanic current known as the Kuro Siwo, or Japan stream, the limits and velocity of which are considerably influenced by the Monsoons of the China sea and the prevailing winds in the Yellow and Japan seas.

During the North-East Monsoon a part of the Pacific drift-current continues its course to the Westward through the Bashee and Balingtang channels, joining the Monsoon drift current which sets strong to the South-Westward at this season, through the Formosa channel and China sea.

During the South-West Monsoon the drift-current which flows to the Northward in the China sea and Formosa channel, joins the Kuro Siwo, and extends the Western limit of its stream to a line joining the island of Tung Ying (on the coast of China) to Tsu sima (in the Korea strait) ; this limit is very perceptible, the waters of Japan

---

* Black stream or tide, so called from its black appearance.

stream being of a dark blue colour whilst that of the colder water of the coast of China is of a pale green; the difference in temperature of the two streams is also very marked.

The main body of the Kuro Siwo is joined at this season by the Monsoon drift-current of the China sea, which also sets to the North-East through the Bashee and Balintang channels, and, being augmented in force, flows rapidly to the Northward along the East coast of Formosa, passing between the North end of that island and the Meiaco sima group, as far as the parallel of 26° N., where it turns to the Eastward and continues in a North-Easterly direction to the Westward of the Liu-Kiu islands until it reaches the Southern point of Japan. Here an off-shoot of the main stream passes to the Northward through the Korea strait into the Japan sea; the main body of the stream, however, continues its course, and trending still more to the Eastward, flows in an E.N.E. direction through Van Diemen Strait. the numerous channels between the islands north of the Linschoten group, and along the Southern shores of the islands of Kiusiu and Sikok, forming (especially along its margin near the shore, where it meets with an opposing tide) races and tide rips often resembling heavy breakers on reefs or shoals. The Kuro Siwo attains its greatest velocity when abreast of Sikok, where it has been known to set 100 miles in 24 hours, its usual velocity in this locality, however, being from 2 to 3 knots an hour.

Continuing its course along the South coast of the island of Nipon, past the Gulf of Yedo, the Kuro Siwo flows through the chain of islands lying south of that gulf, in the numerous channels between which its direction is found to be variable and under tidal influence, thus forming many tide rips and whirlpools; still continuing in an E.N.E. direction as far as latitude $36\frac{1}{2}°$ N., the Northern edge of the Japan stream then leaves the coast of Nipon (which here trends to the Northward from about Inaboye saki) and runs N.E., while a cold current, setting to the Southward along the shores of Kamchatka and the Kuril islands intervenes between the Kuro Siwo and the North-Eastern shores of Nipon and Yezo. Continuing to the North-Eastward, the Kuro Siwo begins to expand, and diminishes greatly in velocity after passing the meridian of 145° E.

At about the meridian of 150° E. the Kuro Siwo, divides, one part flowing to the Northward and, taking the name of the Kamchatka current, is felt as far as the Aleutian islands and Behring strait, whilst the greater portion of the stream continues in an Easterly direction, between the parallels of 32° N., and 41° N., as far as the meridian of 160° E., where it is deflected to the South-Eastward as far as the meridian of 180°, there losing its identity in the drift-currents of the North Pacific ocean which flow to the North and North-Eastward.

The maximum temperature of the stream during the South-West Monsoon is 86°, and its North-Western edge is strongly marked by a sudden thermal change in the water of from 12° to 20°; but the Southern and Eastern limits at this season extending to the Bonin islands are much less distinctly marked.

It has been thought that the Kuro Siwo and the Gulf stream present features that are analogous in the constant high temperature of their waters, but from recent numerous observations of temperature in the former it appears they do not.

In the summer months, viz., from May to September inclusive, the mean temperature of the Kuro Siwo is 82°, or 7° higher than the mean temperature of the ocean, in the same latitude; it is during these months that the flow of tropical waters from the equatorial and drift-currents of the Pacific, together with the Monsoon drifts of the China Sea, unite to increase the velocity and raise the temperature of the Kuro Siwo.

In the months of April and October hot and cold belts are found alternating in the stream, the temperature of which belts differ from 7° to 10°; whilst during the winter months, viz., from November to March, the temperature of the stream, attaining in the latitude of Formosa a mean of 74° (which is rather higher than the temperature of the ocean, in the same latitude), falls on reaching the shores of Japan to 63°, or below the temperature of the ocean at the same season: this may be accounted for by the fact that during these months the great body of water which lies to the westward of the Liu-Kiu islands between the parallels of 26° N. and 28½° N., and which during the summer season flowed to the North-East, at the *rate of from* 20 to 30 miles a day, bringing warm water from the

SEC. IV.   CURRENTS.   115

China Sea to join the Kuro Siwo, is now quiescent. Northward of this still water a cold current, forced down by the prevailing Northerly winds in the Japan and Yellow Seas and setting strong to the Eastward between the parallels 29° N. and 31° N., flows through Van Diemen Strait and joins the Kuro Siwo, the temperature of which is thus considerably lowered; at the same time the stream is contracted to an average width of 250 miles in the meridian of 142° E., and its velocity is diminished to a mean rate of two knots an hour, in consequence of the cessation of the supply of tropical waters.

The Kuro Siwo is, however, undoubtedly a warm current, its waters being generally of a higher temperature than that of the surrounding atmosphere. There is a floating seaweed found in the stream resembling the *fucus natans* of the Gulf stream.*

The **Kamchatka Current** is a branch of the Kuro Siwo, from which it separates in the parallel of 40° N. on the meridian of 150° E., and flows at an average rate of 18 miles a day (with a breadth of about 200 miles) to the North-Eastward as far as 51° N., where a branch trends to the Eastward towards the Aleutian islands; between the parallels of 51° N. and 60° N. there is a difficulty in tracing it, but beyond 60° N. it is again found setting to the North-East; passing West of St. Matthew island and East of St. Lawrence island, it flows along the American coast through Behring strait into the Arctic ocean; in the strait the mean temperature of its waters (52°) is 15° higher than that of the waters along the Asiatic shore. The channel between St. Lawrence island and the coast of Asia is hampered with ice until July, while at the same time the coast of America is free from ice in the month of April.†

The **Oya Siwo**.—A counter current of cold water called the Oya Siwo sets to the Southward, along the South-East coast of Kamchatka and the Kuril islands, and, flowing along the East coast of Yezo island and North-East coast of Nipon is felt as far South as Inaboye saki,

---

* The remarks on the temperature of the Kuro Siwo are deduced from the observations taken by the U.S. expedition to Japan, 1854, the remark books of H.M. ships deposited in the Hydrographic Department of the Admiralty, and the logs of Captain J. O. Hopkins, H.M.S. *Liverpool*.

† Remarks of Dr. John Simpson. R.N., H.M.S. *Plover*, 1852.

It varies in velocity and extent in the different seasons, being much stronger in the winter than in the summer. Except between the Kuril islands, past Cape Noyshap, and through the Tsugar strait the force of the cold stream is about 18 miles a day; through these narrows, however, it sweeps occasionally with great speed, particularly during or after a strong North-Easterly wind. Whirls, eddies, and tide rips take place off Cape Noyshap. This counter current is shown in Japanese itineraries, where the junk course is drawn well off shore, so as to avoid it by passing outside and getting into the Kamchatka current.

During the spring (April), great quantities of floating ice are frequently brought down from the Kuril islands and the Northward, and are left grounded along the Yezo coast from Cape Yerimo Eastward to Cape Noyshap.

The temperature of the Oya Siwo is from 10° to 15° lower than that of the Kuro Siwo, varying according to the season. The average temperature in May was 37° (10° below that of the surrounding atmosphere). It was 66° in July between 10 and 30 miles from the shore; when the temperature rose to 70°, where it bordered on the Kamchatka current flowing to the Northward. After passing westward round Inaboye saki, the temperature of the sea water rose from 66° to 79° where the Kuro Siwo was setting N.E. by E. two knots an hour. The Oya Siwo is not felt close to the shore.

The drift of the N.E. trade sets across the North Pacific Ocean, between the parallels of 9° and 22° N. A portion of this drift (to the Northward of the Sandwich Islands), turns to the North-west and Northward.

The Mexican Current is a continuation of the drift setting to the Southward along the Californian Coast; it runs along the Mexican Coast as far as the Gulf of Fonseca, where it is met and recurved to the Westward by the set Northward and Westward from the Pacific counter current.

The Pacific Counter Current.—A large portion of the Equatorial current (see page 117) sets to the W.N.W. along the North coast of New Guinea; on reaching the shores of the Malay Archipelago this current recurves, and flowing to the Eastward right across the Pacific Ocean between the parallels of about 4° to 8° N., forms the Pacific counter current. On reaching the American Coast it appears to divide, the main portion curving to the Northward, meeting and turning the Mexican stream about the Gulf of Fonseca, from which point the streams appear to run together to the Westward; the other part turns to the Southward, meeting the Peruvian stream between the American Coast and the Galapagos Islands.

The **Peruvian Current** is a cold stream setting to the Northward along the Coast of South America; between the Galapagos Island and the Continent it divides, one part running to the Northward, towards and then round the Bay of Panama, the other curving to the Westward and forming the commencement of the Equatorial stream.

The meeting of such powerful streams as the Mexican, Pacific counter, and Peruvian currents in the great bight of Panama, probably accounts for the extraordinary variable weather, with the tropical squalls, and conflicting currents, experienced in that neighbourhood.

The **Equatorial Current**, which may be considered as a continuation of the Peruvian stream, sets fairly across the Pacific, between the parallels of 4° N. and 10° S. There is a general tendency of its waters near the southern edge to turn to the southward; about the meridian of 180° it appears to divide, one portion running to the Northward along the coast of New Guinea, the other trending to the South-West towards the Australian coast. The Equatorial current in the Western part of the Pacific is much affected by the Westerly Monsoon that prevails between the months of November and March.

The **Australian Current** is a branch of the Equatorial setting along the East coast of Australia as far as the parallel of Sydney, where the off shore portion is met and curved back to the East and North-East by the cold Antarctic current setting to the Northward from the Antarctic regions.

The **S.E. trade drift** of the Pacific sets across the ocean, turning to the Southward about the meridian of Tahiti; and to the Southward of lat. 30° S., the general movement of the waters appear to be to the Southward and Eastward towards Cape Horn.

In the Pacific, as in the Indian and Atlantic Oceans, there is a steady set to the Northward from the Antarctic regions, setting to the North-East, in the neighbourhood of New Zealand and Cape Horn.

ICE.

North Atlantic Ocean.—In the frequented parts of the North Atlantic, the limits of field-ice in March extend from Newfoundland to the Southward as far as 42° N. latitude, and to the Eastward of the meridian of 44° W.

During the second quarter of the year these limits are slightly contracted, but icebergs may be met with within an area reaching out to the point where the meridian of 40° W. crosses the parallel of 40° N., and even to the East and South of this.

By August the field-ice has disappeared, but icebergs are still met with when Westward of the meridian of 38° W. or Northward of the parallel of 41° N. During the winter months the seas are comparatively clear.

Southern Hemisphere.—In the high Southern routes, adopted of late years by navigators in the voyages to and from Australia and New Zealand, the greatest number of icebergs have been seen in the summer season, or in November, December, and January, and the smallest number in June and July. It has also been observed that more icebergs are seen in March. During February, the limit of the icebergs in the South Atlantic extends as far North as 39° S., while in August they are rarely found to the Northward of the 45° parallel.

This difference is specially noticeable in that portion of the ocean to the Eastward of Cape Horn, a district to be navigated with great caution with regard to ice.

Numerous icebergs have also been fallen in with Southward of Cape Leeuwin during January and February, in about the parallel of 44° S. and between the meridians of 105° and 140° E.

PASSAGE TABLES.

The indications of the Thermometer should not be neglected in these seas, as there is generally a diminution of temperature of the air and sea on approaching ice; this, however, must not be assumed as an infallible guide. Icebergs should, if possible, be passed to windward, to avoid the loose ice floating to leeward.

PASSAGE TABLES.

THE average rate of a *Sailing* Vessel in making general ocean passages may be taken as 120 nautical miles a day, and of a *Steam* Vessel as 200 nautical miles a day.

4 miles an hour	=	96 miles a day,	or	672 miles a week.	
5	..	=120	,,	840	..
6	,,	=144	,,	1008	,,
7	..	=168	,,	1176	,,
8	..	=192	,,	1344	,,
9	.	=216	.,	1512	,,
10	..	=240	..	1680	..
11		=264	..	1848	..
12		=288		2016	
13	..	=312	,,	2184	.,
14	..	=336	,,	2352	..
15	,,	=360	,,	2520	..

In the following tables the miles noted under the head of "Steam," show the distance to be traversed by a ship making a direct course, on the most favorable track for steam vessels, between port and port.

Passages noted under the head of "Sail," show the distances that will probably have to be run by sailing vessels; the prevailing winds and currents being taken into consideration.

Navigable Distances in Nautical Miles between some of the principal Ports of the Globe.

ENGLAND TO PORTS IN INDIA, CHINA, AND AUSTRALIA.

Ports.	By Cape of Good Hope.			By Suez Canal.
	Steam	Sail.	Sail N.E. Monsoon.	
	Miles	Miles	Miles	Miles
Plymouth to Bombay	10417	11727	13290	6000
,, Galle	10160	11470	12200	6499
,, Madras	10680	11990	14200	7020
,, Calcutta	11300	12610	14200	7639
,, Singapore	11490	13160	..	8020
,, Hong Kong	12930	14600	15140	9450
,, Shanghai	13730	15400	16140	10249
,, Yokohama	14490	16160	..	11010
,, Melbourne	11890	13200	..	11039
,, Sydney	12440	13750	..	11590
,, Wellington	13360	14670	..	12510

ENGLAND TO PORTS IN THE PACIFIC AND BACK.

	By Cape Horn and Valparaiso.	
	Steam.	Sail.
	Miles	Miles
Plymouth to Valparaiso	8600	9120
,, San Francisco	13700	15360
,, Vancouver	14400	16110
,, Honolulu	14660	15180
Valparaiso to Plymouth	..	9800
San Francisco	14520	15900
Vancouver	15220	16600
Honolulu	15100	17100

AUSTRALIA TO ENGLAND BY CAPE HORN.

	Steam.	Sail.
Melbourne to Plymouth	13060	13940
Sydney to Plymouth	12870	13750
Wellington to Plymouth	11670	12550

ATLANTIC.

Ports.	Steam.	Sail.
Plymouth to Halifax	2434	
,, ,, Northern Passage	—	2400
,, ,, Southern Passage	—	4200
,, Bermuda	2918	3780†
,, St. Thomas	3505	3700
,, Colon	4535	4740
,, Demerara	—.	4080
,, Maranham	—	3980
,, Madeira	1210	1210
,, Gibraltar	1050	1050
,, St. Vincent	2257	2257
,, Sierra Leone	2700	2700
,, Ascension	3700	4200
,, Cape of Good Hope	5890	7200
,, Rio de Janeiro	4945	4945
,, Monte Video	5965	5965
,, Cape Horn	7120	7120
Gibraltar to Madeira	600	600
,, Bermuda	2880	3240
,, Halifax	2670	3840
,, St. Vincent	1560	1560
Halifax to Plymouth	2434	2434
,, New York	580	580
,, Gibraltar	2670	2670
,, Bermuda	750	750
Bermuda to Plymouth	2918	2918
,, Jamaica	1140	—
,, St. Thomas	850	850
,, Demerara	1590	1590
,, Havanna	1140	1140
,, Cape of Good Hope	6140	—
,, New York	690	690
,, Gibraltar	2880	2880
New York to Havanna	1260	1260
Havanna to New Orleans	590	590
,, Vera Cruz	840	840
Jamaica to Colon	560	560

† By Southern Route.

ATLANTIC.—Continued.

Ports.	Steam.	Sail.
Sierra Leone to Cape Coast Castle	900	900
,, Plymouth	2700	4290
,, Ascension	1000	1500
Cape Coast Castle to Fernando Po	600	600
,, Plymouth	3600	5700
Fernando Po to Sierra Leone	1500	1800
,, St. Helena	—	1980
Ascension to Plymouth	3907	4760
,, Sierra Leone	1000	1000
,, Rio de Janerio	1890	1890
,, St. Helera	680	2460
,, Cape of Good Hope	2380	3420
St. Helena to Cape of Good Hope	1700	2700
,, Plymouth	4587	5440
,, Rio de Janeiro	2140	2140
Rio de Janeiro to Plymouth	4945	5600
,, ,, Nov. to Feb.	—	6300
,, Ascension	1890	2760
,, St. Helena	2140	2760
,, Bahia	—	710
,, Pemambuco	—	1080
,, Monte Video	—	1020
,, Cape Horn	2300	2300
,, Valparaiso	3680†	4180
,, Cape of Good Hope	3270	3270
Monte Video to Plymouth	5965	6620
,, ,, Oct. to April	—	7020
,, Cape of Good Hope	3600	3600
Cape of Good Hope to Plymouth	5890	7140
,, Bermuda	6140	6140
,, St. Helena	1700	1700
,, Rio de Janerio	3600	3600
Cape Horn to Plymouth	7120	8000
,, St. Helena	—	3720
Cape Virgins to Cape Pillar	325	—

† Via Magellan Straits.

INDIA, CHINA, AUSTRALIA.

Ports.	Steam	Sail
Plymouth to Gibraltar	1050	—
Gibraltar to Malta	980	—
Malta to Port Said	940	—
,, Alexandria	820	—
Alexandria to Port Said	140	—
Brindisi ,,	930	—
,, to Malta	360	—
Trieste to Port Said	1287	—
Odessa ,,	1233	—
Port Said to Suez	87	—
Suez to Aden	1308	—
Aden to Bombay	1637	1637*
Bombay to Aden	2180*	3975*
Aden to Point de Galle	2134	2134*
Point de Galle to Aden	2480*	3255*
Bombay to Point de Galle	960	—
Calcutta to Aden	3400*	4395*
,, Bombay	2100*	5000*
Aden to Seychelles	1040	3130*
,, Zanzibar	1720	1720†
,, Mauritius	2830*	3610*
,, Batavia	—	3960*
,, ,,	—	4090†
Batavia to Aden	—	4090*
,,	—	3960†
Seychelles to Zanzibar	990	990
,, Mauritius	930	—
,, Bombay	1770	1770*
Zanzibar to Cape of Good Hope	2460	—
,, Bombay	2520	2520*
Bombay to Cape of Good Hope	4527	4527†
,,	—	5820*
Mauritius to Point de Galle	2100	2100*
,, Cape of Good Hope	2340	2340
Natal to Mauritius	—	2400

* In S.W. *Monsoon.*   † In N.E. *Monsoon.*

India, China, Australia.—*Continued*.

Ports.	Steam	Sail
Point de Galle to Calcutta	1140	1140*
,, Singapore	1510	—
,, Cape Leeuwin	3130	—
,, Melbourne	4570	—
,, Cape of Good Hope	4440	4440†
,, ,,	—	4860*
Cape of Good Hope to Bombay	4527	6090†
,, Calcutta	—	7000†
,, Madras	—	7000†
,, Straits of Sunda	—	5460
,, Singapore	—	5960
,, Melbourne	—	6000
Cape of Good Hope to Mauritius	2340	3120
Calcutta to Cape of Good Hope	5580	5580†
,, ,,	—	6600*
Madras ,,	4960	4960†
,, ,,	—	5520*
,, Bombay	1480	4320*
Singapore to Melbourne, by Strait of Sunda	—	5100
,, ,, by Baly Strait	—	4800
,, Sydney (N.W. monsoon)	4380	4380
Singapore to Batavia	550	—
Hong Kong to Sydney (N.W. Monsoon), Torres Strait	4620	4620
,, ,, ,, N. of New Guinea	5040	5040
Singapore to Hong Kong	1440	1980†
Hong Kong to Shanghai	800	1000†
,, Singapore	1440	1980*
,, Yokohama	1560	—
,, River Amur	2400	—
River Amur to Petropaulski	840	—
King George's Sound to Adelaide	1016	—
Adelaide to Melbourne	505	—
Melbourne to Sydney	550	—
Sydney to Brisbane	540	—
Brisbane to Port Denison	591	—
Port Denison to Cape York	652	—

* *In* S.W. Monsoon.   † *In* N.E. Monsoon.

PACIFIC.

Ports.	Steam.	Sail.
Rio de Janeiro to Valparaiso	3680*	4180
Cape Horn to San Francisco	—	7700
Valparaiso to Panama	2610	2610
,, San Francisco	5100	6240
,, Vancouver's Island	5800	6990
,, Honolulu	6060	6060
,, Tahiti	4470	4470
,, Callao	1292	1292
,, Sydney	—	7950
Callao to Valparaiso	1292	2400
,, Payta	500	—
,, Guayaquil	700	—
Guayaquil to Panama	780	—
Panama to Valparaiso	2610	4320
,, San Francisco	3240	5200
,, Vancouver's Island	3940	5940
,, Honolulu	—	~~5500~~ 521
,, Tahiti	4474	4474
,, Sydney	—	7954
,, Wellington	6790	6874
,, Acapulco	1410	---
,, Realejo	680	..
Acapulco to San Blas	490	---
,, San Francisco	1850	—
San Blas to Mazatlan	120	
San Francisco to Valparaiso	5100	6100
,, Honolulu	2080	2080
,, Yokohama	4500	5680
,, Vancouver's Island	700	..
,, Sydney	—	6460
,, Wellington	5820	5820
,, Fiji Islands	4740	
,, Cape Horn	—	6500
Vancouver's Island to Honolulu	2400	2400

* By Magellan

Pacific.—*Continued.*

Ports.	Steam	Sail.
Vancouver's Island to Valparaiso	5800	6820
Petropaulski to Vancouver	3000	3000
,, Honolulu	2820	2820
Honolulu to Petropaulski	—	3450
,, Vancouver's Island	2400	2580
,, San Francisco	2080	2280
,, Yokohama	3600	3600
,, Hong Kong	4950	4950
,, Sydney	4380	4380
,, Tahiti	2380	—
,, Valparaiso	—	7300
,, Fiji Islands	2780	2780
,, Cape Horn	—	7100
Yokohama to Vancouver's Island	—	4300
,, San Francisco	—	4500
,, Shanghae	1035	—
Nagasaki to ,,	454	—
Shanghae to Tientsin	735	—
Yokohama to Hongkong	1560	—
Hongkong to Manila	650	—
Tahiti to Wellington	2400	2400
,, Sydney	3480	3480
,, Fiji Islands	1800	1800
Sydney to Wellington	1200	1200
,, Valparaiso	—	6240
,, Cape Horn	—	5750
,, Hong Kong	—	5820
,, Yokohama	—	5280
,, Fiji Islands	1665	—
Melbourne to Wellington	1470	1470
,, Cape Horn	—	5940
Auckland to Fiji Islands	1080	—
Auckland to Wellington	540	—
Wellington to Lyttleton	200	—
Lyttleton to Otago	230	—

Sec. IV   PASSAGE TABLES.   127

Table, shewing whether a vessel gains or loses by keeping away a given number of Points from the True Course, thereby enabling her to economise fuel.

The upper line shows the rate in knots per hour steaming head to wind. The other lines show the rate in knots a ship must go, if she keeps away, to be in as good a position as she would have been had she held on her course.

Points from the wind.	RATE IN KNOTS PER HOUR.								
0	1	2	3	4	5	6	7	8	9
1	1·02	2·04	3·06	4·08	5·1	6·12	7·14	8·16	9·17
2	1·08	2·16	3·25	4·33	5.41	6·49	7.58	8.66	9·74
3	1·2	2·4	3.61	4.81	6·0	7·21	8·42	9.6	10·82
4	1·4	2·83	4.24	5.65	7·07	8.48	9.9	11.3	12.73
5	1·8	3.6	5.4	7.2	9.0	10.8	12.6	14.4	16.2
6	2.6	5.2	7.84	10.45	13.06	15.68	18.3	20.9	23.52

Example.—A ship is able to steam 4 knots an hour head to wind; by keeping away four points, and setting fore-and-aft sails, she makes 6 knots an hour. The table shows that if she made 5·65 knots, she would be in as good a position as if she had remained steering head to wind; therefore, if she can steam 6 knots, she gains nearly half a knot an hour; but if she can only steam and sail at the rate of 5 knots, she is losing.

If the vessel keeps away 5 points she must steam or sail at the rate of 7·2 knots, to be in an equally good position.

Section 5.

LIGHTS; BUOYS;

HYDROGRAPHICAL ABBREVIATIONS; SOUNDING;

DETERMINING POSITIONS;

MEASURING DISTANCES AND HEIGHTS, &c.

LIGHTS.

Two systems only are in general use, viz., the catoptric and the dioptric; the former being produced by an Argand oil burner, shown in the focus of a paraboloidal silvered metal reflector, any number of which, from 1 to 10, may be placed on the same plane, as the power and range may require.

In the other, the dioptric system, the light is produced from one central oil burner placed in the focus of a glass instrument, by which all the rays emanating from the sections of the flame best suited for the purpose, are bent, so as to be sent nearly horizontally, and only to the sea's surface.

The catadioptric system is a combination of the two preceding systems. Both of these systems admit of lights being shown as fixed, revolving or flashing, and of various orders, determined in the catoptric by the number of lamps and reflectors, and in the dioptric by the size of the instrument, with a corresponding central flame.

In addition to these two well-known systems is that of the magneto-electric light; it belongs to the dioptric class, the light being shown from an apparatus of that character.

Each of these systems presents some conditions which give it a superiority over the other, and these conditions may be briefly enumerated as follows:—

A catoptric revolving light may, according to the number of burners and metallic reflectors used, be deemed equal to a dioptric lens. Seven reflectors on a face revolving are generally used, and are worked at about the same cost as a first order dioptric light. Where, however, there are 10 reflectors on a face, as at Beachy Head, Scilly, &c., the great volume of light, with a slow revolution, shows a superiority over the intensified flame of the lens, and coming to the eye with a great body of light, it illumes to a greater extent the atmosphere, and thus has an advantage, especially where there is a haze.

The catoptric system has the further advantage over the dioptric of facility in erection, at a distance from where skilled mechanics are obtainable, requiring less delicacy in putting up and focussing the light, and being less liable to be put out of adjustment by volcanic or other disturbances. It has also the advantage that its first cost is not more than half that of a dioptric light, and, with care in the use and cleaning of the reflectors, its consumption of oil and stores, except in the case of 10 reflectors on a face, is not more than that of the lenticular light.

The power of a reflector is much increased by what is termed the holophotal arrangement, where an annular lens is placed in front of the frame, while all the back rays of light which are otherwise lost, are thrown back into the flame by a hemispherical mirror. Three reflectors of this kind to a face make an excellent revolving light.

For a *fixed* light, however, the catoptric is far inferior to the dioptric system, and is now only used for lights of that character in positions where want of importance does not warrant the use of a lens; but it possesses one advantage, that it is not in the same degree subject to be deteriorated for want of care and skill on the part of the attendants.

In a Catoptric Fixed Light,—the reflectors (with an Argand lamp in the focus of each) are arranged round a circular frame, and if intended to illuminate the whole horizon, there are generally 13 in each tier; those in the second tier being placed over the interstices of the other, so as to produce as nearly as possible an equal distribution of light.

The reflectors generally used are 21 inches in diameter at the lips, and the Argand burner $\frac{7}{8}$ of an inch.

In a Catoptric Revolving Light—the reflectors are grouped together with their axes parallel to each other on three or more faces, so as to throw their combined light in one direction.

The number of faces depend upon the rapidity of revolution, and shortness of flash intended to be given; but where the interval between the appearance of each face amounts to half a minute or *more, there are* usually only three faces; which may contain one or more *reflectors each.*

Dioptric lights, whether fixed, revolving, or flashing, vary in the size of the lamp and apparatus according to the order required, and are numbered from 1 to 6. The 1st, 2nd, and 3rd orders are all used for coast illumination; the 4th, 5th, and 6th order lights are generally employed at the entrance of harbours, or as lower leading lights. The 1st order is that chiefly in use at the English lighthouses. As a fixed light the dioptric possesses great advantages over the catoptric, especially where it is required to illumine the whole circle of 360 degrees. Over an arc too, according to its area, it has the power of intensifying the light, by utilizing it where not required, and returning it, from where it would otherwise be wasted by reflectors and refractors, to strengthen the illuminated arc; and it has also a great advantage over the reflector in its facilities for marking channels or outlying dangers, which is done by placing a sector of coloured glass vertically on the lens, in conjunction with one on the lantern glass, in line with the danger to be marked; and this can be done with an accuracy impossible in a catoptric light.

In a Dioptric Revolving Light,—the panels of glass usually range from 8 to 16 in number; but occasionally there are only six sides or panels, which produces a very powerful flash, by concentrating a larger arc of light.

Until within a few years the illuminants or combustibles have consisted of animal or vegetable oils, chiefly rape oil, or what is termed Colza. In the United States, lard oil is used; in the tropics, cocoa nut oil is used with advantage; it is cheaper than Colza, and is said to attain a higher degreee of illumination.

Petroleum, Revosine, Paraffin, and other mineral oils have been burnt in the smaller order lamps in France, Australia, and elsewhere, since about 1856; but it was not until about 1869, that Captain Dory succeeded in using mineral oils in the higher order lamps with great advantage, both in economy and power of illumination. They have further the advantage of requiring no attention, as they burn steadily through the whole night without trimming; this is not the case with the other oils that char the wicks.

About 1865, Mr. J. WIGHAM, of Dublin, succeeded in introducing gas illumination for lighthouse purposes, and several of the Irish lights now use it successfully. The gas is manufactured at the Light Station.

At first oil gas from shale was used, latterly it has been made from Cannel coal. By an ingenious arrangement of concentric circles of jets, either one or more of these rings can be lighted, (according to the state of the atmosphere) by which the number of jets can be increased from 28 to 108. It is capable of exhibiting by automatic action, a variety of characteristics, the gas being alternately turned on and off, so as to produce an intermittent light of different intervals in a fixed apparatus; or flashes, when the apparatus is revolving.

By this contrivance much gas is saved as compared with the ordinary intermittent light in which the flame remains at full power; the intermissions being produced by screens alternately lowered and raised.

Annual Consumption of Oil.

THE burners consist of one or more concentric wicks, alternating with concentric air passages.

The Argand or single wick, used in the 4th order dioptric lights and in reflectors, consume about 42 gallons per lamp.

2 wick burners, small size, 112 gallons per lamp.
2 ,, ,, large ,, 214 ,,
3 ,, ,, ,, ,, 412 ,,
4 ,, ,, ,, ,, 828 ,,

Recently five wick, and even six wick burners have been introduced in special cases.

Apparent Lights.

THE apparent light is an ingenious and very useful contrivance, by which beacons marking sunken rocks lying off lighthouses may be illuminated by reflection from an apparatus hermetically sealed, placed on top of the beacon.

This apparatus consists of a plane mirror, upon which the rays of light from the lighthouse impinge; the reflected rays being strengthened by passing through glass prisms. This contrivance was first introduced by Mr. T. STEVENSON, C.E., at Stornoway, in the Hebrides, to illuminate the beacon on Arnish Rocks. They are distant, 530 feet from the special light placed in a window in the lower part of the lighthouse, on a level with the beacon apparatus.

It has since been applied at Odessa, and also at Port Curtis in Queensland.

BUOY SYSTEMS.

System of Buoyage, in use by the Corporation of Trinity House, in Buoying New Channels.

The side of the channel is to be considered starboard, or port, with reference to the *entrance* to any port from seaward.

The entrances of channels, or turning points, shall be marked by spiral buoys with or without staff and globe, or triangle, cage, &c.

Single-coloured can buoys, either black or red, will mark the starboard side, and buoys of the same shape and colour, either chequered or vertically striped with white, will mark the port side; further distinction will be given, when required, by the use of spiral buoys with or without staff and globe, or cage; globes being on the starboard hand and cages on the port hand.

Where a middle ground exists in a channel, each end of it will be marked by a buoy of the colour in use in that channel, but with horizontal rings of white, and with or without staff and diamond, or triangle, as may be desirable; in case of its being of such extent as to require intermediate buoys, they will be coloured as if on the sides of a channel. When required, the outer buoy will be marked by a staff and diamond, and the inner one by a staff and triangle.

Wrecks will still continue to be marked by green nun-buoys.

Regulations for the Colouring of Buoys.

As regards chequered buoys, each buoy, exclusive of the nozzle, is to be divided *horizontally* into four, and *vertically* into eight equal parts; but the white squares are then to be further reduced by one inch all round, being coloured either in red or black, as the case may be.

As regards vertically striped buoys, each buoy is to be divided into *eight parts,* and each division is to be alternately coloured red and white, or black and white, but the white stripes are to be one third narrower than the black or red.

As regards buoys coloured in horizontal bands or stripes, each buoy is to be divided into five parts, which are to be coloured red and white, or black and white alternately, the white bands being one third narrower than the black or red.

All buoys have their names painted on them in conspicuous letters.

SCOTLAND.
System adopted by the Commissioners of Northern Lighthouses.

Entering port, &c., from seaward, *red buoys* must be left on the *starboard* hand in passing in.

Entering port, &c., from seaward, *black buoys* must be left on the *port* hand in passing in.

Buoys painted *red and black* are placed on detached dangers and may be passed on either hand.

All buoys have their names painted on them.

Fairway buoys are plainly marked. Wreck buoys are painted green.

Liverpool is buoyed on the same system.

Buoys, &c.—FRANCE.

Shipmasters frequenting the ports and harbours on the coast of France must observe the following rules:—

On entering a channel from seaward, all buoys and beacons painted *red* with a *white* band near the summit, must be left to starboard: those painted *black* must be left to port: buoys that can be passed on either side are coloured *red with black horizontal bands*. That part of the beacon *below the level of high water*, and all warping buoys, are coloured *white*. The small rocky heads in frequented channels are coloured in the same way as the beacons, when they have a surface sufficiently conspicuous.

Each beacon or buoy has upon it, either at full length, or abbreviated, the name of the danger it is meant to distinguish: likewise, its number, commencing from seaward, and thus showing its numerical order in the same channel.

The *even* numbers are on the *red* buoys, and the *odd* numbers on the *black* buoys; the buoys and beacons coloured *red with black horizontal bands* are named but not numbered

The letters and numbers are painted in *white* on the most prominent part of the buoys, and from 10 to 12 inches in length.

The masts of the beacons which do not present sufficient surface are surmounted for this purpose by a small board. All the jetty heads and turrets are coloured above the half-tide level, and on the former a scale of mètres is marked, commencing from the same level.

TIDE SIGNALS.

In French ports, flood and ebb, and the height of the tide, are signalled at intervals by means of black balls, and by flags; these are hoisted on a mast crossed by a yard.

A ball at the intersection of the mast and yard indicates a depth of 3 mètres or $9\frac{3}{4}$ feet. Each ball *below* this, and in the line of the mast, represents an additional height of 1 mètre or $3\frac{1}{4}$ feet,—but each ball *above* it, an aditional height of 2 mètres or $6\frac{1}{2}$ feet. A ball hoisted at the yard-arm and seen to the left of the mast indicates 0·25 mètre or $\frac{3}{4}$ of a foot additional; but seen to the right of the mast, 0·50 mètre, or $1\frac{3}{4}$ feet additional.

In order to show the state of the tide in respect to flood and ebb, a white flag crossed with black from corner to corner, and a black pendant will be used. One or both of these will be flying at the masthead, so long as there are two mètres or $6\frac{1}{2}$ feet water in the channel; thus, the pendant above the flag indicates flood,—the flag alone, high water,—and the pendant below the flag, ebb.

A red flag at the mast-head indicates that the state of the tide is such that a vessel cannot enter.

Buoys, &c.—AMERICA.
Directions.

In approaching the channel, &c., from seaward, *Red Buoys* with *even numbers*, will be found on the *starboard* side of the channel, and must be left on the *starboard* hand in passing in.

In approaching the channel, &c., from seaward, *Black Buoys*, with *odd numbers* will be found on the *port* side of the channel, and must be left on the *port* hand in passing in.

Buoys painted with *red* and *black horizontal stripes* will be found on obstructions, with channel ways on either side of them, and may be left on either hand in passing in.

Buoys painted with *white* and *black perpendicular stripes* will be found in *Mid-Channel*, and must be passed close-to, to avoid danger

All other distinguishing marks to buoys will be in addition to the foregoing, and may be employed to mark particular spo s; a particular description of which will be given in the printed list of buoys.

Perches with balls, cages, &c., will, when placed, be at turning points, the colour and number indicating on what side they shall be passed.

Vessels approaching or passing Light-Vessels of the United States, *in foggy or thick weather*, will be warned of their proximity by the alternate ringing of a Bell and sounding of a Fog-horn on board of the Light-Vessel, at intervals not exceeding five minutes.

Canada.—Is buoyed on the same system.

HOLLAND.

On entering the channel, &c., from seaward, *White Buoys* must be left on the starboard hand. *Black Buoys* on the port hand.

Belgium.—Same as Holland.

Signs and Abbreviations adopted by the Hydrographic Office, Admiralty.

Quality of the bottom.		General Abbreviations.		General Abbreviations.	
b.	— blue	Alt. —	Altitude	Lt. —	Light
blk.	— black	Anch$^{ge.}$ —	Anchorage	Lt. F. —	Light Fixed
br.	— brown	B. —	Bay	Lt. Fl. —	Light Flashing
brk.	— broken	B_k (against a buoy) Black.		Lt. Int. —	Light Intermittent
c.	— coarse	B^k. —	Bank	Lt. Rev. —	Light Revolving
cl.	— clay	C. —	Cape	L.W$_z$ —	Low Water
crl.	— coral	C. G. —	Coast Guard	Mag^z. —	Magazine
d.	— dark	Cath. —	Cathedral	Mag^c. —	Magnetic
f.	— fine	Ch. —	Church	Min. —	Minutes
				(against a light.)	
g.	— gravel	Chan. —	Channel	Mt. —	Mountain
gn.	— green	Cheq. —	Chequered	Np. —	Neaps
grd.	— ground	Cold. —	Coloured	Obsn. Spot —	Observation Spot
gy.	— gray	Cr. —	Creek	P. —	Port
h.	— hard	E.D —	Existence Doubtful	P.D. —	Position Doubtful
m.	— mud	Flg. Lt. —	Floating Light	Pk. —	Peak
oys.	— oysters	Fms. —	Fathoms	Pt. —	Point
oz.	— ooze	Ft. —	Feet or foot	R. —	River
peb.	— pebbles	G. —	Gulf	R. —	Red
				(against a buoy.)	
r.	— rock	Gt. —	Great	Rf. —	Reef
rot.	— rotten	H. —	Hour	Rk. —	Rock
s.	— sand	Hd. —	Head	Sd. —	Sound
sft.	— soft	Ho. —	House	Sec. —	Seconds
				(against a light.)	
sh.	— shells	H^r. —	Harbour	Sh. —	Shoal
spk.	—speckled	H.S. —	Horizontal Stripes	Sp. —	Springs
		(against a buoy.)			
st.	— stones	H.W. —	High Water.	Str. —	Strait
stf.	— stiff	H.W.F.&C. {	High Water Full and change }	Tel. —	Telegraph
w.	— white	I. —	Island	Var^n. —	Variation
wd.	— weed	I^s. —	Islands	Vil. —	Village
y.	— yellow	Kn. —	Knots	Vis. —	Visible
				(against a light.)	
		L. —	Lake	V.S. —	Vertical Stripes.
		Lat. —	Latitude	W. —	White
				(against a buoy.)	
		Long. —	Longitude	W. Pl. —	Watering Place

CONVENTIONAL SIGNS.

Rocks with less than six feet water upon them......	⊕	⊕
Rocks awash at low water	⊛	⊛
Signifies no bottom found at the depth expressed...	𝟙𝟘	𝟝𝟟𝟘
Anchorage for large vessels........................	⚓	
Anchorage for small vessels........................	⚓	⚓
Currents are represented by	⇒	
Flood tide stream	→	
Ebb tide stream	→	

All charts and plans are, where practicable, constructed upon the True Meridian, *i.e.*, the East and West marginal lines are drawn parallel to the true meridian. Soundings are reduced to mean low water of ordinary spring tides, and are expressed in fathoms and fractions of a fathom, or in feet and fractions of a foot, such being denoted on the chart.

The underlined figures on the dry banks represent in feet or fathoms the depth over them at high water, or the heights of the banks above low water.

The velocity of tide is expressed in knots and fractions of a knot. The period of the tide being shown thus :—1st Qr., 2nd Qr., 3rd Qr., 4th Qr., for 1st, 2nd, 3rd, and 4th quarters. The rise of tide is measured from the mean low water level of ordinary springs. The range of tide is measured from the low water of one tide to the high water of the following tide. All heights are given in feet above high water ordinary springs, and in places where there is no tide, above the level of the sea. All bearings, including the direction of winds and currents, are magnetic, except when otherwise expressed. Bearings of lights are given as seen from seaward, and not from the lights.

The natural scale, or the proportion which the chart bears to the earth (obtained by reducing the number of feet in the minute of latitude to inches, and dividing the product by the scale), is represented thus, $\frac{1}{12150}$.

A cable's length is assumed to be the 10th part of a sea mile.

Meridians adopted in the Construction of Foreign Charts.

Russia, Germany, Sweden, Denmark, Norway, Holland, Austria, and United States of America, adopt the meridian of Greenwich.

France adopts the meridian of Paris, assumed to be in longitude 2° 20′ 9·4″ E. of Greenwich.

Spain adopts the meridian of San Fernando, Cadiz, assumed to be in longitude 6° 12′ 16″ W. of Greenwich.

The Pulkowa Observatory of St. Petersburg (sometimes referred to in Russian Charts), is assumed to be in longitude 30° 19′ 40″ E. of Greenwich.

The Royal Observatory of Naples (sometimes referred to in Italian Charts), is assumed to be in longitude 14° 14′ 31·3″ E. of Greenwich.

Soundings upon Foreign Charts are expressed thus:

			English feet:	English fathoms.
Danish and Norwegian	fathom		(Favn)=6·175 or	1·029
Dutch		,,	(Vaden)=5·575 ,,	0·929
French		,,	(Brasse)=5·329 ,,	0·888
French	Mètre		(Mètre)=3·281 ,,	0·547
Portuguese	Fathom		(Braca)=6·004 ,,	1·000
Prussian		,,	(Faden)=5·906 ,,	0·984
Russian		,,	(Сажень)=6·000 ,,	1·000
Spanish		,,	(Braza)=5·492 ,,	0·915
Swedish		,,	(Famn)=5.843 ,,	0·974

The Dutch "*Elle*," the Spanish, Portuguese, and Italian "*Metro*," and the French "*Mètre*," are identical.

Sometimes upon French plans of harbours the soundings are given in feet.

A *pié́d* usual = 13·124 inches, or 1·094 feet. A *mètre* is 3 *pié́ds*; a *pié́d du roi* = 12·7896 inches. *Brasse*, = 5·329 feet, is used upon old French charts instead of mètre.

Upon some Italian charts the soundings are in French *pié́d*.

ADMIRALTY CHARTS.

THE sailor's attention is here called to the vast amount of practical information that may be found upon the charts published by the Hydrographical Office of the Admiralty. These works of art, from the simplicity of their style, their number and large circulation, are unfortunately for the general interests of the community, not studied with the attention they deserve. As a very slight error in the position of a light or buoy; dot, cross, or figure; might lead to the gravest disasters, every mark upon an Admiralty chart, has been delineated by the Hydrographic Draughtsmen with the greatest care and consideration, and no pains are spared by these gentlemen in their endeavour to attain, where it is possible, even mathematical accuracy, and to lay before the public the labours of the Nautical Surveyors, Explorers, and Amateurs, not only of England, but of the civilized world; reducing their various styles into a comprehensive system, and thus furnishing the intelligent seaman with a powerful guide, which common industry will soon enable him to thoroughly appreciate and take advantage of.

QUALITY OF BOTTOM
As expressed on Charts of different Nations.

English.		French.		Italian.	Spanish.		German.	
Clay	cl.	Argile	A.	Argilla.	Arcilla	Arc.	Lehm	L
Coral	crl.	Corail	Cor.	Corállo.	Coral	cl.	Korallen	K
Gravel	g.	Gravier	Gr.	Rena or Ghiaja	Cascájo	Co.	Grob sand	g.s.
Mud	m.	Vase	V.	Fango.	Fango or Limo	F.	Schlamm	Sch.
Rock	r.	Roche	R.	Roccia.	Roca	R.	Fels	F.
Sand	s.	Sable	S.	Sábbia or Aréna.	Arena	A.	Sand	S.
Shells	sh.	Coquilles	Coq.	Conchiglie.	Conchuela	ca.	Muschel	M.
Stones	st.	Pierre	P.	Pietre.	Piedra	P.	Stein	st.
Weed	wd.	Goêmon or Herb	H.	Alga.	Alga	Al.	Gras	G.
Fine	f.	Fin	fin.	Fino.	Fina	f.	Fein	f.
Coarse	c.	Gros	g.	Grosso.	Grueso.	g.	Grob	g.
Stiff	stf.	Dure	d.	Tenace.	Tenaz		Zäh	Z.
Soft	sft.	Molle	m.	Molle.	Muelle.		Weich	W.
Black	blk.	Noire	n	Nero.	Negro.		Schwarz	schw.
Red	rd.	Rouge	r.	Rosse.	Roxez.		Roth.	
Yellow	y.	Jaune	j.	Giallo.	Amarillo.		Gelb.	

GENERAL HYDROGRAPHIC INFORMATION.

The sailor has often the opportunity of giving considerable assistance to his fellow seamen, by forwarding to the Admiralty, or Lloyds, information from which the existing charts and sailing directions may be added to, or corrected. The following suggestions of "What to observe," and "How to report," may be found useful.

The language of such communications should be simple, concise, seamanlike, and descriptive of such matters as the wants of a sailor are likely to require; as the appearance of the coast on making the land, with its prominent headlands and mountain peaks, giving their estimated height, and the distance from which they can be seen; or if the coast be low, describing remarkable trees, conspicuous sand hills, or any natural or artificial beacons that may help to distinguish the locality: the outlying dangers, with means of clearing them, and the general character of the soundings, with the nature of the bottom.

Give information as to where pilots are to be obtained; the special signals, pilot regulations and charges. Where to get steam tugs. In case of bad weather, the best anchorage or the nearest harbour of refuge to run for; or in an extreme case, as leak, &c., describe the best place to lay the ship aground, and the best time of tide for doing so. Describe all lights, buoys, and beacons, and notice any convenient creeks or rivers for wooding or watering.

Give general directions for sailing, steaming, or working, by day, by night, and in a fog. State the prevailing winds and their seasons; land and sea breezes where they exist. Time balls, tide signals, and fog signals, are to be described. Currents to be noted, and whether permanent, or affected by periodical winds. When rivers of any importance are visited, give the points to which the tide reaches, and to which they are navigable; strength before and after rains.

In describing seaports, especially those subject to changes from increased commerce, give so much of nautical statistics as will enable seamen to judge of their importance, and probable supplies that can be obtained; as—Where fresh water is to be found, in what manner, with its quality. Facilities for coaling a steamer, and price of coal. How coals are obtained, and if from a near or distant locality. Whether wood can be procured in quantity sufficient for steaming purposes or to combine with coal. Any special custom house regulations; holidays, &c. Dock accommodation, both floating and graving. Width of entrance gates. Patent slip? Gridiron? Facilities for repair of a vessel, in wood and iron; or the machinery of a steamer. Conveniences for heaving down. Power of largest crane or sheers. Quarantine regulations. Hospitals or homes for sailors. Shipping Office for Seamen. Prevailing diseases, if of a virulent character; seasons at which they may be expected, and brief precautions for guarding against them. Diet, Clothing, &c. National prejudices, in cases where serious offence might be given unintentionally. Current money and its usual value in English money. Population; number of fishermen, and seafaring men. Foreign consuls or vice-consuls. Means of communication by steam vessels, by rail, by electric telegraph, &c.*

Tides.—For both springs and neaps give the height of high water above the level of low water ordinary springs. The times and heights of high and low water should be taken when the moon passes the meridian of the place between 11^h and 2^h, and also between 6^h and 8^h. In describing tide-streams in the offing, caution must be observed in not confusing the "flood" and "ebb" streams; it is rather desirable to use the terms "east" or "west" going stream, as the case may be.—See page 167.

* **Currents.**—In general, the daily error of the reckoning at each following noon has been adopted as the effect of a single current acting throughout the whole interval; whereas it is probable that in some parts of the ocean where the vessel may be crossing the boundaries of opposing currents, during the twenty-four hours the ship will have passed through different streams, and have been affected by different impulses. To form consistent views on a subject so necessary in every

---

* Sailors and others, wishing to convey useful information in any charts or sketches they may be forwarding, are recommended to imitate the style of a modern Admiralty chart, of some coast or harbour similar in character to that which they may be reporting upon; so as to delineate their labours on the same system as that adopted by the *Hydrographic Office* of the Admiralty.

voyage, and so deeply interesting when considered on a more extensive scale, it is desirable that in such cases, or where ripplings, discoloured water, flocks of birds, or shoals of fish are observed, systematic observations for latitude and time should be made at different hours of the day or night, taking advantage especially of the clear horizon an hour before sunrise and after sunset, for the express purpose of apportioning the whole error of the day's work into its most likely periods. In determining the amount and velocity of an ocean current, due consideration should be given to the probable effects of any bad steerage; to the action of the swell and sea upon the vessel, and also to any changes in the variation and deviation of the compass.

Trade winds and Monsoons.—The annual periods and geographical limits of the trade winds, monsoons and rains, and the times and local peculiarities of the daily sea and land breezes, afford ample scope for observation on many stations to the zealous officer. Observations, especially on the hurricanes or revolving storms of the Atlantic and Indian Oceans, and the typhoons in the China Sea, are also much required.

Lights.—Lights and lighthouses claim particular attention. Every detail that can be useful to the mariner, should be obtained; and in describing a new light, give its character and brilliancy; the intervals of its flashes, if revolving; whether the illuminating apparatus be dioptric, or catoptric, and its order or class; likewise the bearings from seaward on which the light is either visible or obscured, the height of the centre of the lantern above the high water of ordinary spring tides, the height also of the building from the base to vane, its form, colour, or other peculiarity.

In case of a new light, or error in description of an old light, it should be immediately reported to the Admiralty or Lloyds by special letter, in order that the correction may be made in the published List of Lighthouses.

Vigias.—Numerous imaginary dangers are traditionally inserted in all Ocean Charts, and from which they might be expunged if ample evidence of their non-existence within wide limits of their alleged positions could be obtained. Pieces of wreck, fish, and sundry floating substances account for a large proportion of these reported rocks. In the neighbourhood of such imagined dangers, a few deep-sea casts of the lead would be well bestowed, and the observant seamen will keep his eyes open to every unusual appearance in the sea, such as partial ripplings, discoloured water, flocks of birds, or shoals of fish as they may, if not caused by the meeting of opposing currents, be indications of some change in the nature or depth of the bottom; in all such cases a deep-sea cast should be obtained.

Variation of Compass.—As the lines of equal variation traced on a chart are valuable to seamen, not only for the ordinary purposes of navigation, but also to determine by comparison the varying error of their compasses, pains should be bestowed in multiplying observations at sea for the variation of the compass, more especially when the track of the vessel passes through those spaces where the variation changes rapidly. These observations should be tabulated, accompanied with the date, ship's position, direction of ship's head, and tables of deviation of the compass as determined wholly or partially at sea or in port.

Meridian Distances.—In determining the longitudes of imperfectly known places, meridian distances well measured will be very valuable. All chronometers should, if possible, be rated by equal altitudes of the sun; and the meridian distances should depend upon those places the positions of which rest on good authority. A meridian distance, where the interval elapsed between the times of rating the chronometers exceeds eight or ten days, can rarely be depended upon for accuracy.—See page 193. When a meridian distance is measured to a point or island on an unsurveyed coast, a rough plan of the immediate neighbourhood should also be provided.

Courses and Bearings are invariably to be given by compass, corrected for the deviation. All dangers, especially those on which a vessel has struck or grounded, are always to be determined by two or more horizontal sextant angles of well selected objects, see page 178. A careful magnetic bearing should also be given of one of the objects used.

All soundings to be reduced to the low water level of ordinary spring tides.

When an Admiralty chart is mentioned, give its number and the date to which it was last corrected. If an Admiralty Pilot,

Sec. V.   HYDROGRAPHIC INFORMATION.   149

or Directory, give page and edition. If the Admiralty Light list, give date of list, and number of the light.

Sailing along a coast of which there are no reliable charts; angles, bearings, and true bearings, with astronomical observations, should be taken to acquire some knowledge of the nature of the coast, and to correct the errors of the existing chart.

On moonlight nights latitudes from altitudes of stars North and South of the Zenith should be obtained, with frequent observations in the Northern hemisphere of the Pole star. Time, or latitude, may also be found, by observing stars just before sunrise or after sunset.

The dead reckoning should be carefully kept, and carefully worked, and frequent comparisons made with the astronomical positions to determine the amount of current that may be affecting the vessel.

A True bearing from amplitude may be obtained at sunrise, with bearings of conspicuous objects, noting the direction of the ship's head, the course to be steered, the patent log, if in use, and the time. Another set of bearings to the points taken at sunrise, and *true bearing** from Azimuth, may be taken when the forenoon observations are obtained, other objects ahead being selected to continue the work by; a third set may be taken at noon, and the observations continued on similar principles in the afternoon and evening. Bearings of all objects in transit should be taken, and the direction of the ship's head and her courses carefully noted. If no current has been experienced, the courses steered and distances run, will form bases from which an attempt at projection may be made; or from the true positions determined by observations, the bearings may be laid off, thus seeing how nearly conspicuous points could be recognized, and compared with the same points as marked on the faulty chart.

* Although this problem of *True bearing* is fully treated upon by INMAN & RAPER in their Navigations, it is not unfortunately in the common use that it should be among sailors. It consists in simply measuring the angular distance between the sun's limb and the object of which the true bearing is required. This angle reduced to the horizontal angle by right angled spherics, and applied to the true bearing of the sun will give the true bearing of the object. The sun should be in a favourable position for taking an azimuth, and the angular distance measured should always be double the altitude of the sun.

By following the above simple directions intelligent seamen have frequently been able to make valuable additions to existing charts, and rough work of this description is always most acceptable to the Hydrographic Office of the Admiralty, and infinitely more valuable than finished drawings, that may, or may not, be founded on facts.

SOUNDING.

A VESSEL may, at times, be anchored in a bay or off a coast of which no chart exists; in this case it may be considered necessary, not only to sound round the ship, but to be able to make some report upon the anchorage. If, therefore, time, opportunity, or inability, does not admit of making a survey of a harbour, an idea of its form, extent, and depth of water, may be arrived at by simply running lines of soundings from the ship, anchored near the centre, to the prominent points, noting at these points the masthead angle, taken both on and off the arc, and the compass bearing of the ship.*

In such a case a *flag or black ball* should be hoisted close up to the ship's main truck, to facilitate the taking of the masthead angle, and the following practical hints on sounding may be found useful.

Sounding is especially work for a sailor, requiring all the ready wit and tact of his profession. In sounding, the sailor has to manage air and water, the rise and fall of the tides, the velocity of currents, and to fit in his work to suit wind and weather; these forces becoming firm allies to the man who studies them, and foes only to him who knows not how to use them.

The instruments, by the help of which the important operation of sounding is carried on, are the sextant and lead line.

---

* If shoals are discovered in harbours or roadsteads, or upon coasts where the chart furnishes the sailor with reliable points upon the shore, the soundings should, in such cases, be fixed by sextant angles taken to such points. Sounding by masthead angle should only be used if no reliable positions can be found, or if time does not permit of making a rough triangulation of the place. The necessary geometry for determining positions by sextant angles will be found at page 178. Sailors who may wish to gain a further insight into Practical Nautical Surveying, are referred to a *small work on that important subject*, by Staff-Commander Thomas A. Hull, R.N., Superintendent of Admiralty Charts; published at 2s., by J. D. Potter, Poultry, E.C.

The Sextant, a now common but no less invaluable instrument, the inventor of which may be looked upon as one of the greatest benefactors to mankind, might be considered by the sailor as the tongue of his profession, without which he would be, to a certain extent, dumb, and able only indistinctly to give utterance to his discoveries.

The sailor should therefore lose no opportunity of making himself thoroughly conversant with this useful instrument, understanding all its adjustments, peculiarities, causes of error, the size of angles to be safely measured with it, and the means of re-quicksilvering its reflectors. With the assistance of the sextant, Captain Moriarty found the lost end of the Atlantic cable, and without it no ocean can be properly sounded or telegraph cable fairly laid.

When in port, lying at anchor, a simple exercise in using the sextant is to take a round of angles, consisting of six or more, between distinctly seen objects lying nearly in the same horizontal plane, and see how near in addition they may be brought to 360°. Handle the sextant both ways, inverted as well as direct; many objects from being indistinct cannot be reflected from right to left, and the left hand object that can be reflected, must in such case be used.

Every opportunity should be taken of fixing the ship on the chart by bearing and angle. Taking it for granted that the harbour in which the vessel is anchored is well surveyed, it is good practice to take angles to the different points, peaks, and islets, and laying the same off from the ship's position, to see if they agree with those on the chart. True bearings of objects, both right and left of the sun, should be frequently observed, taking care that the angle measured between the sun and the object, is always double the altitude of the sun.

If angles are taken between two objects close to each other, care should be taken that they are on the same level, or that the angle measured is nearly horizontal; with large angles, it is of little consequence that the objects used are not on the same level, as the error caused by obliquity is in these cases small.

The Arc of Excess.—It is good practice to measure small angles off as well as on the arc; the arc of excess in most sextants is 5°. Observed masthead angles will generally be less than 5°, and angles of elevation taken to peaks on making the land will always be less than 5°. Taking the angles off and on the arc, adding them together, and dividing by 2, gives an angle free of index error. At sea, measuring the sun's semi-diameter, and comparing that obtained with that given in the Nautical Almanac, renders the observer dexterous in observing small angles with the sextant. If the object is clear and well defined, the inverting tube will be found of great assistance.

A man-of-war's man should lose no opportunity of volunteering to lay out targets in well-known harbours or roadsteads by the masthead angle, taking the angle both off and on the arc of the sextant, that being one of the methods used in nautical surveying for determining an approximate base. When the target is moored, if time permits, get a round of angles from it; these plotted on the plan of the harbour will afford an opportunity of testing the accuracy of the base obtained from the masthead angle, taking care to fix the ship's position for the way in which she may be swung, either on leaving or returning on board.

Admiral Sir E. Belcher gives the following detailed directions for re-silvering sextant glasses when injured by damp or wet :—

"The requisites are clean tinfoil and mercury, a hares' foot is handy, lay the tinfoil, which should exceed the surface of the glass by a quarter of an inch on each side, on a smooth surface (the back of a book), rub it out smooth with the finger, add a bubble of mercury about the size of a small shot, which rub gently over the tinfoil until it spreads itself and shows a silvered surface, gently add sufficient mercury to cover the leaf, so that its surface is fluid. Prepare a slip of clean paper the size of the tinfoil. Take the glass in the left hand, previously well cleaned, and the paper in the right. Brush the surface of the mercury gently to free it from dross. Lay the paper on the mercury, and the glass on it. Pressing gently on the glass, withdraw the paper. Turn the glass on its face, and leave it on an inclined plane to allow the mercury to flow off, which is accelerated by laying a strip of tinfoil as a *conductor* to its lower edge. The edges may, after twelve hours' rest, be removed. In twenty-four give it a coat of varnish, made from spirits of wine and red sealing-wax."

Sec. V. SOUNDING. 153

The Lead line, by means of which the soundings are obtained, should be looked upon, and treated, as a valuable instrument, and therefore to be used for no other purpose but sounding, except in the absence of a chain for measuring a base line, or ascertaining the height of a cliff.

A good plan is to take for this purpose a ship's line that has been some little time in use, and is, therefore, well stretched, marking it when wet, to feet as far as 5 fathoms, on the same principle that the ordinary line is marked to fathoms, with two knots at 20 feet, and a bit of leather at 4 fathoms or 24 feet. The line should always be measured on the nails driven into the quarter-deck for that purpose, both when leaving and returning to the ship, taking care when measured that the line is always thoroughly wet. If obliged to use a new line, and shoal water is found in sounding out a bay or inland, the line should be re-measured in the boat, on the boat-hook marked to feet, for that purpose. The line should always be marked from the heel of the lead.

Men should be accustomed to heave the lead from the boats, and should be narrowly watched when first at work, in order to ensure their giving the correct soundings. The foremast awning stanchion, armed, forms a good support for the leadsman's breast-rope. Some little practice and method is required on the part of the man heaving the lead from the boat, and also on the part of the officer, to determine, first when an "up and down" cast can be obtained without laying on the oars, and secondly, when it may be necessary to stop the boat altogether.

In the boat, as in the ship, the end of the lead line should always be secured before beginning to sound, and care should be taken that a "dry lead" is taken from the ship for this work, with the necessary tallow for arming, as the nature of the bottom should always be noted, especially marking any changes that may occur, such changes often giving warning of neighbouring shoals.

Sounding in a tide-way it may be necessary to anchor the boat to obtain the masthead, or other angles, to fix the boat's position.

Noting.—The soundings should be noted at equal intervals and distances. If a patent log cannot be obtained, this must be done by the judgment of the officer, by time or strokes of the oars. Under 5 fathoms the lead should be hove continously, and the boats position fixed by masthead angle of ship near the 3 and 5 fathom line at low water on approaching and leaving the land.

A space must be left underneath the figures of the soundings in the note book, in which they be reduced to low water by a table derived from observations made on the tides.—See page 167. The reduced soundings should be noted in red ink.

The time should be noted whenever soundings are taken, or at least every half-hour. In the absence of a watch the signal man on board the ship should be directed to hoist a numeral flag every hour, showing the number of the bells struck, keeping the same flying for five minutes, and hoisting the dinner pendant at noon. Attention in the notation of the time is not only a great assistance to the memory when plotting the soundings on a rough plan, but without it the reduction of the soundings to low water, cannot be honestly carried out.

The greatest care should be taken in entering the soundings, bearings, masthead angles, time, and other matters in the note book in such an intelligible manner, that another sailor could use them, so that in case of accidents, common to seamen, if the note book were saved, the work would not be lost. On return to the ship the work should be plotted as soon as possible, while the memory is green. Such precautions, adapted to the simple capacity of the industrious, enable the sailor to prove beyond a doubt the correctness of his work, and the existence of rocks, or errors in the chart, which his useful labours may have detected.

To ensure the lines sounded upon being straight lines, the ship should either be kept in transit with some object beyond her, or some well-marked object should be found on the shore in line or transit with the point or mark for which it is intended to steer, this second mark lying sufficiently far behind the point steered for to show any deviation from the straight line required. If pulling or sailing to seaward, the same should be looked for astern; such objects kept in transit *must* insure a straight line being followed; on nearing the shore *a good* look-out should be kept for a third object in the same line, in case of the second mark used dippping behind the foreland. Keeping the boat on known straight lines is the only sure method of sounding in a tide way. As there is some difficulty in steering by stern marks, the use of them should be made a point for practice. In all cases the compass-bearing of the points steered for or from, should be noted.

In sounding under sail 5 fathoms may be obtained without lowering the sail, but if the water is deeper, the sail should be lowered if running free, or the sheet eased off when sailing near the wind, to insure getting an "up and down cast."

A Bareca for Beacon should be fitted as a buoy, with a buoy rope of about ten fathoms made of lead-line, and a pig of ballast to moor it by. This will be found useful to drop upon any sunken rock that may be discovered, the boat pulling and sounding round it in all directions to find out the extent and nature of the danger, carefully noting the courses steered and, in the absence of a patent log, estimating the distances pulled from and towards the Baréca.

Boat's Anchor.—In using the boat's anchor in sounding, where the boat may frequently be anchored on rocky ground, it is advisable to bend the cable like a buoy-rope to the crown of the anchor, stopping it with spun-yarn to the ring; then if the flukes jamb in the rock the stop will carry away, and the anchor may be weighed with ease. This precaution has saved many a boat's anchor, prevented a return to the ship, and the disarrangement of a well contemplated scheme.

Gaining Local Information.—If the bay or coast to be sounded is inhabited, take pains to gain the friendship of the fishermen, as these men making their living by their knowledge of localities useful to themselves, but dangerous to shipping, are often able to point out the positions of rocks or foul ground. Boats or canoes fishing should be piously avoided by a vessel coasting or entering a strange port or roadstead, as they are liable to be working in the vicinity of rocks or shoals,

Shoals out of sight of Land.—If a shoal is fallen in with out of sight of land, or if a ship is sent to search for and examine a danger so placed, the vessel should, on finding the shoal water, it

possible, anchor, mooring if convenient. Four boats might then be
sent away to sound on north, east, south, and west lines from the ship
for 2½ miles, hoisting a flag every half hour, and taking a mast-head

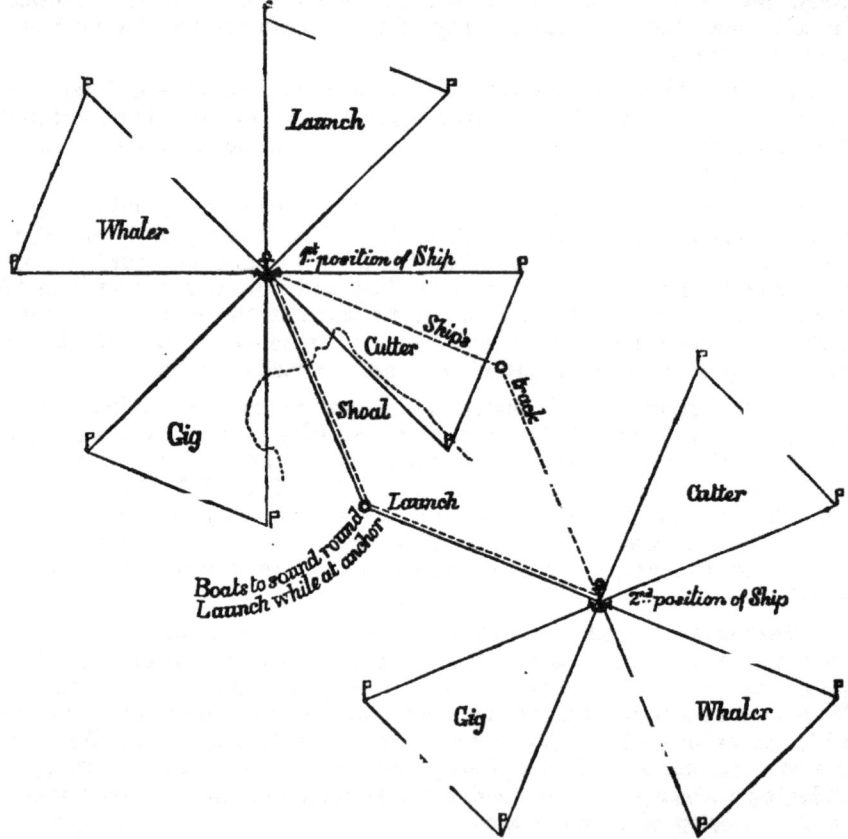

angle of the foremast, at which truck a black ball should be hoisted
an observer at the ship at the same time taking a bearing of the
respective boats. At the end of the three miles, distance to be

determined by mast-head angle, the boats will pull respectively E.S.E., S.S.W., W.N.W., and N.N.E. for 1·9 miles, fixing their position by mast head angle, and hoisting their flags to allow the ship to obtain bearings of them. They will then return to the ship on S.W., N.W., N.E., and S.E., courses, sounding and hoisting their flags as before.

If shoaler water be found, requiring further or more extended examination, the launch should be sent in the direction of the foul ground to anchor on a given bearing at a distance of not over three miles from the ship; a mast-head angle will be taken from the launch to determine the distance. Ship will weigh at signal from launch, and take up a second position as convenient not more than three miles from the launch, and another mast-head angle will be obtained from launch when ship has anchored, and hoisted her mast-head flag or ball. Launch will return to ship sounding, and the three remaining boats sound round the ship as before.

SOUNDING.

Table showing Time Occupied in Deep Sea Sounding, from 400 to 1,400 Fathoms, with different Weights and Lines.

Lines Employed.	Weight of Sinker.		Mean of time in sinking from 400 to 1400 fathoms.
	Description.	In lbs.	min. sec.
Seine twine.	Iron and lead.	13	36 . 03
Log line.	Shot (iron).	68	29 . 47
Deep sea line.	Iron.	96	26 . 53
Small line (American).	Shot.	32	26 . 14
Ditto ditto	2 Shot.	64	26 . 16
Ditto ditto	Ditto.	64	25 . 37
Contract log line.	Lead.	100	18 . 32
Albicore line.	Ditto	188	17 . 34
Hawser laid deep sea line, No. 1 used for first 100 Fathoms, after that No. 2 line.	Iron	224	16 . 18
	Iron	336	14 . 17
	Iron	448	12 . 26
	Iron	560	11 . 06

This table is compiled from the mean of a number of soundings taken by Captain Spratt, Commander Mansell, Commander Dayman, R.N., and Lieutenants Lee and Berryman, U.S. Navy. The last four are by Staff-Commander Johnson, *Instructor of* Nautical Surveying, at the Royal Naval College, Greenwich.

LEAD LINE.
Weight of Lead 7 to 14 lbs.

Marks. Fathoms.	Deeps. Fathoms.
	1
2—A piece of leather with 2 strips	
3—A piece of leather with 3 strips	
	4
5—White	
	6
7—Red	
	8
	9
10—A piece of leather with a hole in it.	
	11
	12
13—Blue	
	14
15—White	
	16
17—Red	
	18
	19
20—Two knots	*

* For surveying, the lead line is marked in feet up to 5 fathoms.

DEEP SEA LEAD LINE.
Weight of Lead 28 to 30 lbs.

Is marked the same as a hand line up to 20 fathoms, then with one knot at 25 fathoms, three knots at 30, one knot at 35, four knots at 40, and so on to 95.

Then at 100 fathoms a piece of bunting, at 105 one knot, at 110 a piece of leather, at 115 one knot, at 120 two knots, and so on as in the first 100 fathoms.

Massey's Sounding Machine.

The register wheels are set at the starting points, the small one at the 10 fathom mark; and the large one on the opposite side at the 150 fathom mark, the catch is then shut down on the fan. On entering the water the catch is lifted and the fan revolves until bottom is reached, when the fan stops and the catch falls on it. It registers the depth in fathoms.

Walker's Harpoon Sounding Machine.

THE first wheel marks one revolution in 30 fathoms; and the second, one in 150 fathoms.

The dials are set by turning the button on the first dial until the pointer is opposite the number 30, and the index of the wheel on the other side of the machine points to 150.

Burt's Nipper and Buoy.

Is a small metal apparatus, similar to a snatch block, attached to a canvas conical bag which acts as a buoy; the line is snatched in the block, the whole forming a sort of floating pulley. As long as the strain of the descending lead is on, the lead line runs freely, at the same time the bag is, to a certain extent, submerged. On the strain being removed, by the lead reaching the bottom, the line is nipped by a spring, and the bag rising to the surface gives notice that the *sounding* has been obtained, and that the line may be run in.

Log Line

Is marked with a piece of bunting at the end of the stray line.
One knot at every half-knot.
A piece of leather at the first knot.
Two knots at two knots.
Three knots at three knots, and so on.

The log line is measured to correspond with a glass running 28 seconds. When the 14 second glass is used, the distance on the log-line must be doubled.

To calculate the length of a knot on the log line.

$3600^s : 28^s :: 6080$-ft. : required length $= 47$-ft. 3in.

The log line should be repeatedly examined, by comparing each knot with the distance between the nails, which are (or should be) placed on the quarter-deck for this purpose, at the proper distance.

The line should be wet whenever it is required to re-measure it, or to verify the marks.

When heaving the log, the log ship should be thrown out well to leeward to clear the eddies near the wake, and in such a manner that it may enter the water perpendicularly, and not fall flat upon it - before a heavy sea the line should be paid out rapidly when the stern is rising, but when the stern is falling, as this motion slacks the line, the reel should be retarded.

The sand-glasses should frequently be tested by a good watch, as in damp weather the sand does not run freely, and sometimes hangs in the neck of the glass.

Massey's Patent Log is a piece of mechanism which being towed from the quarter, with line enough to clear the eddy in the wake of the ship, shows the distance actually gone through the water by means of the revolutions of a fly, which are registered upon a dial plate to knots and tenths. Patent logs should be tested when ships are running the measured mile.

Walker's Harpoon Ship Log is on the same principle, but the rotator is a continuation of the part that holds the wheelwork. The dials are the same as in Massey's.

Table for converting French Mètres and Décimètres into English Feet and Fathoms.

By Robert C. Carrington, F.R.G.S., Hydrographic Office, Admiralty.

1 French Mètre = 39·3708 English Inches.

Mètres.	Feet.	Fathoms.	Mètres.	Feet.	Fathoms.	Mètres.	Feet.	Fathoms.	Mètres.	Feet.	Fathoms.	Mètres.	Feet.	Fathoms.
0·			2·	6½	1	4·	13	2⅛	6·	19½	3¼	8·	26¼	4⅜
·1	⅜	—	·1	6⅝	1⅛	·1	13⅜	2¼	·1	20	3⅜	·1	26⅜	4⅜
·2	⅝	—	·2	7⅛	1⅛	·2	13¾	2¼	·2	20⅜	3⅜	·2	26⅞	4½
·3	1	⅛	·3	7½	1¼	·3	14	2⅜	·3	20⅞	3½	·3	27	4½
·4	1¼	¼	·4	7⅞	1⅜	·4	14⅜	2⅜	·4	21	3½	·4	27½	4⅝
·5	1⅝	¼	·5	8¼	1⅜	·5	14¾	2½	·5	21½	3⅝	·5	27¾	4⅝
·6	2	⅜	·6	8½	1½	·6	15	2½	·6	21⅞	3⅝	·6	28	4⅔
·7	2¼	⅜	·7	8⅞	1⅝	·7	15⅜	2⅝	·7	22	3⅔	·7	28⅜	4⅔
·8	2⅝	½	·8	9	1⅝	·8	15¾	2⅝	·8	22½	3¾	·8	28¾	4¾
·9	3	½	·9	9½	1½	·9	16	2¾	·9	22¾	3¾	·9	29	4¾
1·	3¼	½	3·	9¾	1½	5·	16¼	2¾	7·	23	3¾	9·	29½	4⅞
·1	3⅜	⅝	·1	10⅛	1⅝	·1	16⅝	2¾	·1	23¼	3⅞	·1	29¾	5
·2	4	⅝	·2	10½	1¾	·2	17	2⅞	·2	23½	3⅞	·2	30	5
·3	4¼	¾	·3	10¾	1¾	·3	17⅜	2⅞	·3	23¾	3⅞	·3	30½	5
·4	4⅔	¾	·4	11	1⅞	·4	17¾	3	·4	24	4	·4	30¾	5
·5	5	⅞	·5	11¼	1⅞	·5	18	3	·5	24⅜	4	·5	31	5⅛
·6	5¼	⅞	·6	11¾	1⅞	·6	18⅜	3	·6	24¾	4	·6	31⅜	5¼
·7	5½	⅞	·7	12	2	·7	18¾	3	·7	25	4⅛	·7	31¾	5¼
·8	5¾	1	·8	12½	2	·8	19	3⅛	·8	25⅜	4⅛	·8	32	5⅜
·9	6¼	1	·9	12¾	2	·9	19½	3⅛	·9	25⅝	4¼	·9	32¼	5⅜

Table for converting French Mètres and Décimètres into English Feet and Fathoms.—*(continued.)*

Mètres.	Feet.	Fathoms.	Mètres.	Feet.	Fathoms.	Mètres.	Feet.	Fathoms.	Mètres.	Feet.	Fathoms.	Mètres.	Feet.	Fathoms.
10	32¾	5⅜	13	42¾	7	16	52¼	8¾	19	62⅓	10⅓	22	72	12
·1	33	5⅜	·1	43	7⅛	·1	52¾	8¾	·1	62⅔	10⅜	·1	72¼	12
·2	33¼	5½	·2	43⅓	7¼	·2	53	8¾	·2	63	10½	·2	72¾	12
·3	33¾	5½	·3	43⅔	7¼	·3	53⅓	9	·3	63⅓	10½	·3	73	12¼
·4	34	5⅝	·4	44	7⅜	·4	53¾	9	·4	63⅔	10½	·4	73¼	12¼
·5	34¼	5⅝	·5	44¼	7⅜	·5	54	9	·5	64	10⅔	·5	73¾	12¼
·6	34¾	5¾	·6	44¾	7½	·6	54¼	9	·6	64¼	10⅔	·6	74	12⅓
·7	35	5¾	·7	45	7½	·7	54¾	9	·7	64¾	10¾	·7	74¼	12⅓
·8	35¼	5⅞	·8	45¼	7½	·8	55	9⅛	·8	65	10¾	·8	74¾	12⅓
·9	35¾	4⅞	·9	45¾	7⅝	·9	55¼	9¼	·9	65¼	11	·9	75	12½
11	36	6	14	45¾	7⅝	17	55¾	9¼	20	65½	11	23	75¼	12½
·1	36¼	6	·1	46¼	7¾	·1	56	9¼	·1	66	11	·1	75¾	12⅔
·2	36¾	6	·2	46½	7¾	·2	56¼	9⅜	·2	66¼	11	·2	76	12⅔
·3	37	6⅛	·3	47	7¾	·3	56¾	9⅜	·3	66⅔	11	·3	76¼	12¾
·4	37¼	6⅛	·4	47¼	7¾	·4	57	9½	·4	67	11¼	·4	76¾	12¾
·5	37¾	6¼	·5	47½	8	·5	57¼	9½	·5	67¼	11¼	·5	77	12⅞
·6	38	6¼	·6	47¾	8	·6	57¾	9⅝	·6	67½	11¼	·6	77¼	12⅞
·7	38⅓	6⅜	·7	48¼	8	·7	58	9⅝	·7	67¾	11⅓	·7	77¾	13
·8	38¾	6⅜	·8	48½	8	·8	58¼	9¾	·8	68¼	11⅓	·8	78	13
·9	39	6½	·9	48¾	8	·9	58¾	9¾	·9	68½	11½	·9	78¼	13
12	39¼	6½	15	49	8⅛	18	59	9¾	21	68¾	11½	24	78¾	13
·1	39¾	6⅝	·1	49¼	8¼	·1	59½	10	·1	69	11½	·1	79	13⅛
·2	40	6⅝	·2	49¾	8¼	·2	59¾	10	·2	69¼	11½	·2	79¼	13⅛
·3	40¼	6⅔	·3	50	8¼	·3	60	10	·3	69¾	11⅝	·3	79¾	13⅓
·4	40¾	6⅔	·4	50¼	8⅜	·4	60¼	10	·4	70	11⅝	·4	80	13⅓
·5	41	6¾	·5	50¾	8⅜	·5	60¾	10	·5	70¼	11⅔	·5	80¼	13⅜
·6	41⅓	6¾	·6	51	8½	·6	61	10⅛	·6	70½	11⅔	·6	80¾	13⅜
·7	41¾	7	·7	51¼	8½	·7	61¼	10¼	·7	71	11¾	·7	81	13½
·8	42	7	·8	51½	8½	·8	61¾	10¼	·8	71¼	11¾	·8	81¼	13½
·9	42¼	7	·9	52	8⅝	·9	62	10⅜	·9	71½	12	·9	81½	13¾

SPHEROIDAL TABLES.

SHOWING the length in feet of a degree, minute, and second, of latitude and longitude, with the corresponding number of statute miles in each degree of latitude, and the number of Minutes of latitude, or Nautical miles, contained in a degree of longitude, under each parallel of latitude.—For every degree.

From the Tables by R. C. CARRINGTON, F.R.G.S.

	Latitude.					Longitude.			
Latitude.	Length of one degree in statute miles.	Length in feet of a			Latitude.	Length of one degree in minutes of latitude or nautical miles.*	Length in feet of a		
		Degree.	Minute.	Second.			Degree.	Minute.	Second.
0°	68·704	362755·6	6045·93	100·77	0°	60·410	365233·7	6087·23	101·454
1°	68·704	362756·7	6045·94	100·77	1°	60·400	365178·4	6086·31	101·438
2°	68·704	362760·1	6046·00	100·77	2°	60·373	365012·7	6083·54	101·392
3°	68·706	362765·7	6046·09	100·77	3°	60·326	364736·5	6078·94	101·316
4°	68·707	362773·6	6046·23	100·77	4°	60·261	364350·0	6072·50	101·208
5°	68·709	362783·6	6046·39	100·77	5°	60·177	363853·2	6064·22	101·070
6°	68·711	362795·9	6046·60	100·78	6°	60·074	363246·3	6054·11	100·902
7°	68·714	362810·4	6046·84	100·78	7°	59·954	362529·5	6042·16	100·703
8°	68·717	362827·1	6047·12	100·79	8°	59·814	361703·0	6028·38	100·473
9°	68·721	362846·0	6047·43	100·79	9°	59·656	360767·0	6012·78	100·213
10°	68·725	362866·9	6047·78	100·80	10°	59·480	359721·7	5995·36	99·923
11°	68·729	362890·1	6048·17	100·80	11°	59·285	358567·6	5976·13	99·602
12°	68·734	362915·2	6048·59	100·81	12°	59·072	357304·8	5955·08	99·251
13°	68·739	362942·5	6049·04	100·82	13°	58·841	355933·9	5932·23	98·871
14°	68·745	362971·8	6049·53	100·83	14°	58·592	354455·1	5907·59	98·460
15°	68·751	363003·1	6050·05	100·83	15°	58·325	352868·8	5881·15	98·019
16°	68·757	363036·3	6050·61	100·84	16°	58·040	351175·7	5852·93	97·549
17°	68·764	363071·4	6051·19	100·85	17°	57·737	349376·0	5822·93	97·049
18°	68·771	363108·4	6051·81	100·86	18°	57·416	347470·5	5791·18	96·520
19°	68·778	363147·3	6052·46	100·87	19°	57·077	345459·5	5757·66	95·961
20°	68·786	363187·9	6053·13	100·89	20°	56·722	343343·7	5722·40	95·373
21°	68·794	363230·2	6053·84	100·90	21°	56·348	341123·7	5685·40	94·756
22°	68·801	363274·3	6054·57	100·91	22°	55·958	338800·1	5646·67	94·111
23°	68·810	363320·0	6055·33	100·92	23°	55·550	336373·6	5606·23	93·437

* The figures in this column, divided by 6, will give the length, in cables, of a *minute of longitude* in its corresponding latitude, thus: in latitude 51°=37·861÷6 =6·31 cables in a minute of longitude.

SPHEROIDAL TABLES.

Latitude.					Latitude.	Length of one degree in minutes of latitude or nautical miles	Longitude.		
	Length of one degree in statute miles.	Length in feet of a					Length in feet of a		
Latitude.		Degree.	Minute.	Second			Degree.	Minute.	Second
24°	68.819	363367.2	6056.12	100.94	24°	55.125	333845.0	5564.08	92.735
25°	68.828	363416.0	6056.93	100.95	25°	54.684	331214.9	5520.25	92.004
26°	68.838	363466.2	6057.77	100.96	26°	54.225	328484.1	5474.74	91.245
27°	68.848	363517.9	6058.63	100.98	27°	53.751	325653.4	5427.56	90.459
28°	68.858	363570.8	6059.51	100.99	28°	53.259	322723.6	5378.73	89.645
29°	68.868	363625.0	6060.42	101.01	29°	52.751	319695.6	5328.26	88.804
30°	68.879	363680.5	6061.34	101.02	30°	52.228	316570.3	5276.17	87.936
31°	68.889	363737.1	6062.29	101.04	31°	51.688	313348.5	5222.48	87.041
32°	68.900	363794.8	6063.25	101.05	32°	51.133	310031.2	5167.19	86.119
33°	68.912	363853.5	6064.23	101.07	33°	50.562	306619.5	5110.33	85.172
34°	68.923	363913.1	6065.22	101.09	34°	49.976	303114.2	5051.90	84.198
35°	68.934	363973.6	6066.23	101.10	35°	49.375	299516.4	4991.94	83.199
36°	68.946	364034.9	6067.25	101.12	36°	48.758	295827.2	4930.45	82.174
37°	68.958	364096.8	6068.28	101.14	37°	48.127	292047.7	4867.46	81.124
38°	68.970	364159.5	6069.33	101.16	38°	47.481	288178.9	4802.98	80.050
39°	68.982	364222.6	6070.38	101.17	39°	46.821	284222.0	4737.03	78.951
40°	68.994	364286.3	6071.44	101.19	40°	46.146	280178.2	4669.64	77.827
41°	69.006	364350.4	6072.51	101.21	41°	45.459	276048.7	4600.81	76.680
42°	69.018	364414.9	6073.58	101.23	42°	44.757	271834.7	4530.58	75.509
43°	69.030	364479.6	6074.66	101.24	43°	44.042	267537.5	4458.96	74.316
44°	69.042	364544.4	6075.74	101.26	44°	43.313	263158.3	4385.97	73.108
45°	69.055	364609.4	6076.82	101.28	45°	42.571	258698.4	4311.64	71.861
46°	69.067	364674.4	6077.91	101.30	46°	41.817	254159.2	4235.99	70.600
47°	69.079	364739.3	6078.99	101.32	47°	41.050	249541.9	4159.03	69.317
48°	69.092	364804.1	6080.07	101.33	48°	40.270	244848.2	4080.80	68.013
49°	69.104	364868.6	6081.14	101.35	49°	39.479	240079.2	4001.32	66.689
50°	69.116	364932.9	6082.22	101.37	50°	38.676	235236.5	3920.61	65.343
51°	69.128	364996.8	6083.28	101.39	51°	37.861	230321.4	3838.69	63.978
52°	69.140	365060.2	6084.34	101.41	52°	37.035	225335.5	3755.59	62.593
53°	69.152	365123.1	6085.39	101.42	53°	36.198	220280.3	3671.34	61.189
54°	69.164	365185.4	6086.42	101.44	54°	35.350	215157.2	3585.95	59.766
55°	69.176	365247.0	6087.45	101.46	55°	34.492	209968.0	3499.47	58.324
56°	69.187	365307.9	6088.47	101.47	56°	33.623	204714.0	3411.90	56.865
57°	69.198	365367.9	6089.47	101.49	57°	32.745	199396.9	3323.28	55.388

SPHEROIDAL TABLES.

Latitude					Latitude				
	Length of one degree in statute miles.	Length in feet of a				Length of one degree in minutes of latitude or nautical miles.	Length in feet of a		
Latitude		Degree.	Minute.	Second	Latitude		Degree.	Minute.	Second
58°	69·210	365427·0	6090·45	101·51	58°	31·856	194018·3	3233·64	53·894
59°	69·221	365485·1	6091·42	101·52	59°	30·958	188579·9	3143·00	52·383
60°	69·231	365542·2	6092·37	101·54	60°	30·051	183083·3	3051·39	50·856
61°	69·242	365598·1	6093·30	101·56	61°	29·135	177530·1	2958·84	49·314
62°	69·252	365652·9	6094·22	101·57	62°	28·211	171922·1	2865·37	47·756
63°	69·263	365706·4	6095·11	101·59	63°	27·278	166261·0	2771·01	46·184
64°	69·272	365758·5	6095·98	101·60	64°	26·337	160548·6	2675·81	44·597
65°	69·282	365809·3	6096·82	101·61	65°	25·388	154780·3	2579·77	42·996
66°	69·291	365858·6	6097·64	101·63	66°	24·432	148976·3	2482·94	41·382
67°	69·300	365906·4	6098·44	101·64	67°	23·468	143120·2	2385·34	39·756
68°	69·309	365952·7	6099·21	101·65	68°	22·498	137219·7	2287·00	38·116
69°	69·318	365997·3	6099·96	101·67	69°	21·521	131276·7	2187·95	36·466
70°	69·326	366040·2	6100·67	101·68	70°	20·538	125293·2	2088·22	34·804
71°	69·333	366081·3	6101·36	101·69	71°	19·548	119270·7	1987·85	33·131
72°	69·341	366120·7	6102·01	101·70	72°	18·553	113211·4	1886·86	31·448
73°	69·348	366158·2	6102·64	101·71	73°	17·553	107116·9	1785·28	29·755
74°	69·355	366193·9	6103·23	101·72	74°	16·547	100989·1	1683·15	28·053
75°	69·361	366227·6	6103·79	101·73	75°	15·536	94830·1	1580·50	26·342
76°	69·367	366259·6	6104·32	101·74	76°	14·521	88641·6	1477·36	24·623
77°	69·373	366289·1	6104·82	101·75	77°	13·502	82425·6	1373·76	22·896
78°	69·378	366316·7	6105·28	101·75	78°	12·478	76184·0	1269·73	21·162
79°	69·383	366342·1	6105·71	101·76	79°	11·451	69918·8	1165·31	19·422
80°	69·387	366365·8	6106·10	101·77	80°	10·421	63631·8	1060·53	17·676
81°	69·391	366387·1	6106·45	101·77	81°	9·388	57325·2	955·42	15·924
82°	69·395	366406·3	6106·77	101·78	82°	8·352	51000·6	850·01	14·167
83°	69·398	366423·2	6107·05	101·78	83°	7·313	44660·3	744·34	12·406
84°	69·401	366438·0	6107·30	101·79	84°	6·272	38506·1	638·44	10·641
85°	69·403	366450·5	6107·51	101·79	85°	5·230	31939·9	532·33	8·872
86°	69·405	366460·7	6107·68	101·79	86°	4·186	25563·9	426·07	7·101
87°	69·407	366468·7	6107·81	101·80	87°	3·140	19179·8	319·66	5·328
88°	69·408	366474·4	6107·91	101·80	88°	2·094	12789·9	213·17	3·553
89°	69·409	366477·9	6107·97	101·80	89°	1·047	6395·9	106·60	1·777
90°	69·409	366479·0	6107·98	101·80	90°	0·0	0·0	0·0	0·0

NOTE.—These tables have been calculated for every ten minutes of latitude, and are published by J. D. Potter, 31, Poultry, E.C.

TIDES.

The success of the sailor and pilot depends very much upon a thorough and ready knowledge of the many *important simplicities* of his calling: such knowledge necessarily endows him with that promptness in action, and fearlessness in conduct, for which members of his profession have ever been pre-eminently distinguished. There are few points of his education which require more attention, than the study of that important movement of the waters, known as Tides. Although his numerous duties will not permit of an attempt at an investigation of the bewildering theories that surround the subject, still he may easily acquire the requisite knowledge to enable him to discover the rise and fall of the tides, with the direction and velocity of the tidal streams, with sufficient accuracy for all practical purposes, as the knowledge of the rise and fall of a foot, or the set of half a knot per hour, is often of invaluable service to the sailor in difficulties.

The Admiralty charts, Tide tables and Sailing directions, give considerable information upon this point, and on entering a new chart the sailor's early attention should be given to the tides thereon delineated, remembering, that the soundings on Admiralty charts are all given for the *mean low water spring tides.* *

The Tide is the rising and falling of the sea, which takes place about twice a day, becoming however each day later by half an hour, to an hour. The rising of the sea is called FLOOD TIDE, the falling EBB TIDE.

The tides do not always rise to the same height, but every fortnight, after the new and full moon, they become much higher than they were in the alternate weeks, or after the first and last quarters of the moon. These high tides are called SPRING TIDES, and the low ones, NEAP TIDES.

---

* The Admiralty Manual of Scientific Inquiry furnishes a paper on the Tides, by the Rev. Dr. Whewell, in which the subject is fully treated. There is also a small work, entitled "The Tides," published under the direction of the Committee of General Literature and Education, appointed by the Society for promoting Christian Knowledge, from which much useful information may be obtained. See also, the *Tide tables* published yearly, by the Hydrographic Office.

Those spring tides are the highest which follow the day of New or Full Moon by one, two, or three days, depending on the locality: for instance, the highest tides take place on the West Coast of Ireland and on the South Coast of England, three transits after the New and Full Moon, unless diverted by gales of wind or other extraordinary causes.

Along the East Coast of England, the highest tides take place four transits after the New and Full Moon. In the river Thames they occur five transits after the same epoch. These differences arise from the fact, that the same tide-wave, which produces high water on the West Coast of Ireland, takes half a day in its progress thence to the East Coast of England, and a whole day before it arrives in the river Thames.

When the moon is in perigee, or nearest to the earth, the rise and fall is sensibly increased.

The Spring tides are greater about the times of the Equinoxes, i.e., about the latter end of March and September, than at other times of the year; and the neap tides then are less.

The height from low water to high water is called *the range of the tide*. On the coasts of England it is greatest at the head of the Bristol Channel where spring tides sometimes rise 47 feet. In some parts of the coast the range does not exceed 4 feet. The times at which high water takes place are equally various. Thus the same day that it is high water at Dover, at Noon, it is not high water at London until 3h. p.m., and at Hull until 7h. 30m. p.m.

The time of high water at any particular place is the same on the days both of New and Full Moon, and is termed the "Establishment of the Port;" this being ascertained by observation, the intermediate tides can be deduced from it.

The Diurnal Inequality is a feature in tidal phenomena, which being peculiarly small in British waters, has not received the attention it merits from the English sailor, for in the Indian seas, and indeed in most other parts of the globe, this diurnal inequality is a regular change, considerable in amount, and almost universal in *prevalence; this* change depends upon the moon being north or south of the equator; its maximum corresponds to, but is not necessarily simultaneous with, the moon's greatest declination, and the period of its vanishing corresponds in like manner with the time of the moon passing the equator.

In consequence of the diurnal inequality, it sometimes happens that the afternoon tides are higher than the forenoon tides, or the reverse, for many weeks together. And hence it has sometimes been stated as a rule, at such places, that the afternoon tides are always the highest, or the reverse. But this is not the rule. The rule of the diurnal inequality depends on the moon's declination. If the afternoon tides are the highest at one time of the year, they are the lowest at another.

The diurnal inequality sometimes affects the time of high water as much as two hours, that of low water about 40 minutes; at the same time a variation of 12 inches may be observed in the height of high water, and of 36 inches in that of low water. Such effects are far too great to be neglected either in the prediction of tides or the reduction of soundings.

The following notes, taken from the Admiralty Tide Tables, will be found useful to the sailor when navigating the seas to which they refer.

English Channel.—In the Solent, and as far to the Westward as Portland, there are what are termed the *first* and *second* high waters. This double high water is probably caused by the tidal stream at Spithead, for, as long as that stream runs strong to the Westward the tide is kept up in Southampton water, and there is no fall, of consequence, until the stream begins to slack at Spithead, but when the stream makes to the Eastward at Spithead, the water falls at Southampton. After low water, the tide rises there steadily for 7 hours, which may be considered as the *first* or proper high water; it then ebbs for an hour about 9 inches, at the end of which time it again commences to rise, and in about 1¼ hours reaches its former level, and sometimes higher; this is called the *second* high water. To the mariner, the knowledge that the high water at Southampton remains nearly stationary *for rather more than 2 hours* may, in some cases, be important. Similar *first* and *second* high waters occur on either shore of the Solent, as shown in the times of high water at full and change, in the tide tables.

At Havre, on the French coast, the high water remains stationary for one hour, with a rise and fall of 3 or 4 inches for another hour, and only rises and falls 13 inches for the space of 3 hours; this long period of nearly slack water is very valuable to the traffic of the port, and allows from 15 to 16 vessels to enter or leave the docks on the same tide.

Rio de la Plata.—After heavy gales from S.E. to S.W., the water in the Rio de la Plata may rise 8 feet above the soundings shown in the Admiralty Chart; and continued winds from N.N.E. to N.N.W., may cause the water to fall to 4 feet less than the soundings.

Gulf of Mexico.—The tides of ports in the Gulf of Mexico, westward of Cape St. George on the North East shore of the Gulf, are usually single day tides, the rise and fall increasing or decreasing with the moon's declination. The rise and fall being so small, the times and heights are both much influenced by the winds. Between Cape St. George and Cape Florida there are two tides during the twenty-four hours, subject to a large diurnal inequality.

Indian Ocean.—The tides at Karàchi, Bombay, and probably the other ports in India, are subject to a large diurnal inequality, which may either accelerate or retard the times of high water, sometimes to the amount of an hour and a half, or two hours, and increase or diminish the rise by a foot or more. *See Tide tables for Karàchi.*

In the River Hoogly the night tides are highest from November to February; the day tides highest from March to October.

China Seas.—The low water at Singapore is affected by a large diurnal inequality, amounting at times to six feet.

At Sarawak River the highest tides occur at the change of the monsoons, viz., May and November. In the N.E. monsoon the higher tides occur at the new moon, and those of the day are higher and more regular than those of the night; while during the S.W. *monsoon* the contrary takes place, and the higher tides are then at *full moon.*

At Whampoa Docks, near Canton, in March, the day and night tides rise to the same level. From April to October, the day tides are the higher, and from November to February the lower. In May and June the level of spring tides is 4 feet, and the neaps 2 feet higher than in March.

From tidal observations made at Shanghae by the engineers to the Customs for the last six months of 1872, the night tides in July and in the following three months, average considerably higher than the day ones. The reverse occurs in the months of November and December. At the Langshan Crossing the tide rises for 3 hours only, and falls for 9 hours.

Australia.—The low water of Port Essington, on the north coast of Australia, is affected by diurnal inequality.

At Port Adelaide, South Australia, between September and February, the a.m. tides were found to rise higher than the p.m. Between March and August the reverse is said to be the case. About the neaps the tides are very irregular.

At Port Augusta, when the wind veers round to west and south, and blows strong, the rise has been as much as 16 feet.

At King George Sound there is a large diurnal inequality of the tides, which sometimes reduces the two daily tides to one.

California and Oregon.—The tides on these coasts are of so complicated a character that the following general explanation is considered necessary :—There are generally in each twenty-four hours, or rather in each lunar day of 24 h. 50 m., two high and two low waters, which are unequal in height and in time in proportion to the moon's declination, differing most from each other when the moon's declination is greatest, and least when the moon is on the equator. The high and low waters generally follow each other thus: starting from the lowest low water, the tide rises to the lower of the two high waters (sometimes improperly called "half-tide"), then falls slightly to a low water (which is sometimes merely indicated by a long stand); then rises to the highest high water, whence it falls again to the lowest low water.

The tides at Sitka, on the West coast of North America, are affected by diurnal inequality.

Wind.—It remains to be noticed that the directions of strong winds, as well as the varying pressure of the atmosphere, considerably affect both the times and the heights of high water. Thus, in the North Sea, a strong N.N.W. gale and a low barometer raise the surface 2 or 3 feet higher than the predicted heights, and cause the tide to flow all along the coast from the Pentland Firth to London half an hour longer than the times predicted in the Tables. Easterly, S.E., and S.W. winds produce opposite effects, which will be felt as far down the Channel as Dungeness. On the contrary, at the entrance of the Channel, at Plymouth, and as far up as Portland, south-westerly winds, with a low barometer, raise the surface of the water; and north-easterly winds and a high barometer always lower it.

Tidal Streams.—Besides the acquaintance with the periods of high and low water, and the comparative height of the day and night tides, it will be necessary to observe the *direction* of the stream of flood and of ebb, and the *time* at which the stream turns; but care must be taken not to confound the time of the *turn of the tide-stream* with the time of high water. Mistakes and errors have often occurred by supposing that the turn of the tide-stream is the time of high water. But this is not so. The turn of the stream generally takes place at a different time from high water, except at the head of a bay or creek. The stream of flood commonly runs for some time, often for hours, after the time of high water. In the same way, the stream of ebb runs for some time after low water.

Tide and Half Tide.—The sailor should, therefore, be careful not to confuse himself by the words "flood" and "ebb stream," but rather to use the terms "east" or "west" going stream, as the case may be. In the English Channel, for example, the eastern stream runs up 3 and 4 hours after high water by the shore abreast, and the water is falling; or it is ebb tide in the harbours, while the eastern, or flood stream, as it is called, is still running up, forming what is known to Pilots as "Tide and half Tide." In the same manner, in the North Sea, the north-east going stream (or ebb as it is called) makes to the eastward 2 hours before it is high water at Dunkerque, Ostend, &c.

The time at which the stream turns is often different at different *distances* from the shore; but the time of high water is not *necessarily* different at these points. The time of *slack water* is not wanted for a theory of tides, though its knowledge is otherwise of considerable importance to seamen.

With regard to the streams of flow and ebb, they are often not merely two streams in opposite directions at different times of the tide; they generally turn successively into several directions, so as to go quite round the compass in one complete tide, either in the direction N., E., S., W. (with the sun), or N., W., S., E. (against the sun).

The tides which take place far up deep bays, sounds, and rivers, are *later* than the tides at the entrance of such inlets, but they are not more irregular; on the contrary, the tides in such situations are often remarkably *regular*.

The tide in its progress up inlets and rivers is often much magnified and modified by local circumstances. Sometimes it is magnified so that the wave which brings the tide at one period of its rise advances with an abrupt front of broken water. This is called a *bore* (as in the Severn, the Garonne, the Amazon). Sometimes the tide is divided into two half-day tides in its progress up a river, as in the Forth in Scotland. In all cases, after a certain point, the tide dies away in ascending a river.

The above information can generally be obtained from the charts and sailing directions, which should therefore be always carefully studied for that purpose.

Instructions for the use of the Tide Pole.

In all localities where operations for Sounding are carried on, a sheltered corner, should, if possible, be found, in which the Tide pole may be erected. A broken oar is driven into the sand, and supported by guys of spun yarn, or secured by stones or shot among the rocks, (the place chosen being in about two feet water), at low water, protected as much as possible, and yet with the sea having free access to it. To this oar a painted batten is lashed, conspicuously marked, the alternate feet being painted black and white, the figures 6 inches long, painted black on the white ground, and *v.v.*; such a batten can be read from some distance with a telescope. The time is noted for the rise and fall of *every 3 inches*, or the state of the pole marked every

15 minutes, great care being taken to determine as near as possible the times of high and low water, by taking a mean of the times at which the water was at equal heights before or after high or low water, on the same principle that apparent noon is found, by adding the times of the equal altitudes of the sun together, and dividing by 2. The state of the wind should be always noted.

If in port at the time of the full and change of the moon, it is advisable to especially observe the high and low water on that day, remembering also that the greatest rise and fall occurs from the third to the fifth tide after the full or change.

The times and heights of high and low water should, if possible, be observed successively both by day and night; and the watch or clock by which the times are taken should always show *mean time at the place.* In order to determine accurately the times and heights of high and low water, the method recommended and generally adopted is, from half an hour before to half an hour after high water (or low water), to note every ten minutes the place of the tide on the gauge, and then by interpolation ascertain the time and height of high water, or low water, as the case may be, for insertion in the Register.

The greater the number of daily results thus obtained the more accurate will be the value deduced from them; if, however, circumstances will not admit of a continuous series of tidal observations being made, the high and low water that take place on those days when the moon's transit occurs between 11h. and 2h., and 6h. and 8h., should be preferred, as from the former (11h. and 2h.) the Tide Hour or Establishment, and the Mean Spring Range are obtained, and from the latter, (6h. and 8h.) the Mean Neap Range is known.*

If there should be no available spot for the erection of the Tide pole, or if the Tide pole is not near enough to be noted from the ship (which may often be done by means of a telescope), supposing the ship to be moored. and the nature of the bottom at the anchorage level, a good idea of the daily rise and fall may be obtained by noting the

---

* See " Directions for reducing Tidal Observations," by the late Staff-Commander *John Burdwood,* R.N., published by the Hydrographic Office.

depth alongside, by the method above described, with a well stretched lead line carefully marked in feet and inches.

From one or other of these methods of observations, a daily tide table may be made, showing, for every hour of the day, the number of feet to be subtracted from the soundings taken, to reduce them to low-water springs.

The soundings should be reduced very shortly after the return to the ship, by the officer who obtained them, and not entered even on the rough before they have been so reduced.

If the ship's stay is not long enough to get the times of high water at the full and change, that fact should be especially noted on the chart, and the rise and fall termed *approximate;* in this case, a comparison of the times and heights obtained, with those of some of the ports given in the Admiralty tide tables, that at the same age of the moon have a similar rise and fall, will enable an observer to form a tolerably correct idea of the necessary amount of reduction.

To determine the rise and fall of the tide, and to reduce the soundings accurately, require very considerable tact, judgment, experience, and patience; and these must not be given grudgingly, as the knowledge of the rise and fall of a foot is often of invaluable service to a sailor in difficulties.

For all ordinary purposes the following Table, for Reducing Soundings to the Mean Low Water Spring Tides, will be found sufficiently correct.

At Spring Tides.

At the 1st hour, before and after high-water, deduct $1\frac{1}{2}$ ⎫

,, 2nd ,, ,, ,, ,, ,, $\frac{3}{4}$

,, 3rd ,, ,, ,, $\frac{1}{2}$ ⎬ Of the full rise at Springs.

,, 4th ,, ,, ,, $\frac{1}{4}$

,, 5th ,, ,, ,, $\frac{1}{12}$

,, 6th ,, ,, 0 ⎭

At Neaps.

At the 1st hour, before and after high-water, deduct $\frac{1}{6}$ ⎫
,, 2nd ,, ,, ,, ,, ,, $\frac{2}{6}$ ⎪ Of
,, 3rd ,, ,, ,, $\frac{3}{6}$ ⎬ ordinary
,, 4th ,, ,, ,, $\frac{1}{6}$ ⎪ rise at
,, 5th ,, ,, ,, $\frac{1}{6}$ ⎪ springs.
,, 6th ,, ,, ,. .. .. 0 ⎭

The following diagram is intended to explain the terms Spring Rise, Neap Rise, and Neap Range, as made use of on the Admiralty Charts and in the Sailing Directions published by the Admiralty.

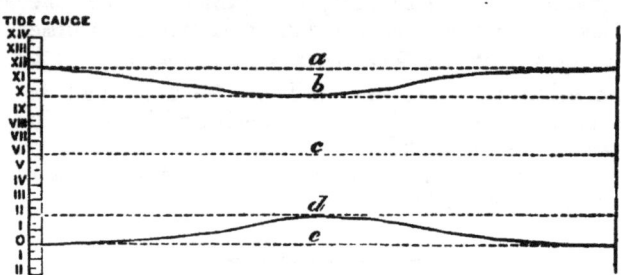

$a=$ Mean Level of High Water Ordinary Springs.
$b=$,,   ,,   ,,   Neaps.
$c=$ Half Tide or Mean Level of the sea both at Springs and Neaps.
$d=$ Mean Level of Low Water Ordinary Neaps.
$e=$ ..   ..   ,,   Springs.

Example.

Spring Rise (or Mean Spring Range) $=e$ to $a=$ 12ft.
Neap Rise . . . . $=e$ to $b=$ 10ft.
Neap Range . . . . $=d$ to $b=$ 8ft.

DETERMINING POSITIONS, &c.

When in sight of land a common method of fixing a ship or boat is by observing two objects that are in transit, *i.e.*, situated in the same line as the ship or boat, and taking an angle between them and some well known object to the right or left, taking care that the angle measured is not less than 25°. A line is then ruled through the objects in transit, any point is taken in it, and from this the angle observed is laid off in the direction of the third object; a line, ruled through that object, parallel to the line forming the angle, will intersect the line ruled through the objects in transit on the position of the ship.

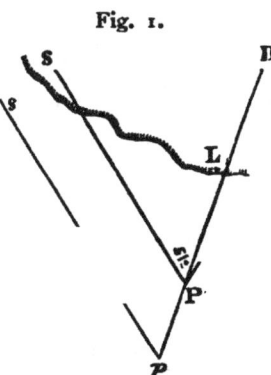

Fig. 1.

Thus in fig. 1, two points, D and L, whose positions are known, are observed to be in transit from a station P, and an angle LPS of 51° is measured between them and a third known point S; through D and L a line DL*p* is drawn, and at any point *p* in that line, an angle D*ps*, equal to 51°, is laid off towards S, and a line *ps* drawn; then through S a line SP is drawn parallel to *sp*, cutting DL*p* in P, the position required.

The position of the ship may also be fixed by projecting the angles measured from her between three objects on shore, whose bearings and distances from each other are known to the observer by means of a chart.

In the selection of three objects to fix a position by, the middle object should, if possible, be the nearest to the observer, or the three objects should be in or near the same straight line, or the observer's position should be within the triangle formed by the objects. If the objects and position of the observer lie in or near the circumference of the same circle, the position cannot be fixed without the assistance of a true or compass bearing of one of the objects.*

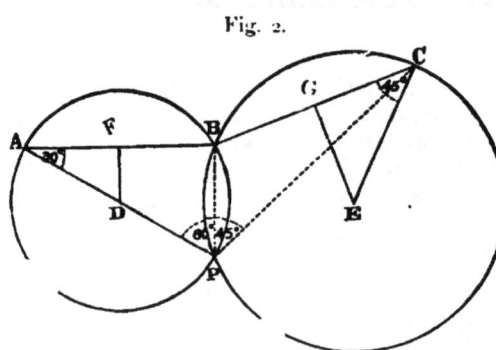

Fig. 2.

**Projecting by Circles.
Case 1.**—Let A, B, and C, be three objects on a chart (see fig. 2): it is required to fix the position of a ship P, by angles; the one measured between C and B being 45°, and that between B and A being 60°.

Join AB, BC, and bisect them in the points F, and G erecting the perpendiculars FD, GE on the same side of the objects as that on which the ship lies. Then at A make the angle FAD=30° the complement of the angle measured between B and A, and at C make the angle GCE=45° the complement of the angle measured between C and B, and draw lines AD, CE, cutting the perpendiculars FD, GE, in D and E. From D and E as centres describe the circles ABP, and CBP, and the intersection of their circumferences at P, will be the position of the ship.

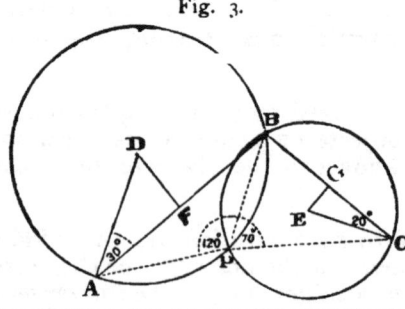

Fig. 3.

Case 2.—Should the angle be more than 90° subtract 90° from it, erect the perpendicular on the opposite side of the line joining the objects to that on which the ship lies, and proceed as before: (see fig. 3,) where the angle measured between C and B is 70°, and that between B and A is 120°.

---

* Where a couple of angles are taken to fix a ship running along a coast, or in a arbour, a compass bearing should always be observed at the same time.

Sec. V. DETERMINING POSITIONS.

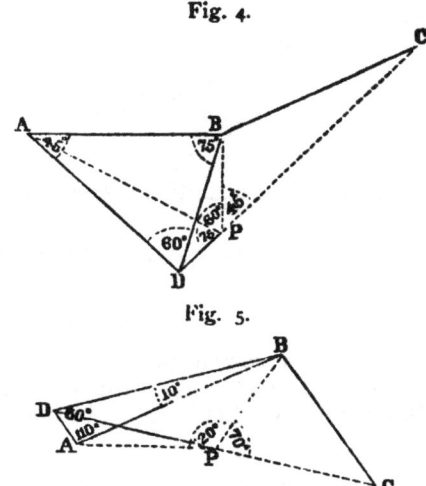

Fig. 4.

Fig. 5.

By the Straight Line Projection. Case 1.—At B, (see fig. 4) on the same side of the objects as that on which the ship lies, make the angle ABD = 75°, or 180° − (60° + 45°): at A make the angle BAD = 45° or the angle subtended by the side BC: join CD, and on CD make the angle BPC = 45°: the point P will be the position required.

Case 2.—When the sum of the angles measured is more than 180° (see fig. 5) at B, on the opposite side of A B to that on which the ship lies, make the angle ABD = 10°, or (120° + 70°) − 180°: at A make the angle BAD = 110°, or 180° − 70°, the angle subtended by BC; join CD; and on CD make the angle BPC = 70°; the point P will be the position required.

A Vessel's position may also be determined by laying off the angles taken between three known objects on a piece of tracing paper with a protractor. Rule upon the tracing paper a straight line representing the line from the ship to the centre object, and from any point on this line set off the angles observed to the right and left, and draw the lines on the tracing paper; lay the tracing paper on the chart, and move the tracing paper over the surface of the chart, until the three lines pass through the respective objects; that done, the point from which the angles where laid off will be the position of the ship.

The following is a useful method of Determining Distance approximately from a Lighthouse, or other object, when running along the land.

Take a bearing of the object, and note the time; see how many points it differs from the course; when this difference is doubled, the ship will be as far from the object as she has run in the interval.

N.B.—Make due allowance for current.

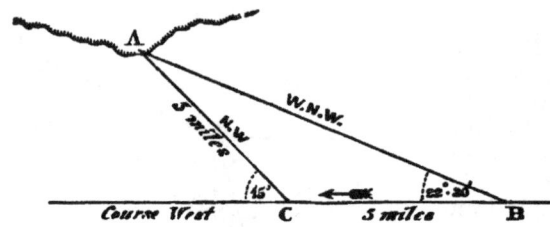

Example.—Steering West, going 10 knots, at 9·45 a.m.; a lighthouse, A, bore W.N.W. Difference between bearing and course = 2 points. Stood on until it bore N.W., (double the difference), time 10.15 a.m.

Distance run = 5 miles = distance of light-house.

The Danger Angle.

The application of this is based on Euclid III., prop. 21, that angles in the same segment of a circle are equal to one another. The Danger Angle should only be used with two well-defined points on a well-constructed chart.

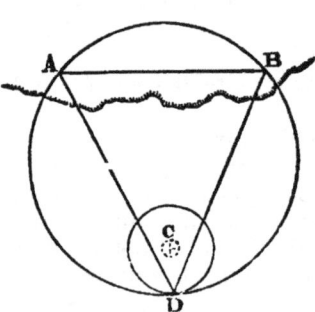

Example.—Suppose C a well-known danger under water, A and B two well-defined points. It is required to pass C at 5 cables distance. Describe a circle round C, with a radius of 5 cables, and on the circumference farthest from A and B, take a point D. Join A B, A D, D B, and describe a circle about the triangle, A D B. The angle A D B = 50°, is the danger angle, which, being set upon a sextant, will, so long as kept on, place the ship on the segment of the circle, A D B.

Should the angle increase, the vessel will be inside the circle, and *vice versa*.

Rule to find the centre of a circle which shall pass through three given points—(Euclid IV., prop. 5):—

Draw straight lines through the points, bisect the lines, and erect *perpendiculars* at the bisecting points: the intersection of these *perpendiculars* will be the required centre.

Finding Distances at Sea.

Commander A. K. Wilson, R.N.

(1) Multiply the height of observer, in feet, from the surface of the water by 573, and divide by the observed angle, in minutes, increased by the dip. The quotient will be the distance in fathoms.

(2) Let d represent distance of the object in miles;

h be the height, in feet, of observer's eye above the water;

h' be the height, in feet, of that portion of the object we are just losing sight of;

Then, $d = \sqrt{\dfrac{3h}{2}} + \sqrt{\dfrac{3h'}{2}}$

Example.—Let observer's eye be 24 feet above the surface of the sea, and let a small vessel be hull down. Assume the hull of this vessel, by her rig, to be 6 feet out of water,

then, $d = \sqrt{\dfrac{3}{2} 24} + \sqrt{\dfrac{3}{2} 6} = 6 + 3 = 9$ miles.

Finding Distance of a Target at Sea.

Measure angle between target and horizon from maintop; add dip for that height, result will be angle subtended at target.

Example.

M S, height of main-top.
T M H, observed angle.
S H M, Dip.
S T M = Dip + observed angle.

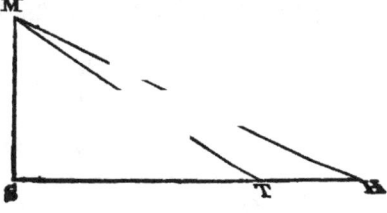

MEASURING DISTANCES.

The distance a vessel is off, when seen from the masthead of another vessel, may be roughly calculated by using the *Table showing the distance of the horizon at different elevations* (see page 197), in the following manner:—

Suppose height of observer's masthead to be 150 feet. A sail is seen from it, with half her topsails visible above the horizon.

An estimate must be formed of the height of half her topsails—say 70 feet.

Height of 70 feet makes the horizon at that height, as 9½ miles distant; and of 150 feet (height of observer's masthead), 14 miles: total distance between the two vessels, 23½ miles.

Raper's method of determining Distances at Sea.

The distance of high land, whose height is known, may be found by measuring its altitude with a sextant.

When angles observed with the sextant are not more than 5°, they should be taken both on and off the arc, added together, and divided by 2. The result is an angle free of index error.

Rule.—To the log. cos. of the dip of the height of the summit observed, add the log. cos. of the difference between the altitude and the dip of the spectator. The sum is the log. cos. of an arc, from which subtract the said difference, and the remainder is the distance in miles, or minutes of a degree.

NOTE.—The altitude must exceed the dip of the spectator.

Example—The summit of Mount Etna is observed to be 1° 30′ above the horizon; the height of the eye 20 feet. Required the distance.

Log. cos. dip of Mount Etna* .... 9·999774
Diff. 1° 30′ − 4′ 17″ = 1° 26′ log. cos. .... 9·999864

    2° 20′   Log. cos. .... 9·999638
  −1 26

54 miles approximate distance.

---

* The log. of cosines of some principal points of high land are here given for

MEASURING DISTANCES.

The summit and the horizon being both raised by refraction in proportion to their distances, the altitude must be corrected by this rule.

Subtract from the altitude observed, one twelfth of the distance, and add to the remainder one twelfth of the dip.

Here one twelfth of 54 miles, or 4′ 30″, being taken from 1° 30′; and one twelfth of 5′, or 25″, being added, gives the true altitude 1° 25′ 55″ ; the sum worked over with this altitude gives the distance 56 miles, which is near enough for ordinary purposes.

But it is near enough, in most cases, to estimate the distance, and correct the altitude at once. Thus, an error of upwards of four leagues in the estimated distance will not, in the present example, produce an error of one mile in the calculated distance.

Measuring Distances by Sound.

Sound travels at the rate of 1090 feet in a second of time, when the air is at freezing point, or 32° of Fahrenheit.

This rate is accelerated by 1·19 feet for each degree of Fahrenheit above that temperature, giving 1123 feet (or 374 yards) velocity per second, for a mean temperature of 60°, which may be used for all ordinary purposes.

convenience, and the log. of any intermediate height may be found by simple proportion.

Peak of Teneriffe (12,172)	9·999747
Mount Etna (10,885)	9·999774
St. Antonio, Cape Verde Is. (7,398)	9·999846
Mount Athos (6,774)	9·999859
Pico Ruivo, Madeira (6,188)	9·999872
Mount Ida (5,800)	9·999879
Black Mountain, Cephalonia (5,382)	9·999888
Volcano, Guadaloupe (5,108)	9·999894
Table Mountain, Cape of Good Hope (3,904)	9·999910
Mount Vesuvius (3,878)	9·999920
Diana Peak, St. Helena (2,687)	9·999944
Mount St. Peter, Ascension (2,217)	9·999954
Rock of Gibraltar (1,437)	9·999970

Example.

	h.	m.	s.
Observed flash of gun at..	8	25	33
Heard report...................	8	25	40
	0	0	7

$374 \times 7 = 2618$ yards.

Using acceleration, Thermometer 80° Fahrenheit.

80°	At 32° sound travels..............	1090 feet (per sec).
32°	1·19 acceleration × 48	57 ,,
Diff. 48°	At 80 degrees, sound travels ...	1147 feet.

$1147 \times 7 = 8029$ feet, or 2676 yards.

Distances measured by sound should be, as long as possible, and if guns are fired from only one end, the distance should be measured on a calm day, or when there is not much wind, as sound is accelerated or retarded by wind. The gun fired should have extreme elevation given to it; and, if possible, guns should be fired from each end of the required distance.

The best method of measuring distance by sound is first to ascertain the number of times that the watch to be used beats in a minute of time. Most pocket chronometers beat 5 times in 2 seconds, or 150 beats to a minute; common watches vary. The number of beats of the watch ascertained, it is easy to find the number of yards that sound travels in that beat, making due allowance for the temperature. —At 60° Fahr. for instance, sound travels 150 yards in one beat of a chronometer beating 150 times in one minute.—That number of yards multiplied by the number of beats observed will give the distance required.

A simple method of counting the beats is by binding the watch close to the ear by means of a pocket handkerchief; this leaves hands and eye free to use glass and note book. The observer, comfortably placed with his eye on the ship or fort, begins to count on seeing the *flash*, finishing with the *report* of the gun.

Measuring Heights.

To find the height of an object, C B, the base of which is accessible, and on the same level as the observer.

Measure the angle B A C, and distance A B, then :—
C B = tan B A C × A B.

To find the height of an inaccessible tower, A B, and the difference of level between B and D, the nearest accessible point to B A.

Erect a stake the height of the eye at D, and measure the angles of elevation A E C, and A D C, as well as the base E D :—

$$A C = \frac{E D}{\cotan A E D - \cotan A D C}$$ The height B C can be ascertained in the same way, which subtracted from A C gives A B.

To find the height, from the angle of elevation or depression to an object, the distance being known.

$$\text{Height} = \frac{\text{Minutes and decimals in angle} \times \text{fathoms in distance}}{573.}$$

To be used for short distances only.

Measuring Heights (Roughly).

If the object be accessible, measure the length of its shadow, and at the same time the shadow of a pole of known length (or your own shadow): the lengths of the shadows are proportional to the heights. This must be quickly done, as the sun is always on the move.

Or,—Recede from an artificial horizon until the reflected image of the summit is seen in the artificial horizon.

Measure the distances between this position and the artificial horizon, also between the artificial horizon and the foot of the object. The distances are in proportion to the height of the observer's eye and the height of the object.

Measuring Distances on Shore.

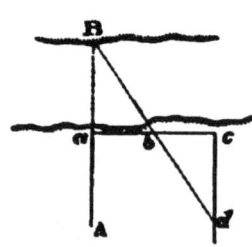

To continue the measurement of the base line A B, or to measure the distance across an intervening river, &c.

Set off two equal lengths in the same straight line at any convenient angle from a to b and from b to c, marking b and c with pickets; measure from c on a line *parallel** to A B the distance c, d, between c and a point d, where the picket at b came in line with the object B.

The length c, d, will be equal to the required distance a B.

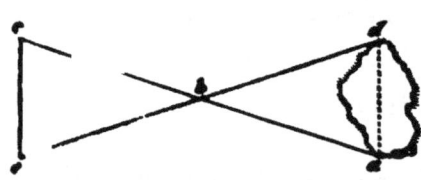

Suppose it is required to ascertain the distance $a\ d$ across an intervening pit or pond.

Measure a straight line from a to c, and also from d to e, making $c\ b = ba$ and $b\ e = b\ d$, then by equal triangles $e\ c$ is equal to $a\ d$.

Whence by measuring $e\ c$ we get the required distance $a\ d$.

Extemporary Measurements.

It is at all times well to know the length of the different joints of the limbs.

Suppose the nail-joint of the forefinger to be one inch, the next joint will be 1½ inches, the next 2 inches, and from the knuckle to the wrist 4 inches; in this case the finger is bent, so that each joint may

---

* This may be done by pocket sextant or compass.

be measured separately, though, when held straight, the distance from the tip of the forefinger to the wrist would be only 7 inches. The span with thumb and forefinger would be 8 inches, and with the thumb and any of the other three, 9 inches, or equal to the length of the foot; from the wrist to the elbow would be 10 inches, and from elbow to forefinger 17 inches, and from collar-bone to forefinger 2 feet 8 inches; height to the middle of the knee-cap 18 inches. From the elbow to the forefinger is usually called a cubit, but it is seldom strictly so, a cubit being 18 inches.

In like manner, the full stretch of the extended arms is called a fathom; but it is generally somewhat less. If a man stands with his back to a flat wall, and extends his arms, his fathom will be scarcely equal to his own height; but if he tries to measure the girt of a tree, by placing his breast against, and as it were embracing it, he will find his fathom many inches short, and on an average, perhaps, not more than 5 feet.

The pulse, when in health, and not excited, beats 72 to 75 times in a minute.

The step is commonly supposed to be 3 feet, and the pace $2\frac{1}{2}$ feet; but this is a most uncertain mode of measurement: very few men can take, with any correctness, a hundred consecutive steps or paces. Practice will determine the amount of ground covered in a certain number of paces, if tried over known distances: it, of course, varies, but from experiment, the mean has been found nearly as follows:—

Pacing, at 30 inches per pace, of 108 in a minute, equals 270 feet, or 3·068 statute, or 2·66 nautical miles per hour.

Pacing quickly, at 30 inches per pace, of 120 in a minute, equals 300 feet, or 3·41 statute, or 2·96 nautical miles per hour.

Pacing slowly, at 36 inches, may average 60 per minute, equals 180 feet, or 2·04 statute, or 1·78 nautical miles per hour.

The speed of track horses, on the sides of a canal, may be assumed to be about 3 miles per hour; that of a fair walking horse, to be from $4\frac{1}{2}$ to 5 miles per hour.

CHRONOMETERS.

The important services rendered by chronometers in the ordinary course of navigation, are well understood and fully appreciated by all intelligent seamen. A few notes therefore on the general treatment of these valuable instruments, extracted principally from that most useful work by Admiral C. F. A. SHADWELL, C.B., on the management of chronometers, are here inserted.*

When chronometers are received on board ship, it is of importance that they should be at once stowed away in the place prepared for their reception, and when once suitably located, they should on no account be subjected to removal or displacement.

The chronometers should be placed low down in the ship, (both because there is less motion, and because the temperature is more equable,) amidships, as far from the extremities, and as near the centre of gravity as convenient; not near the chain cables, or other large masses of iron, so as to ensure freedom from the disturbance of magnetic influence; and not in drawers, where the tremour caused by opening and shutting them acts injuriously on their balances.

The best mode of stowing them seems to be as follows:—A box, divided into as many partitions as there are chronometers to be stowed away, should be securely attached by screws to a solid block of wood, about thirty inches in height, and firmly bolted to the beams of the deck below. Each partition should, in depth, be about equal to that of the largest box of the chronometers to be stowed away, and in length and breadth about two inches longer than the sides of the box it is intended to secure. Great care should be taken that the block and partitioned box, thus prepared, should be entirely detached from all contact with contiguous staunchions or bulkheads; the block and box, moreover, should be surrounded with a strong external casing, the sides and lid of which should on no account be

---

* For further information on this important subject, see Admiral Shadwell's most admirable work. "Notes on the Management of Chronometers and Measurement of Meridian Distances." J. Potter, Poultry.

permitted to touch it, a clear space of at least two inches being left all round. Previous to placing the boxes containing the chronometers within the partitions appropriated for them, it will generally be found convenient to unscrew and altogether detach their lids; because if this be not done, when open they occupy much more room, and will require the partitioned spaces for their reception to be made inconveniently large.

Each chronometer in its box thus prepared, and moving freely in its gimbals, is then to be placed in the space allotted to it on a bed of dry sawdust, horse hair, or cotton, about three inches thick—the spaces around the sides of each box being stuffed, with the same material, to within half-an-inch of the top of the box. Of the three substances named above, dry sawdust is preferred.

The marks on the dial-plate of the chronometers should all occupy the same relative positions: that is, the line joining the XII. and VI. hour marks, should all be parallel to one another, both for the sake of the convenience of comparison, and in order that, retaining the same invariable position with reference to the fore-and-aft line of the ship, they may be similarly affected by the local magnetic attraction of the ship's iron, and that in cases where the balances of the chronometers have accidentally acquired any degree of polarity, their mutual influences on each other may be reciprocal.

Generally speaking, it is not the custom to receive the chronometers on board the ship until a few days before leaving the harbour, the time of the officers, who will subsequently be charged with the duty of superintending their performances, being at that period much occupied with other important matters, connected with the equipment of the ship; but, as from the influences of the new circumstances under which the chronometers are placed, the effects of motion in removing them, change of temperature, and possible action of magnetic causes, their rates, after a while, may differ from their previous rates on shore, it is very advisable, when practicable, that they should be received on board at an earlier period, so that they may become *naturalized* in their new position, and may have settled down to a stability of rate under their new conditions before the ship is called on to proceed to sea.

The chronometers having been received on board, and stowed away in the manner above described, in the place appropriated for their reception, it is of importance at once to commence and adopt an uniform and systematic manner of winding them up, and comparing them daily with one another. Methodical arrangements in these particulars favour the stability of rate of the chronometers, assist in the detection of irregularities, and diminish the probabilities of their being accidentally allowed to run down; while, in the reduction of chronometric observations for the determination of meridian distances, systematic plans of comparison are indispensable to accuracy of computation.

As a rule, 8 a.m. is a time conveniently adapted to this duty, and consistent with the arrangements of a man-of-war.

The chronometers should be wound *fisrt* and compared *afterwards*. In winding them they should be habitually attended to, in the same order, from day to day, one by one, as they lie in their places; so that the mechanical habits of regularity in this particular may be a safeguard against the caprice of memory or accidental distraction from any disturbing cause. From want of system, in this particular, we have known instances where the attention of the officer, engaged in this duty, being accidentally distracted during its performance, the chronometers have been compared, but some or all of them not wound up, that operation being forgotton, and the omission not detected till the chronometer ran down. In winding up chronometers, the turns of the key should always be counted, and the last turn made gently and carefully, *until it is felt to butt.*

In winding up box-chronometers, the chronometer should be inverted carefully in its gimbals, held firmly in the left hand, and the key pressed home with the right; after the operation is performed, and the key withdrawn, care should be taken that the keyhole is again covered with the slider to secure it from dust or damp, and then the chronometers should be *gently cased down* into its natural position without violence or jerk.

In winding up pocket-chronometers, the watch should be held firmly in the left hand, with the wrist pressed gently against the *breast, the* key should be turned equably and steadily with the right;

care being taken to avoid giving the watch any circular motion upon the key. The common but vicious practice of turning the left as well as the right hand is injurious, for two reasons : first, because the circular motion affects the regularity of the balance ; and secondly, because the compound motion of the two hands doubles the velocity of winding, and increases the chances of straining or snapping the spring from the jerk at the conclusion of the operation.

The chronometers should be wound *daily*, whether constructed to go for one or two days ; eight-day chronometers, once a week,—say on Sunday, a day easily remembered. If the number of chronometers is large, the precaution should be taken of examining them after winding, either by looking at the winding index on the dial plate, or if not furnished with that apparatus, by trying the key.

The chronometers having been all wound up, they should next be compared. To facilitate their systematic comparison, and to organize more easily the reduction of chronometric observations, it is customary to select one chronometer as a "*standard*,"* to which all observations, for the determination of the time, are in the first instance referred ; the indication of the other chronometers at the same moment being subsequently obtained by means of the comparisons.

The chronometer selected to perform the duty of the "standard" should be one of first-rate character, and by a maker of established repute ; it will be convenient that it should have a clear and distinct beat to half seconds ; also, that its dial plate be well marked ; and it is advisable, although not indispensable, that its rate should be small. Stability of rate, however, is at all times much more important than the smallness of its amount. It is very desirable, when practicable, that the comparisons of the chronometers with the standard, should all be made by one person. The effects of personal equation are thereby much simplified, and the chances of small contingent errors materially diminished.

The practical seaman will never be content to use the rates he may receive with the chronometers on their leaving the shore, when

---

* For the facility of comparison with the other chronometers it will be found convenient to arrange the chronometers in their box in such a manner that the standard shall occupy a central position among them.

these life-like inventions, to use the Admiral's excellent phrase, are not "naturalized in their new positions." Pressure of business, may possibly, force the sailor to use, for a time, the maker's or observatory rate, but he will take the earliest opportunity of getting his own rate, and watch, with a jealous tenderness, the performances of these delicate instruments, on whose accuracy the honour and safety of his ship are alike dependent.

As a matter of practice, it seems advisable, when circumstances permit, that the rate of a chronometer should not depend on observations made at an interval of *less* than *five* or *more* than *ten* days. Seven days will be found a convenient average interval, and in the case of eight-day chronometers, moreover, it embraces the period affected by the whole length of the chain. With the above limitations, it may be laid down as a maxim, that chronometers cannot be rated too often when time and opportunity permit. The error and rates of the chronometers should, if possible, be obtained by observing equal altitudes of the sun; failing this, by forenoon or afternoon sights, taking care that if the observations on the first day are made in the forenoon, that those on the second day should, if circumstances permit, be also made in the forenoon and *vice versa*.

Practice will soon make the earnest navigator an expert observer, and in addition to the safe conduct of his ship, he will often be able, by measuring Meridian Distances, to be of great assistance to his fellow seamen, by furnishing them with the longitudes of places hitherto unknown. The following table will show by what a simple method chronometers may be rated, and meridian distance measured.

Meridian Distance between Beirut and Sidon.

Losing rates are marked —; gaining rates marked +.

Numbers of Chronometers.*	Dent. 1,793. Standard.	McCabe, 187.	Frodsham, 2,714.
	h. m. s.	h. m. s.	h. m. s.
Standard fast on Beirut mean time, June 17th, 1861	7 15 00·79	7 15 00·79	7 15 00·79
Comparisons (standard being fast on chronometers)	..	9 11 15·00	9 35 15·00
Chronometers fast on Beirut mean time, June 17th, 1861	7 15 00·79	10 03 45·79	9 39 45·79
Ditto Ditto, June 11th, 1861..	7 15 11·68	10 03 43·68	9 39 41·68
Rate in six days	10·89	2·11	4·11
Daily rate*	.. − 1·81	+ ·35	+ ·68
Sidon rates in June	.. − 1·91	+ ·18	+ ·59
Beirut-Sidon rate	.. − 1·86	+ ·26	+ ·63
Chronometers fast on Beirut mean time, June 17th, 1861.. ..	7 15 00 79	10 03 45·79	9 39 45·79
Rate in three days by Beirut-Sidon rate	.. − 5·58	+ ·78	+ 1·89
Chronometers fast on Beirut mean time, June 20th	7 14 55·21	10 03 46·57	9 39 47·68
Chronometers fast on Sidon mean time, June 20th	7 15 25·91	10 04 16·91	9 40 18·41
Meridian distance, Beirut and Sidon	30·70	30·34	30·73

```
        s.
       30·70        Long. of Beirut  ..   35° 29' 04" E.
       30·34                               −    7  39
       30·73                              ─────────────
      ──────        Long. of Sidon   ..   35  21  35  E.
        177
      ──────
    30·59 = 7' 39"
```

* The watch used in rating chronometers, should on the days of observation, ed in a box, and not kept in the pocket, care being taken to compare it with the ·d chronometer on leaving and returning to the ship, both in the forenoon and

Sextant.—A practical acquaintance with the sextant is necessary to enable the navigator thus to rate and utilize his chronometers; he should therefore lose no opportunity of making himself thoroughly conversant with this useful instrument, understanding all its peculiarities, causes of error, etc.

He must remember that even in the best sextants errors may exist through the natural eccentricity of these instruments, or the difficulty of fixing the axis of the index glass concentric with the graduated arc; if equal altitudes cannot be obtained, a mean of a.m. and p.m. sights should be taken; if neither equal altitudes, or sets a.m. and p.m. can be obtained for rating the chronometers, care should be taken to work the a.m. sights of the first day with the a.m. sights of the second day, or p.m. with p.m., making sets of those observations that were taken at about the same altitude; and if possible never to work a.m. sights with p.m., or p.m. with a.m., to obtain a rate.

In making these observations, the eye-piece of the inverting tube should, if possible, be used instead of the shades of the sextant; if shades are used, endeavour always to use the same. The meridian altitude of the sun should be taken with the eye-piece, as the latitude obtained from it can then be meaned more satisfactorily with those determined by the stars.

In observing stars north and south of the zenith for Latitude, similar care should be taken to pair those of about the same altitude; therefore, in the northern hemisphere the best south stars to pair with Polaris are those whose meridian altitudes are about the same as the latitude of the place. To obtain latitudes from sun and stars, circum-meridional altitudes are generally used, taken from 20 minutes before, to 15 minutes after, the meridian passage.

Similarly, in taking Lunars, stars lying at about equal distances, east and west of the moon, should be chosen. The practice of Lunar observations should by no means be neglected by young officers; chronometers may fail; in giving an officer charge of a prize, his captain may not be able to spare him a chronometer; but the Great Greenwich Clock, formed by the moon and stars, is at his disposal for about 25 days in each month.

Artificial **Horizon.**—In using the artificial horizon care must be

taken in pouring out the quicksilver. Sir Edward Belcher, at page 8 of his "Nautical Surveying," gives the following useful hints:—
"Place the finger over the orifice of the bottle, and give it a shake in an inverted position, holding it over the trough previously cleaned. Ease the finger, and allow the mercury to flow gently, keeping the bottle inverted, and taking care to stop the opening of the bottle before the last portion with the dross flows. This will produce a clear brilliant surface."

In pointing the artificial horizon towards the sun, see that the shadow is thrown directly behind it, and not on either side. The spot chosen for making these observations should be sheltered from the wind, and to be free from vibration, removed from traffic. The roof of the artificial horizon should be so placed that the same glass is always towards the observer. A mark may be made on the glass or on the frame work to insure this precaution being carried out.

Observations at Night.—Obtaining latitudes at night with the artificial horizon requires some practice; see that it is placed in the true meridian before the observations are commenced. The approximate mean times of the meridian passages of the stars to be used should be calculated from the "Nautical Almanac," the error on mean time of the watch in use should then be applied to them; these times of the meridian passages as shown by the watch in use, should be entered in the sight book as a guide to the observer, together with the approximate meridian altitudes. The stars selected should, if possible, be of convenient altitudes, neither too high or too low; care should be taken to find the reflected image of the star in the quicksilver some little time before it comes on the meridian. A handy sailor should be trained to hold the dark lanthorn, by which the observations must be first noted by the watch, and then read from the sextant, in such a manner that, while the assistant who is taking time can see the watch, no light is thrown in the direction of the artificial horizon. On the time being secured, the light is taken to the observer with the sextant. Practice is required both in holding the lanthorn and in reading off the sextant.

N.B.—A competent and practical knowledge of what can be done with anchor and compass; log-line and lead-line; sextant and chronometer; chart, scale, and dividers; is the secret of success in a profession, whose members live by utilizing the forces of wind and water.

TABLE OF DISTANCES AT WHICH OBJECTS CAN BE SEEN AT SEA, ACCORDING TO THEIR RESPECTIVE ELEVATIONS AND THE ELEVATION OF THE EYE OF THE OBSERVER.

BY ALAN STEVENSON.

Heights in feet.	Distances in statute or English miles.	Distances in geographical or nautical miles.	Heights in feet.	Distances in statute or English miles.	Distances in geographical or nautical miles.	Heights in feet.	Distances in statute or English miles.	Distances in geographical or nautical miles.
5	2·958	2·565	70	11·067	9·598	250	20·916	18·14
10	4·184	3·628	75	11·456	9·935	300	22·912	19·87
15	5·123	4·443	80	11·832	10·26	350	24·748	21·46
20	5·916	5·130	85	12·196	10·57	400	26·457	22·94
25	6·614	5·736	90	12·549	10·88	450	28·062	24·33
30	7·245	6·283	95	12·893	11·18	500	29·580	25.65
35	7·826	6·787	100	13·228	11·47	550	31·024	26·90
40	8·366	7·255	110	13·874	12·03	600	32·403	28·10
45	8.874	7·696	120	14·490	12·56	650	33·726	29·25
50	9·354	8·112	130	15·083	13·08	700	35·000	30·28
55	9·811	8·509	140	15·652	13·57	800	37·416	32·45
60	10·246	8·886	150	16·201	14·22	900	39·836	34·54
65	10·665	9·249	200	18·708	16·22	1000	41·833	36·28

Example—A tower, 200 feet high, will be visible to an observer whose eye is elevated 15 feet above the water, about 21 nautical miles: thus, from the table:—

15 feet elevation, distance visible 4·44 nautical miles.
200 ,, ,, 16·22 ,,
20.66

Table showing the Distance of the Horizon at Different Elevations.

Height.	Distance of the Horizon	Height.	Distance of the Horizon	Height.	Distance of the Horizon	Height.	Distance of the Horizon	Height.	Distance of the Horizon
Feet	Nautic Miles	Feet	Nautic Miles	Feet	Nautic Miles	Feet	Nautic Miles	Feet	Nautic Miles.
1	1·15	33	6·60	85	10·59	245	17·98	450	24·36
2	1·62	34	6·70	90	10·90	250	18·16	460	24·63
3	1·99	35	6·80	95	11·19	255	18·34	470	24·90
4	2·30	36	6·89	100	11·49	260	18·52	480	25·16
5	2·57	37	6·99	105	11·77	265	18·70	490	25·42
6	2·81	38	7·08	110	12·05	270	18·87	500	25·68
7	3·04	39	7·17	115	12·32	275	19·05	510	25·94
8	3·25	40	7·26	120	12·58	280	19·22	520	26·19
9	3·45	41	7·35	125	12·84	285	19·39	530	26·44
10	3·63	42	7·44	130	13·10	290	19·56	540	26·69
11	3·81	43	7·53	135	13·35	295	19·73	550	26·93
12	3·98	44	7·62	140	13·61	300	19·89	560	27·18
13	4·14	45	7·70	145	13·83	305	20·06	570	27·42
14	4·30	46	7·79	150	14·06	310	20·22	580	27·66
15	4·45	47	7·87	155	14·30	315	20·38	590	27·90
16	4·59	48	7·96	160	14·53	320	20·55	600	28·13
17	4·74	49	8·04	165	14·75	325	20·71	610	28·37
18	4·87	50	8·12	170	14·97	330	20·86	620	28·60
19	5·01	51	8·20	175	15·19	335	21·02	630	28·83
20	5·14	52	8·29	180	15·41	340	21·18	640	29·06
21	5·26	53	8·36	185	15·62	345	21·33	650	29·28
22	5·39	54	8·44	190	15·83	350	21·49	660	29·51
23	5·51	55	8·52	195	16·04	355	21·64	670	29·73
24	5·63	56	8·60	200	16·24	360	21·79	680	29·95
25	5·74	57	8·67	205	16·44	370	22·09	690	30·17
26	5·86	58	8·75	210	16·64	380	22·39	700	30·39
27	5·97	59	8·82	215	16·84	390	22·68	710	30·60
28	6·08	60	8·90	220	17·03	400	22·97	720	30·82
29	6·19	65	9·26	225	17·20	410	23·26	730	31·03
30	6·29	70	9·61	230	17·42	420	23·54	740	31·24
31	6·40	75	9·95	235	17·61	430	23·82	750	31·45
32	6·50	80	10·27	240	17·79	440	24·09	760	31·66

Approximately, correction for curvature, in feet $= \dfrac{2\,D^2}{3}$, D being the distance in *statute* miles.

DIP OF THE HORIZON.

Approximately, the Dip in minutes, is the square root of the Height in feet.

Height.	Dip.	Height.	Dip.	Height.	Dip.
Feet.	′ ″	Feet.	′ ″	Feet.	′ ″
1	0 58	13	3 27	25	4 52
2	1 21	14	3 36	28	5 5
3	1 40	15	3 42	30	5 15
4	1 56	16	3 50	35	5 39
5	2 9	17	3 57	40	6 4
6	2 21	18	4 4	45	6 27
7	2 33	19	4 11	50	6 46
8	2 44	20	4 17	60	7 25
9	2 53	21	4 23	70	8 1
10	3 2	22	4 30	80	8 34
11	3 10	23	4 36	90	9 6
12	3 19	24	4 42	100	9 35

To find the height when the distance is known: To the constant log. 3·78404 add the log. of the distance in miles and the log. tan. of the corrected altitude. The nat. number of the sum of these logs being added to the number from the following table is the height in feet.

Miles.	Feet.	Miles.	Feet.	Miles.	Feet.	Miles.	Feet.	Miles.	Feet.	Miles.	Feet.
1	0·9	11	107	21	390	31	850	41	1487	55	2677
2	3·5	12	127	22	428	32	906	42	1561	60	3186
3	8·0	13	149	23	468	33	964	43	1636	65	3740
4	14·1	14	173	24	510	34	1023	44	1713	70	4337
5	22·1	15	199	25	553	35	1084	45	1792	75	4976
6	31·8	16	226	26	598	36	1147	46	1872	80	5665
7	43·3	17	256	27	645	37	1211	47	1954	85	6394
8	56·6	18	287	28	694	38	1278	48	2039	90	7172
9	71·6	19	319	29	745	39	1346	49	2124	95	7987
10	88·4	20	354	30	797	40	1416	50	2212	100	8852

The Table is to be entered with the distance of the observer, in miles.

Example. The altitude of a summit is (corrected for dip) 2° 48′, the distance 20 miles, the height of the eye 16 feet.

```
         20 log.   ..   ..  1·30103      Height above level of observer   5949
         Const.    ..   ..  3·78404      Cor. for Curvature at 20 miles    354
         Log. tan. ..   ..  8·68938      Height of observer's eye  ..       16
                                         ─────
5949  ..   ..   ..  log. 3·77445         Height required in feet   ..     6319
```

Sec. V.   MEASURING DISTANCES.   199

To find the Distance of an object by Two Bearings, and the Distance run between them.

Rule.—Under the number of points contained between the course and second bearing, and opposite to the difference between the course and first bearing, will be found a number, which multiplied by the miles made good will give the distance of the object (in miles) at the time the last bearing was taken.

N.B.— Due allowance should be made for current.

colspan="14"	Difference between the Course and Second Bearing in Points of the Compass.	
4	4½	5	5½	6	6¼	7	7½	8	8½	9	9½	10	10½	11	11½	12	12½	Points
1·00	0·81	0·69	0·60	0·54	0·49	0·46	0·43	0·41	0·40	0·39	0·38	0·38	0·38	0·39	0·40	0·41	0·43	2
1·23	1·00	0·85	0·74	0·67	0·61	0·57	0·53	0·51	0·49	0·48	0·47	0·47	0·47	0·48	0·49	0·51	2½
	1·45	1·17	1·00	0·88	0·79	0·72	0·67	0·63	0·60	0·58	0·57	0·56	0·56	0·56	0·57	0·58	3
		1·66	1·35	1·14	1·00	0·90	0·82	0·76	0·72	0·69	0·66	0·65	0·64	0·64	0·64	0·65	3½
			1·85	1·50	1·27	1·11	1·00	0·92	0·85	0·80	0·76	0·74	0·72	0·71	0·71	0·71	4
				2·02	1·64	1·39	1·22	1·09	1·00	0·93	0·88	0·84	0·81	0·79	0·78	0·78	4½
					2·17	1·77	1·50	1·31	1·18	1·08	1·00	0·94	0·90	0·87	0·85	0·83	5
						2·30	1·87	1·58	1·39	1·25	1·14	1·06	1·00	0·95	0·92	0·90	5½
							2·41	1·96	1·66	1·46	1·31	1·19	1·11	1·05	1·00	0·97	6
								2·50	2·03	1·72	1·51	1·35	1·24	1·15	1·08	1·03	6½
									2·56	2·08	1·76	1·55	1·39	1·27	1·18	1·11	7
										2·60	2·11	1·79	1·57	1·41	1·29	1·20	7½
											2·61	2·12	1·80	1·58	1·41	1·29	8
												2·60	2·11	1·79	1·57	1·41	8½
													2·56	2·08	1·76	1·55	9
														2·50	2·03	1·72	9½
															2·41	1·96	10
																2·30	10½

Difference between the Course and First Bearing in points of the Compass.

Example.—*A rock bore N.N.W.; after running West 12 miles, it bore N.E. b. N. Required the distance of the ship from the rock when the second bearing was taken.* The number of points between West and N.E. b. N. is 11: between West and N.N.W. is 6. Under 11 at the top and opposite 6 at the side, stands 1·11, which multiplied by 12 gives 13·32 miles, the distance of the rock when the *second bearing was taken.*

Measuring Heights by Barometer.

IN THE FOLLOWING TABLE, THE DIFFERENCE BETWEEN THE NUMBER OF FEET OPPOSITE THE HEIGHT OF A BAROMETER AT ONE STATION AND THAT AT ANOTHER, IS THE APPROXIMATE DIFFERENCE IN HEIGHT.

Barometer in inches.	Height in feet.	Barometer in inches.	Height in feet.	Barometer in inches.	Height in feet.	Barometer in inches.	Height in feet.
31.0	0	27.8	2864	24.6	6083	21.4	9755
30.9	85	27.7	2959	24.5	6190	21.3	9878
30.8	170	27.6	3054	24.4	6297	21.2	10002
30.7	255	27.5	3149	24.3	6405	21.1	10127
30.6	341	27.4	3245	24.2	6514	21.0	10253
30.5	427	27.3	3341	24.1	6623	20.9	10379
30.4	513	27.2	3438	24.0	6733	20.8	10506
30.3	600	27.1	3535	23.9	6843	20.7	10633
30.2	687	27.0	3633	23.8	6953	20.6	10760
30.1	774	26.9	3731	23.7	7064	20.5	10889
30.0	862	26.8	3829	23.6	7175	20.4	11018
29.9	950	26.7	3927	23.5	7287	20.3	11148
29.8	1038	26.6	4025	23.4	7399	20.2	11278
29.7	1126	26.5	4124	23.3	7512	20.1	11409
29.6	1215	26.4	4223	23.2	7625	20.0	11541
29.5	1304	26.3	4323	23.1	7729	19.9	11673
29.4	1393	26.2	4423	23.0	7854	19.8	11805
29.3	1482	26.1	4524	22.9	7969	19.7	11939
29.2	1572	26.0	4625	22.8	8085	19.6	12074
29.1	1662	25.9	4726	22.7	8201	19.5	12210
29.0	1753	25.8	4828	22.6	8317	19.4	12346
28.9	1844	25.7	4930	22.5	8434	19.3	12483
28.8	1935	25.6	5033	22.4	8551	19.2	12620
28.7	2027	25.5	5136	22.3	8669	19.1	12757
28.6	2119	25.4	5240	22.2	8787	19.0	12894
28.5	2211	25.3	5344	22.1	8906	18.9	12942
28.4	2303	25.2	5448	22.0	9025	18.8	13080
28.3	2396	25.1	5553	21.9	9145	18.7	13219
28.2	2489	25.0	5658	21.8	9266	—	—
28.1	2582	24.9	5763	21.7	9388	—	—
28.0	2675	24.8	5869	21.6	9510	—	—
27.9	2769	24.7	5976	21.5	9632	—	—

If possible, simultaneous observations, by signal or time, of two barometers previously compared at the same station, should be taken. But little dependence can be placed on observations beyond 1000 feet, when made with the Aneroid.

Sec. V. ADMIRALTY KNOTS AND STATUTE MILES.

Table for converting Admiralty Knots into Statute Miles.

(THE ADMIRALTY KNOT = 6080 FEET.* 1 STATUTE MILE = 5280 FEET.)

Admiralty Knots.	Statute Miles.	Admiralty Knots.	Statute Miles.	Admiralty Knots.	Statute Miles.	Admiralty Knots.	Statute Miles.
1·00	1·151	7·25	8·348	13·50	15·545	19·75	22·742
1·25	1·439	7·50	8·636	13·75	15·833	20·00	23·030
1·50	1·729	7·75	8·924	14·00	16·121	20·25	23·318
1·75	2·015	8·00	9·212	14·25	16·409	20·50	23·606
2·00	2·303	8·25	9·500	14·50	16·696	20·75	23·893
2·25	2·590	8·50	9·787	14·75	16·984	21·00	24·181
2·50	2·878	8·75	10·075	15·00	17·272	21·25	24·469
2·75	3·166	9·00	10·393	15·25	17·560	21·50	24·757
3·00	3·454	9·25	10·615	15·50	17·848	21·75	25·045
3·25	3·742	9·50	10·939	15·75	18·136	22·00	25·333
3·50	4·030	9·75	11·227	16·00	18·424	22·25	25·621
3·75	4·318	10·00	11·515	16·25	18·712	22·50	25·909
4·00	4·606	10·25	11·803	16·50	18·999	22·75	26·196
4·25	4·893	10·50	12·090	16·75	19·287	23·00	26·484
4·50	5·181	10·75	12·378	17·00	19·575	23·25	26·772
4·75	5·469	11·00	12·666	17·25	19·863	23·50	27·060
5·00	5·757	11·25	12·954	17·50	20·151	23·75	27·348
5·25	6·045	11·50	13·242	17·75	20·439	24·00	27·636
5·50	6·333	11·75	13·530	18·00	20·727	24·25	27·924
5·75	6·621	12·00	13·818	18·25	21·015	24·50	28·212
6·00	6·909	12·25	14·106	18·50	21·303	24·75	28·499
6·25	7·196	12·50	14·393	18·75	21·590	25·00	28·787
6·50	7·484	12·75	14·681	19·00	21·878	--	--
6·75	7·772	13·00	14·969	19·25	22·166	—	--
7·00	8·060	13·25	15·257	19·50	22·454	—	--

* The Geographical mile is generally defined to be the length of a minute of arc in the earth's equator; but the Nautical mile as defined by hydrographers is the length of a minute of the meridian, and is different for every different latitude; (see Carrington's Table, page 164). It is equal to a minute of arc in a circle, whose radius is the radius of the curvature of the meridian, at the latitude of the place.

Table for converting Statute Miles into Admiralty Knots.

(THE ADMIRALTY KNOT = 6080 FEET.* 1 STATUTE MILE = 5280 FEET.)

Statute Miles.	Admiralty Knots.	Statute Miles.	Admiralty Knots.	Statute Miles.	Admiralty Knots.	Statute Miles.	Admiralty Knots.
1·00	0·868	8·25	7·164	15·50	13·460	22·75	19·756
1·25	1·085	8·50	7·381	15·75	13·677	23·00	19·973
1·50	1·302	8·75	7·598	16·00	13·894	23·25	20·190
1·75	1·519	9·00	7·815	16·25	14·111	23·50	20·407
2·00	1·736	9·25	8·032	16·50	14·328	23·75	20·625
2·25	1·953	9·50	8·250	16·75	14·546	24·00	20·842
2·50	2·171	9·75	8·467	17·00	14·763	24·25	21·059
2·75	2·388	10·00	8·684	17·25	14·980	24·50	21·276
3·00	2·605	10·25	8·901	17·50	15·197	24·75	21·493
3·25	2·822	10·50	9·118	17·75	15·414	25·00	21·710
3·50	3·039	10·75	9·335	18·00	15·631	25·25	21·927
3·75	3·256	11·00	9·552	18·25	15·848	25·50	22·114
4·00	3·473	11·25	9·769	18·50	16·065	25·75	22·361
4·25	3·690	11·50	9·986	18·75	16·282	26·00	22·578
4·50	3·907	11·75	10·203	19·00	16·500	26·25	22·796
4·75	4·125	12·00	10·421	19·25	16·717	26·50	23·013
5·00	4·342	12·25	10·638	19·50	16·934	26·75	23·230
5·25	4·559	12·50	10·855	19·75	17·151	27·00	23·447
5·50	4·776	12·75	11·072	20·00	17·368	27·25	23·664
5·75	4·993	13·00	11·289	20·25	17·585	27·50	23·881
6·00	5·210	13·25	11·506	20·50	17·802	27·75	24·098
6·25	5·427	13·50	11·723	20·75	18·019	28·00	24·315
6·50	5·644	13·75	11·940	21·00	18·236	28·25	24·532
6·75	5·861	14·00	12·157	21·25	18·453	28·50	24·750
7·00	6·078	14·25	12·375	21·50	18·671	28·75	24·967
7·25	6·296	14·50	12·592	21·75	18·888	29·00	25·184
7·50	6·513	14·75	12·809	22·00	19·105	—	—
7·75	6·730	15·00	13·026	22·25	19·322	—	—
8·00	6·947	15·25	14·243	22·50	19·539	—	—

* The Geographical mile is generally defined to be the length of a minute of arc in the earth's equator; but the Nautical mile as defined by hydrographers is the length of a minute of the meridian, and is different for every different latitude; (see Carrington's Table, page 164). It is equal to a minute of arc in a circle whose radius the radius of the curvature of the meridian, at the latitude of the place.

Sec. V.   THE MEASURED MILE.   203

TRIAL TRIP TABLE.

The minutes and seconds of time, in which a vessel passes over the measured mile, being known, look for the corresponding number in this table, which will be the rate of the vessel in miles.

Sec.	3m.	4m.	5m.	6m.	7m.	8m.	9m.	10m.	11m.	12m.	13m.	14m.
0	20.000	15.000	12.000	10.000	8.571	7.500	6.666	6.000	5.454	5.000	4.615	4.285
1	19.890	14.938	11.960	9.972	8.551	7.484	6.654	5.990	5.446	4.993	4.609	4.280
2	19.780	14.876	11.920	9.944	8.530	7.468	6.642	5.980	5.438	4.986	4.603	4.275
3	19.672	14.815	11.880	9.917	8.510	7.453	6.629	5.970	5.429	4.979	4.597	4.270
4	19.564	14.754	11.841	9.890	8.490	7.438	6.617	5.960	5.421	4.972	4.591	4.265
5	19.460	14.694	11.803	9.863	8.470	7.422	6.605	4.950	5.413	4.965	4.585	4.260
6	19.355	14.634	11.764	9.830	8.450	7.407	6.593	5.940	5.405	4.958	4.580	4.225
7	19.251	14.575	11.726	9.809	8.430	7.392	6.581	5.930	5.397	4.951	4.574	4.250
8	19.150	14.516	11.688	9.783	8.413	7.377	6.569	5.921	5.389	4.945	4.568	4.245
9	19.047	14.457	11.650	9.756	8.392	7.362	6.557	5.911	5.381	4.938	4.562	4.240
10	18.947	14.400	11.613	9.729	8.372	7.346	6.545	5.901	5.373	4.931	4.556	4.235
11	18.848	14.342	11.575	9.703	8.353	7.331	6.533	5.891	5.365	4.924	4.551	4.230
12	18.750	14.285	11.538	9.677	8.334	7.317	6.521	5.882	5.357	4.918	4.545	4.225
13	18.652	14.220	11.501	9.651	8.315	7.302	6.509	5.872	5.349	4.911	4.539	4.220
14	18.556	14.173	11.465	9.625	8.295	7.287	6.498	5.863	5.341	4.904	4.534	4.215
15	18.461	14.118	11.428	9.600	8.276	7.272	6.486	5.853	5.333	4.897	4.528	4.210
16	18.367	14.063	11.392	9.574	8.257	7.258	6.474	5.844	5.325	4.891	4.522	4.206
17	18.274	14.008	11.356	9.549	8.228	7.243	6.463	5.834	5.317	4.884	4.516	4.201
18	18.181	13.953	11.323	9.524	8.219	7.229	6.451	5.825	8.309	4.878	4.511	4.196
19	18.090	13.900	11.285	9.490	8.200	7.214	6.440	5.815	5.301	4.871	4.505	4.191
20	18.000	13.846	11.250	9.473	8.181	7.200	6.428	5.806	5.294	4.864	4.500	4.186
21	17.910	13.793	11.214	9.448	8.163	7.185	6.417	5.797	5.286	4.858	4.494	4.181
22	17.823	13.740	11.180	9.424	8.144	7.171	6.405	5.787	5.278	4.851	4.488	4.176
23	17.734	13.688	11.146	9.399	8.127	7.157	6.394	5.778	5.270	4.845	4.483	4.171
24	17.647	13.636	11.111	9.375	8.108	7.142	6.383	5.769	5.263	4.838	4.477	4.166
25	17.560	13.584	11.077	9.350	8.090	7.128	6.371	5.760	5.255	4.832	4.472	4.161
26	17.475	13.533	11.043	9.326	8.071	7.114	6.360	5.750	5.247	4.825	4.466	4.157
27	17.391	13.483	11.009	9.302	8.053	7.100	6.349	5.741	5.240	4.819	4.460	4.152
28	17.307	13.432	10.975	9.278	8.035	7.086	6.338	5.732	5.232	4.812	4.455	4.147
29	17.225	13.383	10.942	9.254	8.017	7.072	6.327	5.723	5.224	4.806	4.449	4.142

TRIAL TRIP TABLE—Continued.

Sec.	3m.	4m.	5m.	6m.	7m.	8m.	9m.	10m.	11m.	12m.	13m.	14m.
30	17·143	13·333	10·909	9·230	8·000	7·059	6·315	5·714	5·217	4·800	4·444	4·137
31	17·061	13·284	10·876	9·207	7·982	7·045	6·304	5·705	5·210	4·793	4·438	4·133
32	16·981	13·235	10·843	9·183	7·964	7·031	6·293	5·696	5·202	4·787	4·433	4·128
33	16·901	13·186	10·810	9·160	7·947	7·017	6·282	5·687	5·195	4·780	4·428	4·123
34	16·822	13·138	10·778	9·137	7·929	7·004	6·271	5·678	5·187	4·774	4·422	4·118
35	16·744	13·092	10·764	9·113	7·912	6·990	6·260	5·669	5·179	4·769	4·417	4·114
36	16·667	13·043	10·714	9·090	7·825	6·977	6·250	5·660	5·172	4·761	4·411	4·110
37	16·590	12·926	10·682	9·068	7·877	6·963	6·239	5·651	5·164	4·755	4·406	4·105
38	16·514	12·950	10·651	9·044	7·860	6·950	6·228	5·642	5·157	4·749	4·400	4·100
39	16·438	12·903	10·619	9·022	7·843	6·936	6·217	5·633	5·150	4·743	4·395	4·095
40	16·363	12·857	10·518	9·000	7·826	6·923	6·207	5·625	5·142	4·738	4·390	4·090
41	16·289	12·811	10·557	8·977	7·809	6·909	6·196	5·616	5·135	4·730	4·384	4·085
42	16·216	12·766	10·526	8·955	7·792	6·896	6·185	5·607	5·128	4·724	4·379	4·081
43	16·143	12·711	10·495	8·933	7·775	6·883	6·174	5·598	5·121	4·718	4·374	4·077
44	16·071	12·676	10·465	8·911	7·758	6·870	6·164	5·590	5·114	4·712	4·368	4·072
45	16·000	12·631	10·434	8·889	7·741	6·857	6·153	5·581	5·106	4·706	4·363	4·067
46	15·929	12·587	10·404	8·867	7·725	6·844	6·143	5·572	5·099	4·700	4·358	4·063
47	15·859	12·543	10·375	8·845	7·708	6·831	6·132	5·564	5·091	4·693	4·353	4·058
48	15·789	12·500	10·345	8·823	7·692	6·818	6·122	5·555	5·084	4·687	4·347	4·054
49	15·721	12·456	10·315	8·801	7·675	6·805	6·112	5·547	5·077	4·681	4·342	4·049
50	15·652	12·413	10·286	8·780	7·659	6·792	6·101	5·538	5·070	4·675	4·337	4·044
51	15·584	12·371	10·256	8·759	7·643	6·779	6·091	5·530	5·063	4·669	4·332	4·040
52	15·517	12·329	10·227	8·737	7·627	6·766	6·081	5·521	5·056	4·663	4·326	4·035
53	15·450	12·287	10·198	8·716	7·611	6·754	6·071	5·513	5·049	4·657	4·321	4·031
54	15·384	12·245	10·169	8·695	7·595	6·741	6·060	5·504	5·042	4·651	4·316	4·026
55	15·319	12·203	10·140	8·675	7·579	6·739	6·050	5·496	5·035	4·645	4·311	4·022
56	15·254	12·162	10·112	8·654	7·563	6·716	6·040	5·487	5·028	4·639	4·306	4·017
57	15·190	12·121	10·084	8·633	7·547	6·704	6·030	5·479	5·020	4·633	4·301	4·013
58	15·125	12·080	10·055	8·612	7·531	6·691	6·020	5·471	5·013	4·627	4·295	4·008
59	15·062	12·040	10·027	8·591	7·515	6·679	6·010	5·463	5·006	4·621	4·290	4·004

Table for ascertaining Velocity of Tide by noting the number of seconds a measured knot of 47·3 feet takes to run out.

Time in seconds.	Rate in Knots.	Time in seconds.	Rate in Knots.	Time in seconds.	Rate in Knots.	Time in seconds.	Rate in Knots.
5	5·76	17	1·65	29	0·96	41	0·68
6	4·80	18	1·55	30	0·93	42	0·67
7	3·99	19	1·47	31	0·90	43	0·65
8	3·50	20	1·40	32	0·87	44	0·63
9	3·20	21	1·33	33	0·85	45	0·62
10	2·80	22	1·27	34	0·82	46	0·61
11	2·54	23	1·23	35	0·80	47	0·60
12	2·23	24	1·17	36	0·78	48	0·58
13	2·15	25	1·12	37	0·76	49	0·57
14	1·99	26	1·08	38	0·74	—	—
15	1·87	27	1·04	39	0·72	—	—
16	1·76	28	1·00	40	0·70	—	—

NOTE.—Three or four knots should be timed, and the mean taken.

Table showing the Number of Feet subtending an Angle of 1′ at given distances to 32 Nautical Miles.

One foot subtends an angle of 1′ at a distance of 3437′·75 feet.

Miles.	Feet.	Miles.	Feet.	Miles.	Feet.	Miles.	Feet.
1	1·77	9	15·92	17	30·07	25	44·23
2	3·54	10	17·69	18	31·84	26	46·00
3	5·31	11	19·46	19	33·61	27	47·76
4	7·08	12	21·23	20	35·38	28	49·53
5	8·84	13	23·00	21	37·15	29	51·30
6	10·61	14	24·77	22	38·92	30	53·07
7	12·38	15	26·53	23	40·69	31	54·83
8	14·15	16	28·30	24	42·46	32	56·60

This Table shows the error in height which would result from an error of 1′ (and, by multiplication, of any small number of minutes) in the altitude.

Again, it gives very nearly the magnitude of an object at a given distance, and subtending an angle not exceeding 2° or 3°, by multiplying the number of feet against the distance in the Table by the number of minutes in the angle subtended.

The length of an arc = No. of degrees × radius × ·07145.

Section 6.

BOATS.

**WEIGHTS AND DIMENSIONS;
MANAGEMENT OF STEAM BOATS;
GENERAL HINTS ON MANAGEMENT,
BOARDING, BOAT-CRUISING, &c., &c.**

Particulars of Steam Boats supplied to H.M. Navy.

The details given may be taken as types of the different classes, though of course subject to slight variations.

Description.	Length.	Speed	Indicated Horse Power.	Average Coal Supply. No. of Hours.	Hoisting Weight of Engines and Boiler.		Weight of Boat and Engine.		Total weight fitted with Torpedo gear and shield.	
	feet.	knots.			cwt.	qrs.	cwt.	qrs.	cwt.	qrs.
Launch (engine Penn)	42	8	32·6	16 cwt. 13·7 hours	36	0	148	0	162	0
Launch (engine Rennie)	42	8·1	36·1	16 cwt. 12·38 hours	43	0	155	0	169	0
Pinnace	37	8	28	8 cwt. 9·1 hours	30	2	105	0	117	2
Pinnace	30	7·8	18	5 cwt. 10¾ hours	22	3	60	0	73	0
Cutter	28	6·8	13	4 cwt. 12½ hours	16	1	45	0	—	

The hoisting weight of a 37 feet Pinnace with steam up, water tanks, coal boxes, &c., full, is 139 cwt.

The Pinnaces and Cutters have sufficient buoyancy to float engines, coals, and crew, in event of being filled by a sea.

Instructions for Working the Engines of Steam Launches, etc.

The following Instructions for working the Engines of Steam Launches, &c., are introduced, as it is possible that the Officer in Charge might be thrown entirely on his own resources.

The engine should not be removed from the boat oftener than can be helped. The boiler of steam pinnaces should be lifted, examined at the bottom, and painted every month.

See that the tanks, fitted for the purpose, are properly supplied with coal and fresh water.

The connection with propellers and water-tight joints must be made good before leaving the ship.

Water is run into the boiler through a hose by removing one of the safety-valves. When the water is showing from one-half to three fourths up the gauge glass, remove the hose, and replace the safety-valve. Great care must be taken to see the valve and its seating perfectly clean before the valve is replaced.

To get up Steam.—Put a surface of coal over the fire bars, shut the ash-pit door, and light up with wood and coal at the front until a sufficient body of fire is obtained to ignite the coal on the bars, when the fire may be pushed back, and the ash-pit door opened.

When steam begins to show by the guage, try the safety-valves, and use the blast, (if the steam be required in great haste,) until sufficient pressure be obtained.

Try the small engine to supply feed-water, and make sure of water supply independent of main feeds.

The Boiler will require the most careful and constant attention while steaming. When attainable, fresh water should always be used.

It is fitted with every requisite for efficiency and safety.

From 40 to 50lb. of steam pressure is quite sufficient for all ordinary service. Leaks about tubes and tube-plates are most frequently caused by forced steaming.

The water must never be allowed to go below the mark of low level.

At high speed it is liable to show higher in the gauge glass than it really is.

The gauge glass and gauge cocks must be frequently tried, the one being a check on the other.

The water moving in the glass with the movements of the boat is another proof of gauge-glass cocks being correct.

Care should be taken to prevent spray from striking the gauge glass, as it is very liable to break it.

Maintain a sufficient quantity of water in the boiler, and keep the feed-water supply as nearly constant as possible. In the event of the water getting low the fire must be checked as quickly as possible; to effect this, open the smoke box door, shut the ash-pit door, and throw on wet ashes. In an extreme case, draw the fire.

In the new class of boats, the feed water passes through a coiled pipe in a cistern, secured to the boiler. This cistern receives the exhaust steam from the engines, heating the feed water on its passage to the funnel; doing its work in a noiseless manner, instead of the intermittent puffing, which is the result of direct connection.

Starting the Engine.—Have every fractional part of the engines carefully oiled, especially cylinders, slide-valves, eccentrics, cranks, and thrust: open the small drain cocks in connection with the cylinders and slide-valves, to get rid of condensed water, and let them remain open for a few turns of the engines. The steam cock may be left a little open while steam is getting up, to warm the engine.

Starting ahead or astern is effected by link motion, and requires no consideration after observing the movement of the handle connected with the link.

Great care should be taken to admit the steam to the engines gently at first, and get them up to their full speed gradually.

Running.—Attention to the engines is required in preventing the over-heating of working parts.

Any unusual noise must be quickly attended to, and cause ascertained. (The loosening of pin, key or screw, promptly remedied; always bearing in mind the old adage about "a stitch in time, &c.")

Sea Water.—If obliged to use sea water for the feed, let the process of brining be as constant and continuous as possible.

The density of the water in the boiler must be frequently tested by drawing some off, and applying the hydrometer with the water at the temperature of its graduated scale.

The bringing, or blowing out should maintain the water at $1\frac{3}{4}$ charges of salt, (salt water being 1) or about the mark 18 on the hydrometer in general use.

Stoking.—The stoking must be careful, and frequent, in just sufficient quantity to keep the fire bars properly covered; attention to this will go far to prevent priming.

Keep the steam at a regular pressure, and the fire-bars free from clinker by hooking them out as soon as formed.

The tubes, fire-box, smoke-box, and the space at the back of the fire-bridge should be kept free and clean; this must be done as opportunity offers.

When the screws of a steam launch are taken off for the purpose of her being used as a Sailing boat, the brass bushes, usually provided for the purpose, should be put on the ends of the shafts (first coating them with white lead and tallow), in order to prevent them from the rapid galvanic action which takes place by their close proximity to the copper sheathing on the boat's bottom. If no bushes are provided, then the ends of the shafts should be lapped round with spunyarn well saturated with stiff white lead and tallow.

Weights of Pulling Boats.

Boat.	Length	Weight without gear.			Boat.	Length	Weight without gear.		
		cwt.	qrs.	lbs.			cwt.	qrs.	lbs.
Launch (not sheathed)	42	75	0	0	Gig	32	9	0	0
,, ,,	40	67	1	14	,,	30	8	3	0
Pinnace	32	43	1	21	,,	28	8	0	0
,,	30	40	3	21	,,	26	7	3	0
Cutter	32	19	3	0	,,	24	7	1	0
	30	18	2	0	,,	22	7	0	24
	28	16	3	0	Cutter Gig	20	7	0	0
	26	16	0	0	Whale Gig	27	7	1	22
,,	25	1½	0	0	,,	25	6	3	24
Cutter-Life, cork-lined	32	20	3	0	Whale Gig-Life	27	7	3	0
,, ,,	28	20	0	0	Dingy	14	4	2	0
Jolly boat	18	8	0	0	,,	12	3	1	0
,,	16	6	0	0					

                                                                           cwt. qrs. lbs.

The average weight of a 12-oared cutter's gear is 9   0   23

        ,,   6-oared gig's gear is   4   1   7

Boats that have been in ships are found to be about 1 cwt. heavier for each *year issued*.

Practical hints for the consideration and guidance of Officers having charge of Boats.

Compiled from various sources.

In *shoving off*, when the ship is not head to wind, pull well clear of her before making sail.

Be very particular about the sails being properly set. In boom boats, set the jib before setting the foresail; haul the runners hand taut before hoisting. The jib is the forestay; a fid if the foresail be set first the mast-head is dragged aft, and the after leech hangs slack.

Sling a dipping lug $\frac{1}{3}$ from the foremost yard-arm; standing lug $\frac{1}{4}$.

The rule of the road for boats is the same as for ships.

As a general rule in sailing, insist on the crew sitting down on the bottom boards of the boat.

Keep weights amidships.

Do not allow men to sit on the gunwale, stand on the thwarts, or climb the mast.

As a general rule, a reef should be taken in directly the boat begins to wet.

Before *reefing on a wind*, tell the men off the different duties; the two bowmen to gather down on the luff; the weather hands by the halyards and downhaul; the lee hands to tie the points; one stroke oarsman to attend the sheet, the other to assist the coxswain in reefing the mizen. No person need stand up. Do not luff up; check the sheets; lower enough to shift the tack easily; gather the sheet aft, that the men may reach the foot of the sail without leaning over the lee gunwale; shift the sheet; tie away; slack the sheet; hoist; resume places, and haul aft. Should the mizen be reefed more *quickly than* the foresail, do not haul the sheet aft until the boat has *steerage* way.

Remember, in running, that you cannot carry all the canvas on a wind that you can before it; therefore make ready for rounding to.

Running dead before the wind in a gig is very dangerous. Little time is lost, and it is much safer to run half the distance with the wind well on one quarter; then lower, dip the sail, and bring the wind on the other quarter.

If the men are sitting to windward in a breeze, make them occupy their proper places before passing to leeward of a vessel.

If *caught* in a hard sudden squall, down helm at once, let fly the sheet and lower the sail.

In a moderate squall ease the sheet.

Trimming a boat requires great attention; if properly done, she should almost steer herself. A tendency to carry *lee* helm should be counteracted at once.

If there is any doubt about weathering a ship, or danger, "go about" in time, do not hug the wind, and drift crab-like on top of it. If there is any doubt about her going round, have an oar ready.

Do not put the helm down too suddenly, or too far over.

Water ballast in barricoes is the safest. Iron or sand stows better, but, in the event of a capsize, would sink the boat.

Do not overload a boat, particularly with men or sand. Sand is much lighter when dry than when wet. Remember that a laden boat carries her way longer than a light one.

In a breeze get the mast down before coming alongside.

Keep clear of a vessel with a stern-way on. Keep the boat stem on to a heavy sea.

Boats may ride out a heavy gale in the open sea, by lashing their spars, oars, &c., together and riding to leeward of them, secured to them by a span. The sails should be loosed and attached to the spars; they contribute greatly to breaking the sea, and if weights be fastened to the clews, the boats *drift* will be much retarded.

Towing.—Take the tow-line to the after thwart or foremost stern-sling bolt, and toggle the bight with a stretcher.

In towing short round, do not attempt to turn before your leaders are round.

The heaviest boats should be nearest the tow.

Weighted boats tow best.

Tow a spar by the small end.

When you take in the end of a warp, coil enough of it forward, so as to be able to make a bend the instant your boat reaches the place you wish to make it fast to.

Wet warps require very careful lashings.

If anchoring a boat on rocky ground bend the cable to the crown of the anchor, and stop it to the ring before letting go; then, if the flukes jamb, the stop will carry away, and the anchor may easily be weighed.

Viewed from seaward, a surf has never so formidable an appearance as when seen from the land; persons in a boat outside the broken water are therefore apt to be deceived by it. They should accordingly, if practicable, proceed along the land outside the surf, until abreast of a coastguard or life-boat station, or fishing village, whence they might be seen by those on shore, who would then signalize to them where they might safest attempt to land, or warn them to keep off; or who might proceed in a life-boat or fishing-boat to their aid, the generality of coast fishing-boats being far better able to cope with a surf than a ship's boat, and the coast-boatmen being more skilful in managing boats in a surf than the crews of ships. If in the night, double precaution is necessary—and it will in general be much safer to anchor a boat outside the surf until daylight, than to attempt to land through it in the dark.

Where a surf breaks at only a short distance from the beach, a boat may be veered and backed 'hrough it from another boat anchored outside the surf, when two or more boats are in company; *or she may* be anchored and veered, or backed in from her own *anchor*.

RULES PUBLISHED BY THE ROYAL NATIONAL LIFE BOAT INSTITUTION,

ON THE

**Management of Open Rowing Boats in a Surf;
Beaching them, &c.**

1. In Rowing to Seaward.

As a general rule, speed must be given to a boat rowing against a heavy surf.

Indeed, under some circumstances, her safety will depend on the utmost possible speed being attained on meeting a sea.

For, if the sea be really heavy, and the wind blowing a hard on-shore gale, it can only be by the utmost exertions of the crew that any head-way can be made. The great danger then is, that an approaching heavy sea may carry the boat away on its front, and turn it broadside on, or up-end it, either effect being immediately fatal. A boat's only chance in such a case, is to obtain such way as shall enable her to pass end on, through the crest of the sea, and leave it as soon as possible behind her. Of course if there be a rather heavy surf, but no wind, or the wind off shore, and opposed to the surf, as is often the case, a boat might be propelled so rapidly through it, that her bow would fall more suddenly and heavily after topping the sea, than if her way had been checked; and it may therefore only be when the sea is of such magnitude, and the boat of such a character, that there may be chance of the former carrying her back before it, that full speed should be given to her.

It may also happen that, by careful management under such circumstances, a boat may be able to avoid the sea, so that each wave may break ahead of her, which may be the only chance of safety in a small boat: but if the shore be flat, and the broken water extend to a great distance from it, this will often be impossible.

The following general rules for rowing to seaward may therefore be relied on :—

1. If sufficient command can be kept over a boat by the skill of those on board her, avoid or "dodge" the sea if possible, so as not to meet it at the moment of its breaking or curling over.

2. Against a head gale and heavy surf, get all possible speed on a boat on the approach of every sea which cannot be avoided.

If more speed can be given to a boat than is sufficient to prevent her being carried back by a surf, her way may be checked on its approach, which will give her an easier passage over it.

2. On Running before a Broken Sea, or Surf, to the Shore.

The one great danger, when running before a broken sea, is that of *broaching-to*. To that peculiar effect of the sea so frequently destructive of human life, the utmost attention must be directed.

The cause of a boat's broaching-to, when running before a broken sea or surf, is, that her own motion being in the same direction as that of the sea, whether it be given by the force of oars or sails, or by the force of the sea itself, she opposes no resistance to it, but is carried before it. Thus, if a boat be running with her bow to the shore, and her stern to the sea, the first effect of a surf or roller, on its overtaking her, is to throw up the stern, and as a consequence to depress the bow; if she then has sufficient inertia (which will be proportional to weight) to allow the sea to pass her, she will in succession pass through the descending, the horizontal, and the ascending positions, as the crest of the wave passes successively her stern, her midships, and her bow, in the reverse order in which the same positions occur to a boat propelled to seaward against a surf. This may be defined as the safe mode of running before a broken sea.

But if a boat, on being overtaken by a heavy surf, has not sufficient inertia, to allow it to pass her, the first of the three positions above enumerated alone occurs—her stern is raised high in the air and the wave carries the boat before it, on its front, or unsafe side, sometimes with frightful velocity, the bow all the time being deeply *immersed* in the hollow of the sea, when the water, being stationary, *or comparatively* so, offers a resistance, whilst the crest of the sea, having the actual motion which causes it to break, forces onward the stern, or rear end of the boat.

A boat will, in this position, sometimes aided by careful oar-steerage, run a considerable distance until the wave has broken and expended itself. But it will often happen, that if the bow be low, it will be driven under water, when the buoyancy, being lost forward, whilst the sea presses on the stern, the boat will be thrown (as it is termed) end over end ; or if the bow be high, or it be protected, as in most life-boats, by a bow air chamber. so that it does not become submerged, that the resistance forward, acting on one bow, will slightly turn the boat's head, and the force of the surf being transferred to the opposite quarter, she will in a moment be turned round broadside by the sea and be thrown by it on her beam-ends, or altogether capsized. It is in this manner that most boats are upset in a surf, especially on flat coasts, and in this way many lives are annually lost amongst merchant seamen when attempting to land, after being compelled to desert their vessels.

Hence it follows that the management of a boat, when landing through a heavy surf, must, as far as possible, be assimilated to that when proceeding to seaward against one, at least so far as to stop her progress shoreward at the moment of being overtaken by a heavy sea, and thus enabling it to pass her. There are different ways of effecting this object :—

1. By turning a boat's head to the sea before entering the broken water, and then backing in stern foremost, pulling a few strokes ahead to meet each heavy sea, and then again backing astern. If a sea be really heavy and a boat small, this plan will be generally the safest, as a boat can be kept more under command when the full force of the oars can be used against a heavy surf, than by backing them only.

2. If rowing to shore with the stern to seaward, by backing all the oars on the approach of a heavy sea, and rowing ahead again as soon as it has passed to the bow of the boat, thus rowing in on the back of the wave ; or, as is practised in some life-boats, placing the after-oarsman, with their faces forward, and making them row back at each sea on its approach.

3. If rowed in bow foremost, by towing astern a pig of ballast or large stone, or a large basket, or canvas bag termed a "drogue" or drag, made for the purpose, the object of each being to hold the boat's stern back, and prevent her being turned broadside to the sea or broaching-to.

A boat's sail bent to a yard, and towed astern loosed, the yard being attached to a line capable of being veered, hauled, or let go will act as a drogue, and tend much to break the force of the sea immediately astern of the boat.

Heavy weights should be kept out of the extreme ends of a boat; but when rowing before a heavy sea the best trim is deepest by the stern.

A boat should be steered by an oar over the stern, or on one-quarter when running before a sea, as the rudder will then at times be of no use. If the rudder be shipped, it should be kept amidships on a sea breaking over the stern.

The following rules may therefore be depended on when running before, or attempting to land, through a heavy surf or broken water :—

1. As far as possible avoid each sea by placing the boat where the sea will break ahead or astern of her.

2. If the sea be very heavy, or if the boat be very small, and especially if she have a square stern, bring her bow round to seaward and back her in, rowing ahead against each heavy surf that cannot be avoided sufficiently to allow it to pass the boat.

3. If it be considered safe to proceed to the shore bow foremost, back the oars against each sea on its approach, so as to stop the boat's way through the water as far as possible, and if there is a drogue, or anything in the boat that may be used as one, tow it astern to aid in keeping the boat end on to the sea, which is the chief object in view.

4. Bring the principle weights in the boat towards the end that is to seaward, but not to the extreme end.

5. If a boat, worked by sails or oars, be running under sail for *the land through* a heavy sea, her crew should, under all circumstances, *unless the beach* be quite steep, take down her masts and sails before

entering the broken water, and take her to land under oars alone, as above described.

If she have sails only, her sails should be much reduced, a half-lowered foresail or other small head-sail being sufficient.

3. Beaching or Landing through a Surf.

The running before a surf or broken sea, and the beaching or landing of a boat, are two distinct operations; the management of boats, as above recommended, has exclusive reference to running before a surf where the shore is so flat that the broken water extends to some distance from the beach. Thus, on a very steep beach, the first heavy fall of broken water will be on the beach itself, whilst on some very flat shores there will be broken water as far as the eye can reach, sometimes extending to even four or five miles from the land. The outermost line of broken water, on a **flat shore, where the waves break in three** and four fathoms water, **is the heaviest**, and therefore the most dangerous, and when it has been passed through in safety, the danger lessens as the water shoals, until, on nearing the land, force is spent and its power harmless. As the character of the sea is quite different on steep and flat shores, so is the customary management of boats on landing different in the two situations. On the flat shore, whether a boat be run or backed in, she is kept straight before or end to the sea until she is fairly aground, when each surf takes her further in as it overtakes her, aided by the crew, who will then generally jump out to lighten her, and drag her in by her sides. As above stated, sail will, in this case, have been previously taken in if set, and the boat will have been rowed or backed in by oars alone.

On the other hand, on the *steep* beach, it is the general practice, in a boat of any size, to retain speed right on to the beach, and in the act of landing, whether under oars or sail, **to turn the boat's bow half round towards** the direction from which the surf is running, so that she may be thrown on her broadside up the beach, when abundance of help is usually at hand to haul her as quickly as possible out of the reach of the sea. In such situations, we believe, it is nowhere the practice to back a boat in stern **foremost under oars**, but to row in under full speed as above described.

BOARDING.

Boarding a Wreck, or a Vessel under Sail, or at Anchor in a Heavy Sea.

The circumstances under which life-boats or other boats have to board vessels, whether stranded or at anchor, or under way, are so various that it would be impossible to draw up any general rule for guidance. Nearly everything must depend on the skill, judgment, and presence of mind of the coxswain or officer in charge of the boat, who will often have those qualities taxed to the utmost, as undoubtedly the operation of boarding a vessel in a heavy sea or surf is frequently one of extreme danger.

It will be scarcely necessary to state that, whenever practicable, a vessel, whether stranded or afloat, should be boarded to leeward, as the principle danger to be guarded against must be the violent collision of the boat against the vessel; or her swamping or upsetting by the rebound of the sea: or by its regular direction on coming in contact with the vessel's side; and the greater violence of the sea on the windward side is much more likely to cause such accidents. The danger must of course, also be still further increased when the vessel is aground and the sea breaking over her. The chief danger to be apprehended on boarding a stranded vessel on the lee side, if broadside to the sea, is the falling of the masts; or if they have been previously carried away, the damage or destruction of the boat amongst the floating spars and gear alongside. It may therefore, under such circumstances, be often necessary to take a wrecked crew into a life-boat from the bow or stern; otherwise a rowing boat, proceeding from a lee shore to a wreck, by keeping under the vessel's lee, may use her as a breakwater, and thus go off in comparatively smooth water. This is accordingly the usual practice in the rowing life-boats. The larger sailing life-boats, which go off to wrecks on outlying shoals, *are however*, usually anchored to windward of stranded vessels, and then veered down to 100 or 150 fathoms of cable, until near enough to throw a line on board.

The greatest care under these circumstances has of course to be taken to prevent actual contact between the boat and the ship.

In every case of boarding a wreck or vessel at sea, it is important that the line by which a boat is made fast to the vessel should be of sufficient length to allow of her rising and falling freely with the sea; and every rope should be kept in hand ready to cut or slip it in a moment if necessary.

Before going alongside a vessel underweigh and hove to, observe if she have head or sternway, and in any case get the masts down before closing her. Wait until the ship has gathered headway, and then go alongside.

Do not shove off during sternway, else the ship in setting to leeward and falling off may bury the boat under her bows, or with her stem and dolphin striker cut the boat down.

In being towed by a vessel, if alongside, contrive to have the rope from as far forward as possible, so as to avoid riding at a short stay: never make it fast, but toggle it with a stretcher through the aftermost of the foremost sling bolts, so as to be able to slip in an instant.

In being towed astern, the closer the better.

When boarding foreign men-of-war the boarding book should not be taken on board, but the information gathered should be entered afterwards.

When boarding a vessel in a British port, assistance should be offered if required.

Boarding Merchant Vessels.—Fill up the columns of the printed boarding book now supplied.

SHIP PAPERS.

THE FOLLOWING ARE THE PAPERS THAT MAY BE EXPECTED TO BE FOUND ON BOARD A MERCHANT VESSEL, THOUGH UNDER DIFFERENT NATIONALITIES THEY MAY VARY.

Every Merchant Vessel should carry on board some Official Voucher for her Nationality, issued by the authorities of the Country to which she belongs.

The Official Voucher of a vessel which belongs to a Country possessing a Register of its Mercantile Marine, is a Certificate of her Registry: in other cases its form varies and passes under different names—"Passport," "Sea-brief," &c.

The Certificate of Registry is a document signed by the Registrar of the Port to which the Vessel belongs, and usually specifies the name of the Vessel and of the Port to which she belongs; her tonnage, &c.; the name of her master; particulars as to her origin; the names and description of her registered owners.

The Passport purports to be a requisition on the part of a Sovereign Power or State to suffer the Vessel to pass freely with her company, passengers, goods and merchandise without any hindrance, seizure, or molestation, as being owned by citizens or subjects of such state. It usually contains the name and residence of the Master: the name, description, and destination of the Vessel.

The Sea-letter, or Sea-brief, is issued by the Civil Authorities of the Port from which the Vessel is fitted out; it is the document which entitles the Master to sail under the Flag and Pass of the Nation to which he belongs; and it also specifies the nature and quantity of the cargo, its ownership, and destination.

The Charter-party is the written contract by which a Vessel is let, in whole or in part; the person hiring being called the Charterer. It is executed by the owner or master, and by the Charterer. It *usually* specifies (amongst other things) the name of the Master, the name and description of the Vessel, the port where she was lying at the time of the Charter, the name and residence of the Charterer, the character of the Cargo to be put on board, the port of loading, the port of delivery, and the freight which is to be paid.

The Charter-party is almost invariably on board a Vessel which has been chartered.

The Official Log-book is the Log-book which the Master is compelled to keep in the form prescribed by the municipal law of the Country to which the Vessel belongs.

The Ship's Log is the log kept by the Master for the information of the owners of the Vessel.

The Builder's Contract is to be expected on board a Vessel which has not changed hands since she was built. It is not a necessary document, but it sometimes serves, in the absence of the Pass or Sea-letter or Certificate of Registry, to verify the Nationalty of a Vessel.

The Bill of Sale is the instrument by which a Vessel is transferred to a purchaser. It should be required whenever a sale of a Vessel is alleged to have been made either during the War or just previous to its commencement, and there is any reason to suspect that the Vessel is liable to Detention, either as an Enemy's Vessel or as a British or Allied Vessel trading with the Enemy.

Bills of Lading usually accompany each lot of goods.

A Bill of Lading on board a Vessel is a duplicate of the document given by the Master to the Shipper of goods on the occasion of the shipment; it specifies the name of the Shipper, the date and place of the Shipment, the name and destination of the Vessel, the description, quantity, and destination of the goods, and the freights which are to be paid.

The Invoices should always accompany the Cargo; they contain the particulars and prices of each parcel of goods, with the amount of the freight, duties, and other charges thereon, and specify the names and address of the Shippers and Consignees.

The **Manifest** is a list of the Vessel's Cargo, containing the mark and number of each separate package, the names of the Shippers and Consignees; a specification of the quantity of goods contained in each package, as rum, sugar, &c., and also an account of the freight corresponding with the Bills of Lading.

The Manifest is usually signed by the Ship-broker who clears the Vessel out at the Custom House, and by the Master.

The **Clearance** is the Certificate of the Custom-House authorities of the last port from which the vessel came, to show that the Custom duties have been paid. The Clearance specifies the cargo and its destination.

The **Muster-Roll** contains the name, age, quality, place of residence, and place of birth of every person of the Vessel's company.

Shipping Articles are the agreement for the hiring of seamen. They should be signed by every seaman on board, and should describe accurately the voyage and the terms for which each seaman ships.

The **Bill of Health** is a Certificate that the Vessel comes from a place where no contagious distemper prevails, and that none of her crew at the time of her departure were infected with such distemper.

BOAT CRUISING.
(ESPECIALLY ON THE EAST COAST OF AFRICA.

Equipment.—Ordinary service boats always require to be raised upon to enable them to carry the necessary weight of provisions, water, &c., in addition to their crew, for ten day's or a fortnight's cruise. Cutters and pinnaces should have from 6 to 8 inches added to their gunwale forward, decreasing to about half that height at the stern of the boat. They should have a large locker built in the bows, at least as high as the thwarts, made watertight, and the top of it covered with copper or lead, in which to stow the boat's crew's night and *spare clothing*. A similar locker to be built in the after part of the *stern sheet* for the officers.

A jackstay should be fitted round the boat, underneath the rubbing strake for the rain awning to be laced down to. All boats should be supplied with a rain awning; two spare oars in addition to her complement; boat-hooks fitted with a stout lanyard, ending in an eye, secured to the hook, and seized two-thirds down the staff, to this eye a line should be attached (a boat hook thus fitted will be found useful when boarding a vessel under weigh); a small grappling iron with a fathom and a-half of light chain; a stout line to throw on board vessels under weigh when desirous of boarding; scuttle-butts made to fit under the thwarts as large as the dimensions of the boat will conveniently allow; breakers, sails, boilers, &c., complete. The oars should always be short as it must be remembered that the boats when provisioned will sit deep in the water. A pump should also be fitted under the coxswain's seat to discharge water at the stern.

Rig.—The Slave Dhows on the East Coast of Africa are specially rigged for running with the Monsoons and they carry very large sails in proportion to their tonnage. The object, therefore, should be to rig boats so as to stand some chance of sailing on equal terms with them. The ordinary service rig will hardly allow of this, and after much experience it has been generally found that boats rigged with two sliding gunters are the best suited for cruising purposes, for this rig is not only easily handled and made safe in bad weather, but by means of bamboo yards, light duck square sails, and even topsails can be set when chasing off the wind. Boats rigged as schooners are also very handy but the sliding gunter has great advantage in its short masts and light yards.

Provisioning and Cooking.—All boats should be fitted with scuttle-butts of such dimensions as to stow conveniently under the boats thwarts on each side of the kelson. They will be found very useful for carrying both provisions and water, and stow better than breakers, though no boat should be without a certain proportion of the latter for convenience in watering. Bread and meat being the two most bulky articles of provision, should be reduced as much as possible, and it has been found generally convenient to supply only half allowance of these articles to boats when detached, furnishing the officer in charge with money to purchase fresh meat, vegetables, &c., to make up for it, and which can generally be procured along the coast. On the other hand, in consideration of the duty on which they are employed, double allowance of small stores and spirits is generally issued.

The ordinary service Cooking Stove can be conveniently used in all boats, from a cutter upwards, to be placed on the large locker built in the bows, which should be covered with copper (as before recommended.) In smaller boats a small tub of sand, with three bricks, answers very well for this purpose.

Clothing and Sleeping.—Every man in a boat should be provided with a blanket frock for sleeping in. He should also have two complete shifts of woollen clothing, with a white cover for his cap, to be worn during the day. The rain awnings should always be spread at sunset, and the oars so arranged on the thwarts as to form a platform for the men to sleep on. As it is of great importance that the men's clothing should be kept dry, the lockers in the bow and stern of boat should be made watertight if possible, the bottom being raised considerably above the kelson. As the health of the men much depends on cleanliness, every opportunity should be taken of clearing the boat and thoroughly cleansing her.

Anchors and Cables.—All boats, if possible, should be furnished with two anchors—at any rate, all boats larger than cutters. For launches and pinnaces, one chain cable should, when practicable, be supplied, as, from the nature of the bottom (coral), hemp cables are often cut through. For the same reason, a few fathoms of chain next to the anchor is advisable, even in the smaller boats. A small amount of sounding line and a boat's lead to be furnished to all boats.

Boarding Vessels.—Two boats, as a rule, should cruise in company, such as, for instance, a pinnace and a gig. In boarding dhows or other slave vessels, if not actually resisting, the gig should board, and the pinnace's crew stand to their arms. If boarding a vessel with way on, a grappling, fitted with a fathom and a half of small chain and a stout line (like a gunner's grappling), should be kept handy in the bows of the boat, ready to throw on board, the other bowman having a line also ready for hooking on. If a slave dhow is resisting, and continuing her course under sail, it is often possible to cut her rudder pendants, and so disable her (many of these vessels steering by pendants from the after part of the rudder). Care should be taken for the men, when boarding such a vessel, to carry, if possible, their rifles, or other fire-arms with them, as the Arabs are extremely good swordsmen, and may generally be expected to out-number the boarders.

Crossing Bars.—For the purpose of steering a boat in crossing the numerous bars of the rivers on the East coast, boats should be fitted with a crutch hole on each quarter where an oar could be worked to assist the rudder. Some people like to pull over bars, others prefer, when the wind is free, to sail. The objection to pulling is that at the critical moment, and when the boat is in broken water, it too often happens that some of the men catch crabs with their oars, and broach the boat to. The best plan when pulling is to toss the oars as the wave breaks round the boat, giving way as quickly as possible afterwards, but this requires a very well-trained crew. Under sail with a good commanding breeze, everything depends upon careful steerage. Some object on the plea that the sail get becalmed between the rollers. Such, to a slight extent, is no doubt true, but the face of the advancing roller always raises her in time to gather the way necessary for her to be properly handled. In crossing bars men should be encouraged to relieve themselves of all superfluous garments, oar lanyards taken off oars, and masts and other buoyant articles left free to float in case of the boat being swamped, a contingency which to those who have seen much of such work is always possible, however well-handled the boat may be.

General Remarks.—Each boat should have a boatswain's bag, containing all the small gear likely to be required for repairing sails, rigging, &c., also some fishing lines and hooks; a carpenter's bag, with material for repairing the boat if stove, and a few ordinary carpenter's tools. She should also have signal book, lead and line, junk axe, and a couple of tomahawks will be found useful.

On Stations where Slavers are met with, Special Instructions are issued.

The following extracts may be useful to Boarding Officers.

If in the course of your search, you are satisfied that the vessel is engaged or equipped for the slave trade, and that she is subject to your authority, you will proceed to detain her.

You will be justified in concluding that a vessel is engaged in the slave trade.

If you find slaves on board; or, if you find in her outfit any of the equipments hereinafter mentioned.

1st.—Hatches with open gratings, instead of the close hatches, which are usual in merchant vessels.

2nd.—Divisions or bulkheads, in the hold or on deck, in greater number than are necessary for vessels engaged in lawful trade.

3rd.—Spare planks either actually fitted, or fit for laying down as a slave deck.

4th.—Shackles, bolts, or handcuffs.

5th.—A larger quantity of water in casks or tanks, than is requisite for the crew of a vessel as a merchant vessel.

6th.—An extraordinary number of water casks, or other vessels, for holding liquid; unless the Master shall produce a certificate from the Custom House, at the place from which he cleared outwards, stating that sufficient security had been given by the Owners that such extra quantity of casks, or of other vessels should be used to hold palm oil, or for other purposes of lawful commerce.

7th.—A greater number of mess tubs or kids than is requisite for the use of the crew of the vessel as a merchant vessel.

8th.—A boiler, or other cooking apparatus of an unusual size, and larger, or capable of being made larger, than is requisite for the crew of the vessel as a merchant vessel, or more than one boiler or *other cooking* apparatus of the ordinary size.

9th.—An extraordinary quantity of rice, of the flour of Brazil, of manioc and cassada (commonly called farina), of maize, or of Indian corn, or of any other article of food whatever, beyond the probable wants of the crew; unless such rice, flour, farina, maize, Indian corn, or other article of food be entered on the Manifest as part of the cargo for trade.

10th.—A quantity of mats or matting greater than is necessary for the use of the crew of the vessel as a merchant vessel; unless the mats or Matting be entered on the manifest as part of the cargo for trade.

The following remarks, with regard to the Capture of Slavers on the East India Station, are extracted by permission, from Captain Colomb's book, "Slave Catching."

By the general instructions, the naval Commander is told:—

"The vessels subject to your authority by virtue of British jurisdiction are:—

"In all waters, which are not foreign territorial waters—

"British vessels, and vessels not entitled to claim the protection of the flag of any state or nation.

"In British waters—

"All vessels, *whatever* be their nationality.

"The vessels subject to your authority by treaty, are vessels "belonging to any other state with which Great Britain has a treaty "for the suppression of the slave trade; but only within the limits "prescribed by the treaty; or, if no limits are prescribed, when found "in waters not being territorial waters."

"Territorial water," in its essence means any water over which, or over the entrance to which, the Power possessing the coast can throw shot.

Custom has given an arbitrary range of three miles; hence territorial waters are waters within three miles of any shore, whose owner is strong enough, or important enough to assert his jurisdiction.

Waters cannot be made territorial artificially, but waters naturally territorial may be made the "high seas" by treaty.

In the Zanzibar territory, the Sultan, by a voluntary Act, has rendered his waters, the "high seas," in respect of the slave trade, during the months of January to April inclusive, in each year. He retains his right for the remaining months.*

Egypt and Turkey have conceded nothing. The approach to their territories in the Red Sea must be marked "dangerous" by the naval officer.

The Sultan of Omān has absolutely surrendered his right over Ormani waters, to hamper the naval officer, from Lamoo to the Euphrates.

To the English naval officer, as such, Persia has conceded no powers.

The ships engaged in slave trading, likely to be met with, are, first, the ships of the Arab tribes bordering the Persian Gulf; secondly, the ships of Omān; thirdly, those of the Sultan of Zanzibar.

Ships of Turkey may be met, but if they are engaged in the slave trade, their detention by the naval officer will be at his own risk.

The ships of Persia may be met, but, ordinarily speaking, they must not be interfered with.

The ships of France, possibly engaged in the slave trade, do not, or did not, engage in it north of the equator, and are not likely to be met in the northern slave trade.

The most ordinary slave-trader of all, however, flies no flag, carries no papers, belongs to nowhere, and claims nobody's protection.

Most Arab dhows fly a plain red flag.

---

* Since the above was written, the mission of Sir BARTLE FRERE has resulted in the signature of the treaties for the suppression of the Slave Trade by the Sultans of Zanzibar and Muscat.

BOAT RACING.

The following Rules, drawn up by the Committee, for the Channel Fleet Regatta of 1872, may be useful as a guide on any further occasion of the same kind.

Courses.—1.—The sailing course to be triangular and about 6 miles. A special course (about 4 miles) will be made for single-banked boats. The long rowing course to be about $2\frac{1}{2}$ miles. The middle course about $1\frac{3}{4}$ miles, and short course 1 mile.

2.—All Buoys to be left on Port hand by rowing boats.

3.—The direction in which sailing boats start to be decided by the Committee on the morning of sailing.

4.—The Committee reserve to themselves the power of altering or increasing the course if requisite, or of altering the days for pulling and sailing according to the weather.

5.—In all races—three boats to start, or no race.

6.—In races where a second prize is given—four boats to start, or no second prize. Where a third prize is given—six boats to start, or no third prize.

7.—All objections to be lodged with the Committee within half-an-hour of the objecting boat passing the winning post. The decision of the Committee to be final.

8.—Entries to close at Noon, on the day before the Regatta, and to be addressed to the Secretary of the Regatta Committee.

9.—All boats to carry distinguishing flags 17 + 14.

10.—Any additional funds will be added to the Prizes, or disposed of as the Committee may decide.

11.—In the service races, no boats to be allowed to use more than the regulation number of oars—the oars to be pulled from their proper thwarts, and in sailing races, service rigs to be adhered to;

neither iron ballast, nor false keels to be allowed. Water ballast only permitted. By service rig is to be understood the ordinary working rig of the boat.

12.—In all service races, ballast may be trimmed but not started. Any sailing boat using oars in the water for any purpose to be disqualified.

PROGRAMME.

FIRST DAY.—ROWING. | SECOND DAY.—SAILING.

No.	RACES.	No.	RACES.
1	Pinnaces. Long race.	1	All Launches. One Prize for Steam, and one for other Launches.
2	Gigs, 4 oars. Middle course.		
3	All Launches, 18 oars. Long course.	2	Pinnaces.
4	Dingies, 4 oars. Short course.		
5	Cutters and Barges, 12 oars or less. Long course.	3	Cutters and Barges, over 28 feet.
6	Whalers and 5-oared Gigs. Middle course.	4	Cutters, 28 feet and under.
7	Launches pulled by Marines. Long course.	5	Life Boats.
8	4 and 5-oared Gigs and Whalers. pulled by Boys. Middle course.	6	Single banked Boats.
9	Galleys and 6-oared Gigs. Long course		Champion Copper Punt Race, no sails, to be paddled with shovels.
10	Cutters and Jolly Boats, 8 or 10 oars. Middle course. *No Boat entering for No. 5 can pull in this race.*		Tubs without keels. Duck Hunts, &c. *Fancy Dresses admissable.*
11	Dingies pulled by Marines. Short course.		
12	Cutters, 12 oars, pulled by Stokers. Middle course.		
13	6-oared Galleys, pulled by Officers. Middle course.		
14	Cutters, 12 oars, pulled by Boys. Middle course.		
15	All comers; Boats may be double banked. Long Course.		
16	Consolation Stakes for all boats who have raced but won nothing.		

The following Rules were drawn up by the Regatta in 1874.

1.—Courses. The sailing course to be Triangular and about 8 miles; a special course will be made for single banked boats, the long rowing course to be about 2½ miles; the middle course about 1¾ miles; and short course 1 mile.

2.—The Committee reserve to themselves the power of altering or increasing the course, if requisite, or of altering the days for pulling and sailing, according to the weather.

3.—In all races three boats to start or no race.

4.—In races where a second prize is given, four boats to start or no second prize, where a third prize is given, six boats to start or no third prize.

5.—All objections to be lodged with the Committee within an hour of the objecting boat passing the winning post. The decision of the Committee to be final.

6.—Entries to close at noon on the day before the Regatta, and to be addressed to the Secretary of the Regatta, H.M.S.—

7.—All boats to carry distinguishing flags, 17 × 14 inches.

8.—Any additional funds will be added to the prizes, or disposed of as the Committee may decide.

9.—The Officer who acts as member of the Committee to do his utmost to see that all boats of his own ship abide by the rules.

10.—No member of the Committee is allowed to pull or sail in a racing boat.

ROWING REGULATIONS.

1.—In the service races, no boats to be allowed to use more than the regulation number of oars, the oars to be pulled on their proper thwarts, (except in Nos. 5 and 8 Races.)

2.—The Committee will give directions previous to the race, on which hand the buoys are to be left.

3.—An Officer to be in all double-banked boats,

4.—If the boats crew are Seamen, a Seaman is to steer; if Marines, a Marine is to steer; if composed of Stokers, a Stoker is to steer.

5.—Any boat purposely fouling another, or making use of the vessels or buoys stationed, as turning points for the purpose of rounding more readily, will be disqualified.

SAILING REGULATIONS.

1.—Service Rigs are to be strictly adhered to: by service rigs, is to be understood the rig that boats use when on service.

2.—Iron ballast or false keels are not to be used, water ballast only permitted.

3.—All boats to be sailed by a full boat's crew.

4.—All boats in the proximity of any boat that may capsize, are to render immediate assistance, and boats rendering such assistance, shall have the option of sailing the two winning boats (provided the Committee are of opinion that those boats stood a chance of taking either of the prizes).

5.—In the event of any boat capsizing in the early part of the race, the Committee have the power of recalling all boats and starting them afresh.

6.—Ballast may be trimmed but not started.

7.—Any boat using propelling power, such as oars, balers, hands or feet in the water, to be disqualified.

8.—The rules of the road with regard to meeting or passing or giving way to one another, are to be strictly adhered to, and any boat purposely fouling another, will be disqualified.

9.—Any boat capsizing, both Officer and boat shall be excluded from racing again for that particular prize.

First Day—ROWING RACES.

Number	Class of Boats.	Maximum No. of Oars.	Minimum Length in Feet.	Course to be Pulled.	By whom Pulled.
1	Cutters	12	30	Long	Seamen
2	Whalers	5		Middle	Midshipmen
3	Launches	18	42	Long	Seamen
4	Dingies	4		Short	Boys
5	Cutters	12	28	Long	Marines
6	Galleys	6		Ditto	Seamen
7	Pinnaces	14	32	Ditto	Ditto
8	Cutters	12	28	Ditto	Stokers
9	Whalers and Gigs	5		Middle	Boys
10	Launches	18	42	Long	Marines
11	Cutters	10	28	Ditto	Seamen
12	Whalers and Gigs	5		Middle	Ditto
13	Cutters	8	23	Ditto	Ditto
14	Cutters	12	30	Long	Boys
15	All Comers in Service Boats. Oars may be double banked			Long	Seamen, Stokers, and Marines
16	Consolation Stakes, for Boats that have Raced and won nothing			Middle	Seamen, Stokers, and Marines
17	Service Boats, belonging to ther own Ships			Ditto	Officers

Second Day—The same races as in 1874.

Section 7.

**OPERATIONS ON SHORE;
LANDING A BATTALION;
ATTACK OF A POSITION; BIVOUAC.
DISEMBARKING TROOPS BEFORE AN ENEMY.**

OPERATIONS ON SHORE.

On the Employment of Small-arm Men, when Landed for Service.

Although there are a vast number of manœuvres laid down in the *Field Exercise for the Army* it is notorious that but very few, and those *simple* ones, are used in real warfare. It should be remembered that 'drill in the army is as much a mode of securing discipline as a preparation for war.' As this is not the case with seamen, it is obvious that they need only practise those movements which are required on actual service; and it cannot be too often repeated that they are very few, and very simple. A knowledge of a great many details of drill by no means implies a power of applying them. No officer is fit to command even the smallest body of men whose aptitude for war is determined by the strength of his memory alone. It is incontestible that the grave dislike and distrust with which the practice of Infantry manœuvres has ever been received by Naval officers has been due to the unpractical shape in which they have generally been presented to them. Common sense revolts from the tedious study of mere 'parade' and 'barrack square' formations, which it is foreseen are of no practical value in the field. While the present practice is derisively termed—and to a great extent rightly—mere 'playing at soldiers,' it should be remembered that the English sailor possesses when rightly trained, qualities in the very highest degree valuable for the warfare of the present day; his activity, intelligence, independence, self-reliance, and above all, his almost daily habit of facing danger of some sort, are of the very first importance, and with careful instruction will go far to make him the beau-ideal of a fighting man.

Train him, then, so as to develop these qualities to the utmost, and do not make the grave mistake of trying to drill into him what may be considered the only faults of our soldiers:—a too great stiffness, precision, and regularity of movement. How much of the little time which can be spared to this subject is devoted to an attempt to imitate the wall-like regularity of the British Line—destroying the very essence of a blue jacket—while the elements of fighting remain uncared for and unknown.

To deny the utility of this training for seamen is to ignore the great importance of the tactics which were so signally successful under Lord Dundonald, and the power which a fleet has, with its numerous steam launches, of harassing an enemy's coast, cutting his communications, destroying his depôts, and keeping his coast garrisons in a constant state of alarm. Some of the finest feats in the last war were performed by French sailors, both as Infantry and Artillery: witness the recapture of Le Bourget and the defence of the Paris forts. Admiral Jurien de la Gravière, no mean authority, says on this point; "Though I above all things hold to our proper rôle of "seamen, I have not the less constantly protested against the "opinions of those who despise our landing parties. It would be, "indeed, lamentable if the 5000 to 6000 picked men of a fleet were "not always capable of executing a simple reconnaissance, coup-de-"main, or short expedition."

But to be able to do this we must give up "playing at soldiers" and "turn to" in real earnest. When relieved from the tedious mass of details which at present encumber it, the study of the minor operations of war is as interesting as it is important.

Movements in the Field.—With regard, then, to manœuvres, we may say that the power of skirmishing in a thoroughly efficient manner is of the first importance: in former days the way was *prepared* by the skirmishers, while the real attack was delivered by the line; now the skirmishers do the main work themselves, and afterwards secure it by the final charge. The success of skirmishing depends almost entirely upon the skill with which the sectional commanders handle their men, and all officers should be thoroughly acquainted with their work in this respect. A company which can wheel, or form fours, to either flank, and form to the front and rear, in addition to skirmishing, may be said to be thoroughly efficient for all practical purposes. In Battalion Drill we may dismiss at once all mere *parade* movements; Naval Officers have far too much to learn now a-*days*, to burden their memories with these cumbrous and useless details. For the remainder, Secs. 4; 7; 3 of 10 and 11; 1 of 13; 14; 17; 18; 19; 22; 1 of 25; 2 of 26; 1 of 29; with the skirmishing, contain all that is necessary.

Detailed arrangements for Landing a Battalion of Six Companies (50 Men in each) and 4 Guns for Six Days.

We will now proceed to give a brief idea, first, of the organisation of a battalion for landing, with a few precautions to be observed in conducting the march; and secondly, of the manœuvres which will be required, and the mode of applying them. It should be remembered that the same skill is required, whether dealing with 100 or 10,000 men; and that an action, however small, can only be brought to a successful issue without undue loss of life by acting in accordance with recognised and approved methods.

The above force has been selected as one which could be easily furnished by a squadron of 3 or 4 ships, and the length of time specified allows for a march inland of about two days for the purpose required.

The subjoined strength is that authorized by the Gunnery Manual, 1873, but it should be remarked that the number of spare ammunition-men and stretcher-men is probably double what would be actually required. The spare men could be employed in the transport of baggage and supplies.

The Battalion consists of ... 4 Field Officers.
1 Surgeon.
18 Officers of Companies.
1 Chief Petty Officer.
300 Rank and File.
12 Markers.
4 Field Officers.   6 Pioneers.
18 Company Officers   24 Stretcher-men.
5 Battery Officers   24 Spare-ammunition-men.
72 Men for Field Guns.   6 Medical Attendants.
6 Pioneers.   3 Buglers.
―――   3 Signalmen.
105 are armed with Pistols   3 Armourers.
―――   1 Master-at-arms.

R 2

	1 Ship's Corporal.
	1 Ship's Steward's Assistan
	408
The Battery	72 Men.
	5 Officers.
	1 Ship's Corporal.
TOTAL	486

The markers are omitted from the battery, as they are of no practical value whatever.

Dress.—The Men will wear: Weight.
    Serge jumper and trousers ... ... $1\tfrac{3}{4}$ lbs.
    Flannel ... .   1 ,,
    Comforter ... ... $\tfrac{1}{2}$,,
    Pair of Stockings ... . $\tfrac{3}{4}$,,
    Boots ... ... $2\tfrac{3}{4}$,,
    Cap ... ... $\tfrac{1}{2}$,,
    Shirt ... ...   1 ,,
    Knife and Laniard ... ... $\tfrac{1}{2}$,,
    Handkerchief ... ... $\tfrac{1}{4}$,,
                                                          9 lbs.

And they will carry:
    Rifle · .. ..   9 lbs. 3 ozs.
    Accoutrements ... ...   3 ,, 14 ,,
    Sword and Bayonet, &c. ... ...   3 ,, 2 ,,
    60 rounds of ammunition .. ..   6 ,, 0 ,,
    Blanket ... ...   5 ,, 14 ,,

Rolled up in it: Weight.
    Shirt ... ...   1 lb. 0 ozs.
    Stockings ... ...   0 ,, 12 ,,
    Towel ... ...   0 ,, 7 ,,
                           2 ,, 3 ,,

Haversack	0 lb.	6 ozs.
Spoon	0 ,,	3 ,,
TOTAL ...	39 ,,	13 ,,
Add 4½ lbs. for 2 days' provisions ...	4 ,,	8 ,,
GRAND TOTAL ...	44 lbs.	5 ozs.

Field Guns' crews will carry:

Cutlass and Pistol	5 lbs.	8 ozs.
Accoutrements	2 ,,	1 ,,
60 rounds of Ammunition	2 ,,	9 ,,
Dress, Blanket, &c.	17 ,,	10 ,,
	27 ,,	12 ,,
Two days' provisions	4 ,,	8 ,,
TOTAL ...	32 lbs.	4 ozs.

The blankets are rolled up in the direction of their length, tied by four stops at equal intervals; a fifth stop then connects the two ends, the blanket being worn like a horse-collar over the *left* shoulder.

Although no water-bottle is supplied for small-arm men, it would be necessary to extemporise something of the sort, as no force could operate without them.

Officers, in addition to their ordinary dress, carry clasp-knife, water-bottle, pocket-book, waterproof, haversack, blanket containing a flannel shirt and stockings, towel, spoon, and binoculars or telescope.

Provisions:—The men will carry two days' provisions, cooked, in their haversacks.

Daily Scale:—

Biscuit or Bread	1	lb.
Meat	{ 1 { 1½	,, salt or ,, fresh.
Tea	1	oz.

Sugar	2 ozs.
Salt	½ oz.
Pepper	1/36 ,,
Total	2¼ lbs.

No rum has been allowed; extra tea instead.
For the remaining four days, transport would be required for

4 ×	486 ×	1 lb.	biscuit	= 1944 lbs.	=	20 bags.
4 ×	,, ×	1 ,,	meat	= 1944 ,,	=	10 barrels.
4 ×	,, ×	1 oz.	tea	= 122 ,,	=	3 half chests
4 ×	,, ×	2 ,,	sugar	= 243 ,,	=	2 casks.
4 ×	,, ×	½ ,,	salt	= 61 ,,	=	1 case.
4 ×	,, ×	1/36 ,,	pepper	= 4 ,,	=	1 case.
				4318 lbs.		or 1 ton 18 cwt. 2 qrs.

A set of scales and weights would be required.

Camp Equipment.—Officers supply their own cooking gear.

1 mess kettle to each section of a company	= 24 kettles.
1 ,, ,, field gun	= 4 ,,
For the remainder	= 4 ,,
	32 kettles.

1 pick for each company and gun	= 10
1 spade or shovel for ditto	= 10
2 tomahawks to each company and 1 to each gun, as hand hatchets	= 16

Drinking cups, in the proportion of one to each man, should be taken.

Reserve Ammunition :—

	Weight.
40 rounds for 312 rifles=12,480 rounds = 23 ammunition boxes	1587 lbs. (14 cwt. 19 lbs.)

20 rounds for 105 pistols — 2100 rounds } 116 lbs.
= 4 boxes.

There is no reserve ammunition for the guns, nor does it appear how any could be safely carried.

Carriage for:—
    The provisions ... ... 38 cwt.
    Spare ammunition ... ... 15 cwt.
    Camp equipment, say ... ... 6 ,,
must be provided in some way. It would take about four carts, if procurable.

Looking to the fact that ships are wholly unsupplied with the stores requisite for an expedition of this sort, it is evident that it would be impossible to detach a force of this nature for a longer period than the men could carry provisions for, viz. :—2 or 3 days, unless transport could be procured.

Moreover, the value of our field guns is very questionable, when we consider the fatigue of dragging them for a long march and the extremely small amount of ammunition they are provided with: 78 officers and men being required to handle 4 guns with only 64 rounds of shell between them! It would probably be better to leave half the guns behind and carry ammunition for the others in the spare limbers.

However, supposing that sufficient carriage has been obtained for the baggage and stores, that it has been decided to take four guns, and that the landing has been unopposed, the force will form up into column of route as follows:—

The leading company of the battalion will form an advanced guard, keeping about half-a-mile in front, being pushed farther to the front if the country is close than if it is open and the view extended. Four companies and the left half-company of the rear company will follow, with as large a front as possible; close in rear of these will come the small arm ammunition carts (if a couple of spare limbers are available the S.A.A. might be placed upon them and drawn by the spare-ammunition and stretcher-men), the guns will then follow in the most convenient order. The armourers, medical attendants and others will be in the rear, behind them the provisions and stores.

In rear of the whole, at a distance varying from $\frac{1}{4}$ to $\frac{1}{2}$ a mile, will come the right half company of the rear company, acting as a rear guard.

The pioneers and one signalman will move with the support of the advanced guard, one signalman with the main body, the other with the rear guard.

The precautions to be observed during route marching must be carefully carried out; all places on either flank likely to conceal an enemy must be searched by a patrol (6 men and a petty-officer) sent out by the support of the advanced guard.

Bivouac for the Night.—"In selecting a site for a bivouac," says Sir Garnet Wolseley, "wood and water are the first requisites, " a good supply of the former is essential, as a bivouac is deprived of " half its comfort unless the men have large fires to sleep near.

" This is all the more essential if the nights are cold. In cold " weather woods are the warmest places. In tropical climates it is " pleasant at night to bivouac in the open ; dry and sheltered situa- " tions should be selected."

To take camp furniture in the shape of tents, &c., even if procurable, would be an insuperable encumberance on the free movement of a small force such as we are considering.

As soon as the site for the bivouac has been selected, the Right Leader will go to the front and carefully decide upon the positions to be taken up by the outposts,—the Left Leader doing the same in rear. The companies told off for this duty (which they will take in turn) will be at once inspected, detailed, and marched off to their places ; the Right and Left Leaders going round all the posts with the captains of the companies, and assuring themselves that all is correctly arranged.

As the line of march will most probably have been along a road, it will be sufficient to assume that attack need only be apprehended from the front or rear, and that the flanks will be sufficiently secure by establishing a small piquet at some slight distance on either flank ; *in other* cases more efficient security must be provided laterally.

VII. OPERATIONS ON SHORE.

ORDER OF MARCH.

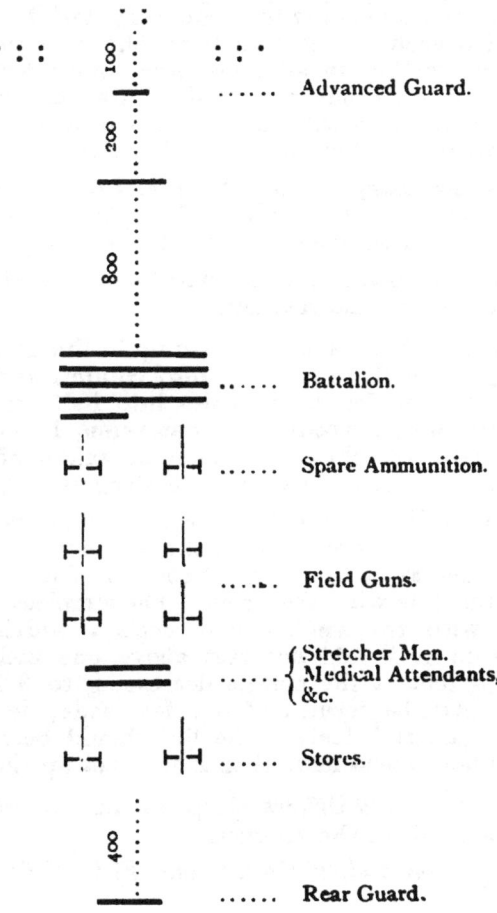

Assuming the strength of a company to be 50 men, 3 P.O.'s. and 3 officers, it will be detailed for outpost duties as follows :—

One half company will be placed about 800 yards in front of the bivouac; the remaining half company will be told off into 2 sections of 3 reliefs each; these will form the two piquets, stationed about 400 yards farther in advance, and about 400 yards apart. Each piquet will send out one-third of its strength as double sentries, to be posted about 400 yards in advance and 100 yards apart, care being taken that they can see each other.

It is necessary to push these well to the front, in order to guard against surprise, and to allow the main body time to prepare for an attack. These sentries will be relieved every hour during the night.

When necessary, the rear will be protected by a company detailed in a precisely similar manner.

The main body will be drawn up in line facing the direction of the enemy, and will be wheeled into column and pile arms; they will then be told off for camp duties into half companies, each half company providing 2 woodmen, 2 watermen, 1 cook, and the rest for duty if required. Blankets will then be taken off, and if safe to do so, accoutrements also; the men will sleep as they stand in the ranks.

The readiest method of cooking the provisions is to suspend the kettles from three sticks over the fire, but where fuel is very scarce it may be necessary to construct cooking places, which should be in rear of, and in line with the arms. The simplest form is a trench dug in a line with the wind, and of such a width that the kettle when placed on it should not rest above one inch at each end; depth, about a foot at the muzzle decreasing to 3 inches at the chimney, which may be formed of a a few sods; length, sufficient for the kettles (about 5 feet). The fire should be lit near the mouth and should not extend more than 2 or 3 feet up the trench.

The Company Officers sleep on the reverse flank of the men, on the other side of the kitchens.

The Commanding Officer and Field Officers together in rear of the centre.

BIVOUAC.

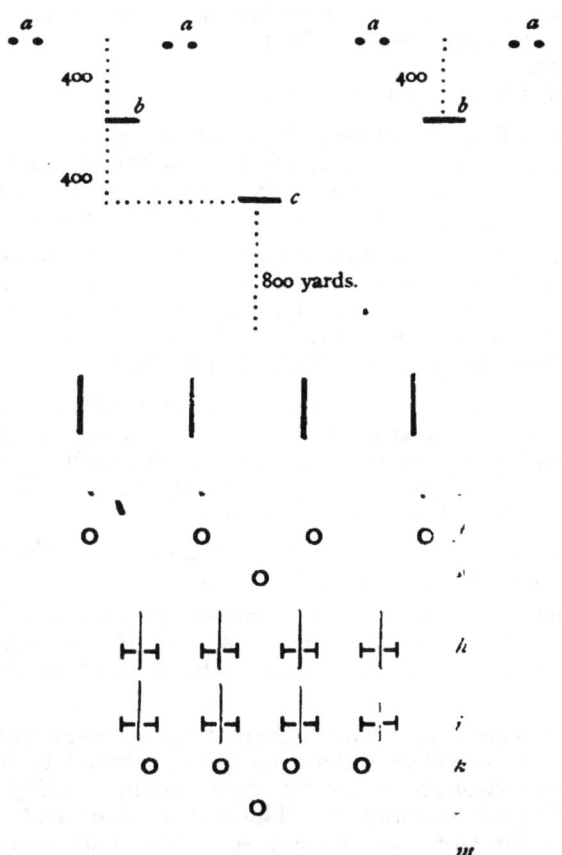

a a....Double Sentries. *b b*....Piquets. *c*....Half Company, Reserve of
 ost. *d*....Main body. *e*....Kitchens. *f*....Company Officers. *g*.. .Field
 ers, Head-quarters. *h*....Stores, &c. *i*....The Guns. *k*....Guns' Crews.
  Battery Officers. *m*....Rear Guard.

The Baggage and Stores will be in rear of the centre.

The Guns will then come in the order directed. Line at close interval will be convenient. The crews and officers in rear, as shown in the sketch.

Latrines will be dug to leeward.

Attack of a Position.—The following may be considered as giving a general idea of the conduct of an attack on an enemy who is waiting to receive it; and let us, for the sake of clearness, assume the small force mentioned above: viz., 300 men and 4 guns.

First, let it be said that no force, however small, should ever march through an enemy's country without an advanced guard of some description, to feel the way, prevent surprises, and give timely notice of the enemy; especial care should be taken to protect the flanks and to search any places which are likely to conceal a foe.

Supposing, then, that you have received intelligence of the vicinity of a body of troops, drawn up to bar your further progress, you will push forward your leading company to skirmish by sections, over a sufficient ground to cover your front, and feel for the enemy. The second company will support them.

As soon as the enemy opens fire upon your skirmishers, they will halt and take cover, and reply to it.

Reconnoitre his position, and make up your mind which is his weak point. Never make a *direct* attack, if you can avoid it, but endeavour to turn a flank, making a feint, if necessary, against the other one.

Never advance your skirmishers in a general line when under fire, as we see sometimes done on parade, but let short rushes be made by the sections from cover to cover as opportunity offers, having due regard to mutual support. The section commanders are the men to look out for this, and should seize any momentary lull in the enemy's fire for a short rush (not a "steady double") to cover in front, whence they can more effectually gall him.

Suppose you determine to push back his right flank, then send out *two* more companies from the reserve to "prolong the line to the left," one as skirmishers, one as support (the Army system of making the old supports prolong the line saves time, but by making the same two companies always work together a feeling of confidence and security is given to the men); keeping your reserve ready to repel any counter attack against your own right. The supports should be well in hand and as much out of fire as possible; with our small companies it is best to keep them in close order, as cover can generally be somewhere found for them.

Press your skirmishers forward against the enemy's flank; seize every stone and blade of grass for cover; if your men suffer much, then move up your supports to *reinforce* them. Get the idea into your men that they must win somehow, the very existence in the drill book of such a manœuvre as 'relieving skirmishers' is sufficient to breed a general self-distrust and faint-heartedness; the practice of it would be fatal to any hard struggle. Let your men run *to* the front as much as they like, but never *from* it.

If the foe has not the wisdom to retire, but holds on with the energy of despair, then having sufficiently pounded him with your guns and small-arms, you must drive him from his position at the point of the bayonet. Pass the word along to fix swords, and to stand by to charge; and when the moment comes "Up Guards and at them"—not a general advance in line at the double, but a rush of every one for the enemy's position. "First in for the Victoria Cross" should be the cry; the sword bayonet used as a cutlass will, in a sailor's hand, do good work on these occasions;—remember the boarding axes at Le Bourget and the work they did.

The more confusion you get into on drill days the better; it will teach you the inestimable art of re-forming with rapidity. War is all confusion to the mere parade officer, 'madness with a method in it' to the real soldier.

Field Guns.—The same rules apply to them as to other artillery. Their office is to prepare the way for the infantry by bringing a heavy fire on the opposing line.

Select as far as possible, commanding positions whence they can fire for a considerable time without need of moving; a gun is of no use whilst *changing position*.

About 2000 yards may be said to be the range at which to open fire if you can get a clear view; and with the naval field guns drawn by men, about 800 yards should certainly be the least distance an approaching enemy should be allowed to close.

Move from position to position as seldom as possible; but when you do, then go as quickly as you can. Keep the limber numbers lying down under cover, and dismount the limber itself when desirable.

With regard to the Field Battery Drill it may be said that the mass of the manœuvres laid down are mere parade formations. In addition to the Gun Drill movements, Sections V., VIII., and IX., contain everything which is of practical value.

Defence of Posts.—Every officer should know the few simple methods of placing a post in a state of defence. How walls, hedges, houses, &c., may be so prepared. Chapter I. of the Sappers' Manual may be studied with advantage touching these points.

ARRANGEMENTS FOR DISEMBARKING TROOPS BEFORE AN ENEMY.

To prepare the necessary orders, the following subjects must be considered and determined.

1. The number of troops to be first thrown on shore. This will depend not only on the amount of force present, and on that opposed to it, but on the number and capacity of suitable boats at command, and on the extent of the shore adapted for the disembarkation.

2. The number and classes of boats available and indispensable to receive troops, to land field guns, to carry guns to cover the operation, steam boats for towing, and ships to provide and man the boats. Also, the boats required for other consequent and simultaneous services; as for Medical Officers, horses, water for immediate use, and reserve ammunition. Each ship of war and transport should be required to furnish a list of her boats ready for service, and their dimensions.

3. The numerical succession of the boats to land the troops from the leader of the 1st Division, as No. I. Boats with guns are to be distinguished by letters. (See Instructions appended.)

4. The succession of troops from the right when landed; naming the corps, numbers, and the ships from which they are to be taken.*

5. The allotment of the troops to the boats; so that, except where boats with skirmishers intervene, boats as numbered may carry troops, in the succession they are to be arranged in on shore. † This is necessary to ensure their rapid formation, and to prevent disorder on the beach. The troops may then be landed simultaneously, or in succession from columns or single boats. In considering the

---

* It is not absolutely necessary that the disopsition of the troops in the boats should be such that the proper right of each battalion should retain its position when landed at the right, yet it is decidedly preferable that it should be so. For although with an inverted line all evolutions are practicable, yet it is better to avoid the inversion.

† See Note above.

sufficiency of the boats, the weight of each soldier with his rifle, ammunition, water and provisions for three days, may be estimated at 200 lbs. If horses are to be landed with an infantry regiment in due proportion for service in the field, from 12 to 15 would be required for mounted Officers, Surgeon's stores, reserve ammunition, fodder for horses, and afterwards for the conveyance of sick, or foot-sore men.

6. To avoid exposing and tiring the men unnecessarily, and losing time, precise arrangements should be made for rapidly and quietly placing the troops in the boats immediately the order is given. With this view a Naval Officer should ascertain and report how many accommodation ladders can be placed over the side of each ship having troops, for different boats to receive them from. He should arrange for having the ladders fixed in proper time, and for their being conspicuously and consecutively numbered from the foremost to the aftermost on each side, so that the boats to receive troops may be "told off" to the ladders. When more than one boat is "told off" to the same ladder, they are to go alongside according to the numerical succession in the order of rowing. To prevent delay, and to ensure the soldiers being landed in their relative positions, he should request that corresponding arrangements may be made on board by having at the appointed time such soldiers at the head of each ladder as are assigned to the boat which is to come alongside it. (See table of details.)

7. Each boat is to be provided with the gear and the other things named in the instructions. (See page 260.)

8. Decide on the place of rendezvous for the boats after receiving troops. Probably it would be at the ship nearest the landing place.

9. Determine on the formation the boats are to assume at the rendezvous; on the Officer to command the flotilla; on the leaders of divisions; how and by what they are to be towed.

10. Make the arrangements for flanking gun boats, or covering ships, and the positions they are to occupy; to do which, the neighbourhood of the place of disembarkation, both afloat and ashore, must be perfectly known or carefully examined.

If the distances from the beach to conspicuous marks on shore in the direction of an unexpected enemy can be given to Officers in the covering boats and covering ships, it will greatly assist them in determining their time fuzes and elevations, and may therefore materially affect the success of the operation.

11. Determine whether the skirmishers and their supports shall land without their valises, leaving them in the boats; if the rest of the landing party shall have their valises on when they get into the boats, but ready for slipping should the boat be much injured by shot; what clothing, cooked provisions, and water each man shall take; what spirits are to be taken with the troops, and to whom the spirits are to be given over; if the Officers are to carry their own things, and Officers' servants to be under arms. Rifles should never be loaded until the men land.

12. Arrange for boats to carry reserve ammunition, (water, of which the supply must be abundant), provisions, medicines and stores, immediately the landing place is secured. These duties are to be assigned to particular Officers.

13. If necessary, arrange for future supply of provisions, stores, and ammunition from the ships, and for conveyance and reception of wounded men.

14. A Beach Master, assisted if necessary by other Officers, should be appointed to control the work on the beach. Under them, a party of seamen should be stationed where the horses and stores are to be landed, with such well stretched rope, slings, purchase-and leading-blocks and spars for light shears, as the case may require.

15. If the landing is to be on an open and ample beach, the boats may probably best approach it in columns in line ahead, the columns being more or less numerous according to the number of boats employed. The fewer boats there are in a column, the less effect will ricochet shot have upon them; but a single line abreast would have a front much too great for the troops when landed, and increase the difficulty of finding a beach adequate for the purpose.

The distance between adjacent columns in feet should be the number of boats in a column $\times 20$; so that when they are beached in line abreast, there may be about 10 feet between adjacent boats.

If the beach be insufficient for the whole force to land and form on at once, probably the disembarkation should be from columns in line abreast. This may enable the 1st line of troops to land and advance, and leave the beach clear for the 2nd. Adjacent lines abreast for beaching should therefore be about a cable's length apart, as a battalion should disembark and form in two minutes in smooth water if the measures for disembarking have been well arranged, and the soldiers and sailors well instructed in the operation.

16. If the landing is to be from lines ahead, the skirmishers may be best placed in the 1st boat of each line; if from lines abreast, the extreme and centre boats; for being separated by equal intervals they may more readily cover the whole front of the troops.

The mode of landing, and the required disposition of the troops on shore, must be arranged by the Military and Naval Officers in concert, that each department may work in accordance with the other.

The foregoing orders having been determined, the necessary orders may be prepared.

Outline of a Scheme of Instructions for the Disembarkation of a Force in the presence of an Enemy, to be issued by the Military and Naval Commanders in conjunction.

A copy of the instructions should be given to each Officer commanding a Corps, or detachment of a Corps, and to the heads of Military departments.

A copy to be given to each Captain of a ship of war. He is to take care that the Officers in charge of boats, or otherwise detached from the ship, have copies of the "tables," and such other parts of the instructions as they are concerned with, and that they perfectly understand the duties entrusted to them.

The Captain of each Transport should have a copy.

The instructions should be issued, if possible, before assembling in the neighbourhood of the place of disembarkation.

TABLE OF DETAILS

For Boats to be Employed in Disembarking the Troops to be Landed First.

Ships which are to provide boats for landing.	Distinguishing No. and Class of Boat.		No. of Soldiers to each Boat.	No. and name of Transport to which each boat is to go.	Side and No. of ladder to which each boat is to go.	From what Corps men are to be taken.	Where the boats are to rendezvous after receiving the troops.
	1	Launch	....	No. (....)	1, 3, 5, Starboard No. 1.	15th	Alongside.
	2	Launch	....				
	3	Pinnace	....				.... there
	4	Cutter	....		2, 4, Starboard No. 2.		
	5	Cutter	....				to form the order
	6	Launch	....	No. (....)	6, 8, 10, Port No. 1.	60th	of Assembly
	7	Launch	....				named.
	8	Pinnace	....				
	9	Cutter	....		7, 9, Port No. 2.		
	10	Cutter	....				(See other table.)
	11	Troop boat	....	No. (....)	11, 13, Starboard No. 1.	82nd	
	12	Troop boat	....				
	13	Troop boat	....		12, 14, Port No. 1.		
	14	Troop boat	....				
	15	Troop boat	....	No. (....)	15, 17, Starboard No. 1.	23rd	
	16	Troop boat	....				
	17	Troop boat	....		16, 18, Starboard No. 2.		
	18	Troop boat	....				

Special arrangements for towing the Divisions of boats will be required, dependent on the number and nature of gun-vessels or steam-boats available.

Instructions for Disembarking, &c.—The ships of war are to prepare the boats named in the annexed "Table of details," and the boats are to be employed in the manner noted against their names in the table, and according to the following instructions.

The ships from which troops are to be disembarked are to have accommodation ladders prepared for different boats to come alongside of at the same time. The ladders are to be conspicuously numbered from the foremost on each side, as No. 1, to the aftermost as the highest number on the same side, so that the boats may readily find the ladders to which they are "told off." A "guess warp" is to be stretched alongside each ship's sides for boats whilst loading, to ride by; and another rope is to be over the stern for loaded boats, waiting for others to proceed with them to the rendezvous. If horses, or other weights have to be disembarked, the purchases in the transports should be rove, and everything prepared for expediting the work. (See page 266.)

Ships to indicate whether they are clear of troops or not by use of the ensigns or other symbols. Each ship from which the troops first to be landed are to be taken, is to wear her ensign at the gaff or ensign-staff until the last boat-load of such troops has left her, when she is immediately to hoist her ensign at the mizen top-gallant mast-head. Those transports which are entirely cleared of troops are to continue to wear the ensigns at the mizen; but those which have still troops to land, are to re-hoist their ensigns at the gaff when the flotilla has quitted the rendezvous for the beach. This must be attended to, to prevent boats returning in quest of remaining troops to empty vessels.

Preparation and fitting of Boats. Each boat is to have the means of stopping shot holes; several buckets for baling; two baracoes of water, and drinking utensils for the use of people in the boats, and a small cask of water to be landed for the immediate use of the troops and horses; her anchor and cable, masts and sails and two spare oars; but she is to be free from all unnecessary lumber. The sails are to be used to reduce labour only when the boats are not loaded with troops. A stout broad gangboard is to be fitted for each *side of the bow*, to hang about two feet below the gunwale, when *fixed for landing.* Whilst proceeding to the shore, the gangboards are

Sec. VII.   DISEMBARKING TROOPS.   261

to be hung along the outer sides of the boat, under the oars, with a line led from the after ends also under the oars and in-board over the bows. When the boat is beached, the stops which hang the gangboards alongside are to be let go; the bowmen jumping on shore place their inner ends, and the foremost men in the boat pull the after ends forward and up by the line, and hang them by slings previously well adjusted.

As the gangboards may fail, or not give so free an exit to the men as is necessary, a grating or stage of plank about two feet wide is to be slung on each bow between the foremost rowlock and the stem, and to be level with the water when the boat has her complement of troops in her. Whilst going to the beach, these stages may be turned back close to the bows, and just before touching the beach, be thrown forward to hang horizontally in the slings. It would be impossible for troops (armed) to jump from a launch's gunwale, a height of nearly seven feet from the ground; but with that height divided by the stage, they can easily get over the bows. This is a most important fitting, for it may be of vital consequence that the men who have to seize the beach should not be detained under a fire they cannot return. The boats' crews should be exercised in placing these gangboards and stages, and if possible the troops in embarking and disembarking. This may be done with the boats on the ship's decks.

If the landing is to be on deep mud or deep sand, the boats which carry the field guns should be provided with plank or mess tables to run the field guns to the hard ground.

Each boat to land troops is to have her number, as shown in the numerical succession for the order of assembly, in white figures eight inches long, on a black ground, on both quarters and both bows; and the boats with guns are to have their distinguishing letter so placed.

Each boat is to have a boat's signal book, and an answering pendant, and a man conversant with the flags specially to look out for signals and to receive orders given by hailing. The boat of the Officer commanding the flotilla, and of each Officer commanding a division, is to have a set of boat's signal flags, a staff

Boats' Signals.

long enough to display four of the flags upon, and a light pole to spread the flags if it should be calm.

Boats' Crews' meals, and the relief of boats' crews. If there is a probability of the boats being absent long after meal hours, the crews are to take a day's provisions cooked, and their spirits; and proper measures are to be taken to have the boats' crews relieved, for the boats may be required night and day.

Boats to be kept in repair. The Officers in charge of boats and their crews should be made aware of the importance of not allowing their boats to be unnecessarily damaged, since injury to the boats may frustrate the intended service, or at any rate increase the difficulty of conducting it. Any injury done to boats is to be repaired at the earliest possible opportunity.

Flotilla to be managed by signal book and by signals appended. The flotilla will be managed by the evolutions in the signal book, with which Officers are presumed to be familiar, and by the signals appended to these instructions. The strictest attention must be paid to signals, and to keeping station.

Time at which boats are to be at the respective transports and proceed to rendezvous. The boats are to be alongside the Transports, and at the ladders to which they are assigned in the table of details at (   ) o'clock, or when the signal is made. If more than one boat is assigned to the same ladder, they are to go alongside according to their numerical succession in the order of assembly. Having embarked the troops, the boats of the same ship of war coming from the same transports are proceeded to the rendezvous, where they are to form in (here state in what order of assembly,) in accordance with the subjoined numerical succession and divisional arrangement, of which an example is shown in the following diagram.

Officers of the Flotilla. Captain............will command the Flotilla. (He is to have three despatch boats in attendance.) Commander....... . will lead the......division; Commander.... ..... will lead the......division; (and so on).

A Lieutenant is to be in the leading boat of the boats of each ship, even when that boat would have in her the Officer commanding the division.

Sec. VII. DISEMBARKING TROOPS.

Distinguishing No. of boat.	If in columns in line ahead, the leading boats are to contain skirmishers and supports.	Or, if in columns in line abreast, skirmishers and supports are to be in the leading column.
1	13	
2	14	
3	15	
4	16	
5	17	
6	18	
7	7	1
8	8	2
9	9	3
10	10	
11	11	
12	12	
13	1	6 5 4 3 2 1
14	2	
15	3	
16	4	12 11 10 9 8 7
17	5	
18		18 17 16 15 14 13

A B C D
E F

The distance between columns for beaching should be (in feet) No of boats × 20.

The distance between columns should be about a cable's length, so that the first line may land and advance before the second is landed.

All Officers appointed to boats of the Flotilla are to be ready to go to the Captain Commanding it, to receive his instructions when the signal is made to do so.

Beach Master and his duties. Commander... ....is to be the Beach Master, assisted by Lieutenant... ....and Lieutenant............

He is to have a beach party of Petty Officers and Seamen, a staff of signal men, and boats to carry messages. These men are to be taken from the ship to which he and the Lieutenants belong.

He is to erect a signal staff in a convenient position, and have a set of signal flags and books with him. He is to cause a strict look out to be kept for signals afloat and on shore; and all ships are to keep a look out for signals made from his signal station. He must, without loss of time, erect such shears or provide such means as may be proper for landing the heaviest weights which have to be brought on shore. The shears should be so placed that a boat drawing four feet water may come under them, or be "plumbed" from their heads when sloped. Several pairs of shears may be required, and at different places. From the shears to hard ground, a road over which weights may be easily moved, should, if necessary, be prepared. (If the operation be on a large scale. See page 274.)

The disembarkation will be conducted as directed under the signal to "beach the boats and disembark." (See page 257.)

If the skirmishers, or any other of the troops are to leave their valises in the boats, the crews are to land the valises, and stack them as near the beach as they can be properly placed, and leave them there with the soldiers appointed to take charge of them.

Boats to succour troop-boats disabled by shot. The following boats, viz. :— ....are to form a line abreast two cables astern of the boats, with troops as at A, B, C, D, ready to go to the succour of any disabled boats, and to land their troops.

Boats for Medical Officers. The Medical Officers are to be in boats with necessary attendants, stores, and appliances, at 3 cables length distance from the flotilla as at E. F.

Sec. VII. DISEMBARKING TROOPS.

The casks of water brought for the troops are to be placed clear of serf and tide, with bungs up; if the sun be powerful, they should be in shade or covered in sand, to prevent leakage. Land casks of water.

At the time stated for the boats to be alongside the transports, the troops, as indicated in the table of details, are to be at their respective gangways, and immediately to go into the boats. They are to be placed in the boats from forward to aft as they are to be ranged from right to left on shore. They are to be cautioned not to stand up in the boats, and not to make a noise. Sidesmen are to be stationed at the ladders to pass rifles into the boats, and assist the soldiers. Details respecting the troops. Getting into boats.

Before quitting the ships, the skirmishers and their supports should be "told off," and if possible the Officers and men comprising these parties should have the ground pointed out to them for which they are to push after leaving the boats. They should have their valises off whilst in the boats, and leave them in the boats, that they may spring instantly on shore and advance. The valises will be landed by the boat's crew, stacked on the beach, and left in charge of the rear guard. The skirmishers will return for them when the landing has been made good.

The rest of the troops are to have their valises on, and to land with them on; but whilst in the boats they are to keep them unbuckled, ready for slipping in case the boats be much injured by shot. In no case are the troops to load whilst in the boats.

The troops first landed should have trenching tools with them, as it may be necessary at once to render buildings defensible, or to throw up cover.

If all the troops cannot be landed at once, a second table of details should be filled in for the guidance of the Officers commanding the boats, after the first party has been landed, and for Military Officers on board each transport. A second table of details to be provided if all the troops cannot be landed at once.

The guns, reserve ammunition, projectiles, provisions, medicines, stores, and water if necessary, which are to be next landed must be ready for leaving the ships as soon as the beach is secured.

As many horse slings must be ready as are equal to one quarter the number of horses, unless it should be necessary to hoist the horses out of the boats on shore, as well as out of the ships, in which case a few more would be wanted. Slings for hoisting out field guns, and ammunition waggons, in such numbers that there shall be no waiting for them.

Topsides of transports should be so cleared away under lower yards or derricks, that little hoisting may be required to get guns and heavy weights over. All the appliances for each battery of artillery should be mustered, and things so disposed of and lists taken that each may leave the ship complete.

Whilst the transports are being cleared, proper Military Officers should be on board to determine in what succession the things are to be despatched, and to see that expedition and care are used in sending them.

In each transport the duties and stations of the soldiers and crew for clearing her should be reduced to writing, and every Officer and man made to understand clearly what he has to do to expedite the disembarkation.

The Naval duties relating to clearing transports of guns and stores, is assigned to Captain................ The boats carrying these things are to have distinguishing flags preconcerted with the Chief of the Military Staff and Beach Master, so that the department to which they are consigned may readily find them.

Landing horses from ships' boats on a beach. When horses are to be landed from ships' boats, it may be the best plan to place the boats' sterns on the beach, and to have inclined planes from inboard to the sterns, and other inclined planes from the sterns to the shore. For those boats which, having landed infantry, are to land horses, the inclined planes should be previously carefully fitted; their component pieces all marked, and men taught to put them rapidly together, so that the boats as they return from one service may be *quickly* prepared for the other. The outboard plane may be made of *mess tables* supported by studding-sail yards and booms cut to the

Sec. VII.   DISEMBARKING TROOPS.   267

required length, and sufficiently numerous to have the plane sustained at very short spaces. The boats should be prepared with dunnage. If a pier of a convenient height can be found, or formed, the horses may walk to it from the boats so fitted. Before sending horses and heavy weights from the ships, care must be taken that the means exist on shore of landing them with safety.

The Captains of H. M. Ships, should, without loss of time, place themselves in communication personally, or otherwise, with the Commanding Officers of troops, the Chiefs of Military departments, and with the Captains of Transports with whom they have to act, so that a complete understanding may exist between all branches of the service with regard to the duties which they are jointly to carry into effect. *Captains of H. M. Ships to communicate with Military Officers with whom they have to act.*

This order to be signed by Military and Naval Commanders.

SIGNAL No.

Beach the Boats and Disembark.

With a Cornet Under. | With a Cornet Over.

Signal to be used when the boats are close to the beach on which the landing is to be.

Execution.

If from lines ahead,

Each boat is to be beached ten feet clear of, and on the Starboard side of the boat preceding it. | Each boat is to be beached ten feet clear of, and on the Port side of the boat preceding it.

Each leader is to preserve exactly her proper distance from the next leader in the direction of the front.

The second boat of each column is to sheer twenty feet out of the wake of her leader, on the side indicated by the position of the cornet, and preserve her proper distance from the leader. She will be steered with sufficient accuracy if her gunwale at her broadest part be kept "on with" the outer ends of the oars of the leader. —Each following boat of every column is to act in the same way with regard to her next ahead, as *the second boat does* with the leader.

DISEMBARKING TROOPS.

If from a single line abreast,

The boats are to be beached at the distance apart at which they are then arranged.

If from lines abreast,

Each boat in the 2nd, 3rd, 4th, 5th, or 6th column is to be beached ten feet clear of, and on the Starboard side of her corresponding boat in the column preceding it. | Each boat in the 2nd, 3rd, 4th, 5th, or 6th column is to be beached ten feet clear of, and on the Port side of her corresponding boat in the column preceding it.

The leader of each succeeding column is to adjust accordingly the position of his column with regard to the one preceding it.

If in two lines, the boats of each column must be forty feet apart; if in three, sixty feet apart.

Note—When the preparative is made with this signal, the bow men are to lay their oars in and prepare gangboards and other appliances for landing the troops. The anchor is to be got ready and the cable seen clear. On the inner end of the cable there is to be a buoy, that it may be slipped if the anchor should have been let go too soon. As soon as the boats touch the beach, their cables are to be set taut and oars to be got ready to keep the boat end on.

SIGNAL No.
Boats are to be prepared to Re-embark Troops.

If the boats have been beached from lines ahead and are lying in line abreast on the beach in numerical succession.

Execution.

If special orders to the contrary have not been given, the following instructions are to be followed:—

The troops as they arrive on the beach, and as they stand in the ranks, are to be embarked, the boats being taken in the numerical succession as they are ranged on the beach. Officers are to be appointed by the beach master to show them the boats they are to occupy, and to direct how many men each boat is to contain.

The boats are to be prepared with gang boards and for being kept afloat by being pulled and pushed out as the troops get into them. If the heavy boats cannot otherwise be kept afloat, gangways are to be prepared over the light boats. Particular boats are to be ready to embark field guns with rapidity. If necessary, arrangements are to be made for boats when full of troops to proceed to co-operate by rifle fire with the covering gun boats on the flanks, so as to facilitate the withdrawal of the other men and the last boats. Such boats must be cautioned not to mask the fire of any boats employed for the same purpose.

Particular boats are to be prepared for wounded men, and when *full* are to quit the beach.

SEC. VII.   DISEMBARKING TROOPS.   269

If the boats have been beached from two, three, or four lines abreast.

The troops as they arrive on the beach, and as they stand in the ranks, are to be embarked in the boats of the 1st division in their numerical succession, omitting, till the boats of the first Division are full, all the boats of the 2nd Division and 3rd Division, which will be intermixed with the 1st Division. When the boats of the 1st Division are full, those of the 2nd Division and 3rd Division are successively to be filled in the same way. Officers are to be appointed by the beach master to show the troops the boats they are to go into, and to direct how many men each is to contain.

The other directions under the first case apply to this.

SIGNAL No.

The boats about to embark troops, or those of the Division or Ship indicated, are, on receiving their complement of troops or material, to leave the beach, and form the order denoted.

SIGNAL No.

The Boats about to embark troops, or those of the Division or Ship indicated, are, on receiving their complement of troops or material, to leave the beach, and proceed to their Ships or on the duty ordered.

If the disembarkation is to be on a very great scale, the following matters must be determined upon in addition to those already mentioned. — *If the disembarkation is on a great scale.*

It may be assumed that a fleet containing a great expeditionary force would rendezvous previous to proceeding to the place of disembarkation. At the rendezvous every detail should be settled that the least possible time may elapse between the arrival of the expedition at its destination, and the disembarkation.

The army and the transports should be "told off" in corresponding divisions; ships of war attached to each division to conduct it; the rest of the fleet being as little crowded as possible, organized as a distinct formation for the protection of the whole.

The ships of each division of transports should be distinguished by wearing under the convoy flag, 1, 2, or 3, to denote that they contain troops of the 1st, 2nd, or 3rd division of the army. — *Transports, how to be distinguished*

Each transport should have special distinguishing pendants assigned to her, to be shown with her divisional distinguishing signal when she alone is addressed, or referred to.

DISEMBARKING TROOPS.   Sec. VII.

Order of sailing and scheme for anchoring to be issued.   The order of sailing for the transports, as well as for the ships of war, must be promulgated, and the necessity for a strict attention to station-keeping enforced.

A scheme for anchoring at the destination must also be issued, and transports required to have all cables bent, and warps, and boats for laying the warps out, ready for instant use.

All ships should leave the rendezvous with storm sails bent, and with the baggage, guns, and horses, well secured in case of bad weather.

Precautions against an attack at sea.   If there is the least prospect of being attacked at sea, that greatest of dangers to a large combined expedition, every ship of war should be absolutely clear for action; at least one watch's hammocks stowed, and the watch at their guns. Arrangements should be made for look-out vessels to encircle the fleet, and night alarm signals established. Before leaving the rendezvous, it may be well to ascertain by signals, repeated by a chain of ships, extending over 25 miles, if the neighbourhood is clear of enemy's vessels.

Small vessels to precede and buoy dangers &c. near intended anchorage.   Small vessels should be selected to precede the fleet a little, and anchor where it may be necessary to indicate dangers, and to buoy shoals. This also may be done somewhat sooner, as a feint on some other part of the enemy's coast than that intended to land on.

If possible, the time of arriving at the rendezvous and the time spent at it, should be so adjusted that there may be a good moon for making the passage.

Hospital ship to be provided.   If no Hospital ship has been provided, it should be determined at the rendezvous which transport when cleared of troops can be best converted into one. Arrangements should be concerted for preparing her without delay for the reception of sick and wounded men, and attaching Medical Officers and Medical stores to her.

At and men work.   At the rendezvous, the expedition should be freed from Officers and men who, from illness, have not a prospect of being soon equal to work. The same rule should be applied to horses.

Sec. VII. DISEMBARKING TROOPS.

As the usual establishment of boats of a ship of the line will not carry more than 300 men in heavy marching order, before a disembarkation on a great scale could take place many additional boats must be procured. As each accoutred infantry soldier weighs about 200lbs., as each horse occupies about 9ft. and weighs about 10 cwt., and as the weight and dimensions of the guns, ammunition, and accompaniments of a battery of Artillery are known, and as the amount of each of the "arms" which are to be simultaneously landed must be decided, the required boat accommodation may be calculated. This is the least boat tonnage that will suffice, for after the landing place is secured, the rest of the troops, the military train, guns, stores, camp equipage, &c., must be sent on shore, whilst probably there will be demands for boats to remove wounded and sick men. *Additional boats to be procured.*

It may be assumed that each large ship of war, by fitting davits along her sides, by placing boats inside her present boom boats, and others within each other on the quarter-deck, might carry boats equal to receiving 600 men, and these boats might be all hoisted out or lowered down in half an hour.

Greatly increased accommodation could be provided if collapsible boats were supplied to the expedition.

The boats for landing Infantry must be prepared with oars, anchors, cables, painters, and the bow fittings as recommended for launches. (See page 261). Others, besides being so prepared, must also have portable but secure decks for Artillery and Cavalry. If it be impossible to procure boats which, taken singly, are wide enough to place horses and guns upon, flats or rafts may be made of two boats. They should be secured together by lashings from the bow of each to the stem of the other, and by athwartship lashings. Beams strong and numerous enough for the proposed work, must be placed athwart both, and be secured to each to carry the deck or planking. Round the deck there must be strong stanchions and ropes. These flats should be steered by oars. *Preparation of boats, flats, &c.*

It will be a great convenience that these flats when loaded should float about the same height as the pier at which they are to land, so that horses may step from one to the other, and things be transferred without a necessity for hoisting; and this should also be borne in mind when constructing the pier.

Construction of jetties or piers. Some flat-bottomed broad boats may be required for forming piers over, and plank to deck these from gunwale to gunwale, and to form a brow to the shore, must be provided. A plan of the intended pier should be formed; indeed, it might be advantageous to put the pier itself together, that no necessary appliance or fitting may be overlooked. When the horse, gun, and pier boats have been fitted, carpenters and others should be instructed in putting the fittings rapidly in place, and the fittings should be all marked and stowed ready for instant use. The boats must be assigned to different ships to take care of, and to Officer and man.

When flat bottomed boats or suitable country craft cannot be procured, a jetty may be speedily constructed from a ship's resources by one of the following plans:—

1st. The outer end of the jetty may rest on ship's iron tanks sunk as a standing buttress, and if necessary, filled with stones, sand, and water, to render them stable. Between the outer end of the jetty and the shore there may be other buttresses, made probably of shallower tanks; the spaces between the clusters of tanks and the shore must be bridged over by spars and plank. Mess tables and stools may be used for the purpose. If the depth of water be greater than the height of a tank, a riding tier of tanks may be added. In this case the lower tier should be placed on a raft, well secured to it and to each other, towed into position, and sunk.

2nd. If the ground be very uneven, the outer end of the jetty may be on as many empty floating tanks as will give the required buoyancy. They must be placed on a raft, be well secured to it, lashed together, and the lashings set taut by wedges between the tanks. The raft must be towed into its position and there secured by several anchors and hawsers, and by quarter guys to the shore. The inner end must not be so rigidly secured that it will not admit of the undulating of the outer end by any trifling swell or by variations of weights that may be upon it.

3rd. Should the features of the shore or other circumstances *render it* desirable, both ends of the pier may be on floating tanks, and be supported intermediately by other tanks. The inner end may be connected with the shore by a "brow," or the weights may be lifted from it by derrick or shears. The outer end must be kept in position by quarter guys, anchors and hawsers. It may be convenient that the hawsers should be "single whips," rove through blocks at the rings of the anchors, the standing parts being made fast to the raft, and the hauling parts taken on shore.

To prevent such a floating pier inclining too much by wind, or by weights placed on one side, spars may be secured under and athwart the floating tanks, projecting several feet on each side, with one or more empty tanks at each extremity. The immersion of the tank on one side, and weight of that on the other, would both resist any tendency of the jetty to incline. Pumps should be fitted to the floating pier.

For the ready management of a vast number of boats, it would be desirable to have them in separate flotillas, which, although acting, if necessary, in concert, should have distinct formations, and have their details managed by their respective Commanding Officers. The Officers should be assigned to the flotillas, and the boats be properly prepared as stated before, so as to be ready to proceed on service immediately the expedition arrives at its destination. *Management of a large number of boats.*

Each boat of the different flotillas, should have below the figure on her bows and quarters showing the number of her numerical succession in her own flotilla, a horizontal line and the number of the flotilla to which she is attached, as

$$\frac{1\quad 2\ \&c.;}{1\quad 1} \quad \frac{1\quad 2\ \&c.;}{2\quad 2} \quad \frac{1\quad 2\ \&c.,\ \&c.}{3\quad 3}$$

When one of these flotillas, or a part of it, as a division or the boats of a ship, is addressed, it will be by the letter B (according to the boats' signal book) under the numeral flag indicative of the particular flotilla; combined in the case of a part being addressed with the appropriate flag.

To conduct the disembarcation, a table of details would be required for each flotilla; and second tables if it had to land a second party of troops, all of which should be arranged before quitting the rendezvous.

When the boats are organized in separate flotillas, the whole of them may be applied to landing one entire division at a time, or, if equal to it, they may land divisions simultaneously; or the first or advanced line of several divisions may be landed together. It should be determined whilst at the rendezvous how the boats are to be applied.

It must be decided at the rendezvous if the landing of the Cavalry and Artillery is to be proceeded with at the same time as that of the Infantry.

As the boats with horses and guns must be towed, and will be slower than the boats which are not towed, it might be well that each flotilla to land Cavalry and Artillery should have a supplementary squadron of boats for the purpose. These boats should not be formed with the other boats so as to expose and detain them longer than necessary under an unreturned fire. If the Artillery and Cavalry are to be landed after the Infantry, as would usually be the case, the boats which landed the Infantry must have been previously "told off" to different transports to return to tow boats with gun and horses. The boats with Cavalry and Artillery must not be overloaded, and as many of their oars as possible are to be used.

Beach Master's duties.   In addition to the duties already named for the guidance of the Beach Master (at pages 264 and 272), he is, when the disembarcation is on a great scale, to ascertain from the Chief of the Military Staff, with what Officers he is to communicate on the beach respecting the landing of guns, ammunition, shot, stores, medical stores, water, provisions, &c., and in whose charge these things are to be left,

As far as the strength of his party will permit, he is to place the things landed in the most convenient position, and to cause due care to be taken of them.

He is to inform the Military departments with whom he is to *co-operate*, how the boats landing things for them will be distinguished, and *where* they will be found.

He is to communicate also with the Naval Officer charged with taking the before-mentioned things from the ships to the shore; and he is, during the operations, to keep up such communications by signal as events may render necessary.

He is to expedite the business on the beach, and to prevent any loitering or delay in the boats, or on the part of Officers or men of the ships; and if necessary, he is to point out to the proper Military Officers that the beach should be kept clear of all unnecessary people. Should he, on any emergency, require more strength than his party afford him, he should apply for the temporary aid of a fatigue party, or sentries, from the Senior Military Officer near him.

The Beach Master is to take care that the piers, jetties, shears, and all appliances for disembarking troops, horses, guns, and stores, are kept in good order, and he is, without loss of time, to apply for such artificers as he may require, to repair any damage that may interfere with the perfect working condition of his department.

The reserve ammunition, intrenching tools, stores, and forage for each division of the army must be assigned, and arrangements for their being landed conveniently for their respective divisions; (see previous remarks on boats carrying these things having distinguishing flags). *Reserve ammunition, stores, intrenching tools, &c.*

The animals, and other means of land transport, should be sought or determined at the rendezvous, and its power estimated, that it may be determined to what purpose it is to be applied; observing that reserve ammunition, provisions, surgeon's instruments, medicine, and money chests must have the preference. *Preparation of land transport.*

At the place of disembarcation how ships containing infantry, cavalry, artillery stores, &c., are to be distinguished from each other. At the place of disembarcation, all ships, whether ships of war, or transports which have on board Infantry who are to be disembarked, are to have the No. or Nos. of the Regiment, or Regiments, and the letter R. in characters two feet long on a board hung over the stern, and on a board in each main rigging.

Ships having Cavalry are to have a letter C with the number of the Regiment in the three places before named.

Ships having ammunition to be landed, are to have the word "Ammunition" in red characters, two feet long, on a board hung over the stern, and on a board in each main rigging.

Ships having Stores to land, are to have the word "Stores" in black characters as before stated.

Every ship, as soon as perfectly cleared, is to have the boards removed.

(The ships will prepare accordingly.)

Arrangements made for re-embarking. — Arrange for a retreat by having piers constructed and boats held ready, so that there may be a speedy re-embarcation. (See signals appended.)

Every detail that it is possible to determine should be settled before leaving the rendezvous.

The troops to be "told off" in divisions, and the transports in corresponding divisions.

Order of sailing for transports, and "order of anchorage."

Ships of war prepared for defence.

Circumspection in quitting the anchorage.

Distinguishing signals for transports.

Ships to be prepared for bad weather.

Arrangements for Hospital Ships.

In what succession the Infantry, Cavalry, Artillery, reserve ammunition, shot, provisions, stores and camp equipage are to be landed.

The procuring and preparation of boats.

The troops to be first thrown on shore, and their succession from the right to left.

The composition of the flotillas of boats.

The allotment of troops to boats, and their preparation for going *into them.*

Skirmishers, their valises, and cooked provisions.

What things Officers are to take.

The order of assembly.

The arrangements for towing the various divisions, and detailing each steamboat for her particular duty.

The ships and boats, with guns to cover the landing.

Boats to land horses, guns, stores, reserve ammunition, &c.

The number and the numbering of the accommodation ladders of transports.

Preparation of transports for rapid disembarcation.

Beach Masters and beach party of seamen and soldiers.

Construction of piers, shears, &c.

Means of land transport, and how to be applied.

Orders for disembarcation prepared and issued.

Section 8.

ROCKET AND MORTAR APPARATUS FOR SAVING LIFE FROM SHIPWRECK;

LIFE-BUOYS, LIFE-BELTS, & CORK MATTRESSES; HINTS TO BATHERS;

INSTRUCTIONS FOR RESTORING THE APPARENTLY DROWNED;

FIRST HELP IN ACCIDENTS, &c., &c.

DISINFECTANTS, &c.

Sec. VIII.   SAVING LIFE FROM SHIPWRECK.

Board of Trade Instructions in respect of the Rocket and Mortar Apparatus for

SAVING LIFE FROM SHIPWRECK.

It is unnecessary to describe minutely the manner in which the Rocket Apparatus is to be arranged for firing, as a knowledge thereof, and of the manner of using the apparatus, can only be obtained by actual practice. It may, however, be observed that an angle of 45 degrees for the mortar, and 35 to 38 degrees of Boxer's Rocket, are the elevations which appear to give the greatest range. The first shot or rocket should always be fired with the rocket line in the box, and the box should be slightly tilted towards the wreck. But for subsequent shots the line may be faked on the beach, care being taken that no impediments are in the way of its running out rapidly when the rocket or shot is fired.

Great care should be taken in arranging the apparatus with precision for firing the first shot or rocket, as after the line becomes wetted and dirty the chances of effecting a communication are more remote.

The rocket line should be fastened to the rocket stick. The line should also have a knot made near the hole at the end of the rocket staff, so that if the line is burnt near the rocket the knot will catch the stick.

When the line has been thrown over the ship, and has been grappled by the crew, a signal will be made in the following manner:—If in the day-time, one of the crew, for this purpose separated from the rest, will wave his hat or his hand, or a flag or handkerchief; or (if at night) a rocket, a blue light, or a gun will be fired, or a light shown over the ship's gunwale for a short time, and will then be concealed.

On this signal being seen on shore, the inshore end of the shot or rocket line should be made fast to the whip, by being bent round both *parts of it* at about two fathoms from the tailed block, and a signal should then be made as follows, for those on the wreck to haul off the line.

One of the men on shore is to be separated from the rest, and in the day-time is to wave a small red flag, or at night is to show a red light for about a minute, and then again conceal it.

The crew of the wreck on seeing this signal will haul on the shot or rocket line till they get the whip and tailed block, when they will make the tail of the block fast to the mast, *as high up as circumstances will permit*, or to the HIGHEST secure part of the vessel, and will cast off the rocket line, and make the signal as before for those on shore to haul off the hawser.

As soon as the signal is perceived by those on shore, the whip (being previously made fast to the hawser at two or three fathoms from its end) will be manned, and the hawser hauled off by it to the wreck, by those on shore.

As soon as the persons on the wreck get hold of the hawser, they will proceed to *make it fast to the wreck at about* 18 *inches ABOVE the place where the tail of the block is fixed;* and when they have secured it, and disconnected the hawser from the whip, they will signal as before to the people on shore.

On perceiving this signal, the hawser is to be set up by means of the double block tackle purchase ; and the breeches buoy (the block of which will have been adjusted on the hawser) is to have the whip secured to it by a clove hitch ; and, by means of the whip, is to be hauled off to the wreck by those stationed for the purpose on the shore; who, also, on the next signal being shown, implying that a person is secured in the sling, will haul him ashore, and repeat the same operation to and fro until all are landed. The parts of the whip line should be kept as far apart as possible.

Circumstances may require some deviation from the above rules. For instance, if the wrecked vessels be subjected to violent motion by the beat of the sea, it will be better not to set up the hawser at all, but to man it with as many hands as can be spared, and reeve it over a triangle, if necessary, when by hauling and veering on it, following the motion of the vessel, a sufficiently uniform strain on it would be obtained without the risk of carrying it away.

Again, circumstances might arise, as they have sometimes done, when the immediate breaking up of the wreck might be imminent, and the delay in getting the hawser on board be of serious moment. In such a case the floating sling buoy should be hauled off by the whip alone, and the wrecked persons brought ashore in it floating in the water.

And again; in cases where the wreck happens on a flat shore, the hawser need not be set up at all, but the whip made to answer for both hawser and whip. When this is the case, the travelling block should be taken from the sling life buoy, and one end of the whip should be run through the thimble attached to the life buoy slings. The ends of the whip should then be made fast to the grummets on the sides of the life buoy.

In all other cases the hawser should be set up when practicable.

LIFE BUOYS.

The life buoys at present in common use in the navy are the SERVICE LIFE BUOY, KISBIE'S, AND CIRCULAR CORK LIFE BUOYS. A life buoy invented by Messrs. WELCH and BOURCHIER is also on trial.

The Service Life Buoy is supposed to be capable of keeping four men afloat. It is fitted with a portfire that burns 20 minutes, for the purpose of denoting its position when let go at night.

Great coolness and caution is requisite to float on this buoy. As soon as you get hold, place your feet on the balancing plate, grasping the pole with your hands. In this position you will float with your head well above water. Some men get frightened and endeavour to raise themselves higher up the buoy, which is certain to overbalance it, and throw them headlong into the water again.

Kisbie's Life Buoy of the ordinary circular form and size will support two men holding on outside. The best position for one man to keep himself afloat is to slip the buoy over his head, and rest his arms over it on either side. These buoys are most frequently stuffed with rushes instead of cork, and should the outside covering not be perfectly watertight they soon get sodden and lose much of their buoyancy.

Welch and Bourchier's Life Buoy consists of an air-tight casing, which may be either tapering in form towards the base, or have perpendicular sides, being either square, rectangular, circular or oval, having a central space for the reception of the person or valuables to be saved. This casing, which may be made in one or more compartments, and of any suitable metal or other material, may be enclosed in an open framework of wood, which serves to protect it, and adds to its floating power. If preferred, it may be enclosed in a framework of metal, having a grating at the base to form a support for the person in it, and open for the free passage of the water.

The apparatus is also provided with one or more hollow tubes or sockets, containing the signal staffs; carrying flags to indicate its position by day, and signal fires, which light by friction tubes, to shew its position by night; also for the purpose of attracting the attention of passing vessels, should it be lost, or of preventing ships running it down.

These signal staffs are automatically raised out of their sockets, when the buoy is dropped into the water by means of a weighted rod sliding down the inside of the fixed staff, which also forms the ballast. The weight is attached to the inside lower part of the second or middle section of the signal staff by a cord or chain passing over a sheave in the head of the lower fixed staff.

Small tanks, containing fresh water for drinking, sufficient to sustain life for a week, are fitted beneath the air casing, with suitable small tubes with mouth-pieces leading to the top of the lower fixed tube for sucking up the water.

Cork or other floats are attached to the upper part of the buoy by cords of sufficient length to go round a man's body, so as to afford a *means of* support in the water for several persons at a time, who can *thus hold* on to the buoy.

LIFE BUOYS.

The apparatus, which can be hoisted by any ordinary tackle, may be suspended from two hollow tubes or davits, projecting horizontally from any part of the ship in which it is considered advisable to place it. These davits may also be fitted to a piece of timber or sheet of metal, fitted at the back with metal clamps, to hook into loops on a ships' side or elsewhere, so that in the event of going alongside a wharf, or another ship, for the purpose of coaling, loading, or unloading, &c., the apparatus, together with the davit fittings, may, by having bolts fitted to the wood or metal sheet carrying the davits, be altogether and at once unshipped, and re-shipped again at pleasure with great ease; or the davits may be fitted with hinges so that the apparatus may be simply turned up inboard. The davits are hollow cylinders with a slot cut in the upper side of each, into which an eye or socket, attached to the upper part of the buoy, is inserted, and there firmly retained in its place by a sliding bolt. A cap is fitted to the davits above each signal staff to keep the latter from rising into position while the buoy is suspended at the davits. Each telescopic staff is provided with a projecting arm, carrying a pan, which contains a signal light or port-fire, to be ignited by means of a trigger line and friction tube, one of the trigger lines being passed through a small hole through the ship's side or stern, between the davits, which at sunset *only*, should be secured to the "letting go" handles, and loosed at sunrise. The buoy is detached by withdrawing the bolts before mentioned, and when so detached falls direct into the water; the firing of the port-fire, when required, and the elevation of the signal staff or staffs being simultaneous with its descent. If the buoy is "let go" at night, one fuze or port-fire always remains unignited and is at the service of the man in the buoy.

The buoy may be readily hoisted without lowering a boat; whilst hoisting, the man in the buoy should draw the mast down and hook the stop chains, then the eyes being inserted in the slots, the bolts must be shot the same, the caps turned over the staffs, the stop chains unhooked, and the buoy is again ready for letting go.

LIFE BELTS.

2   2

SEC. VIII.   LIFE BELTS.   287

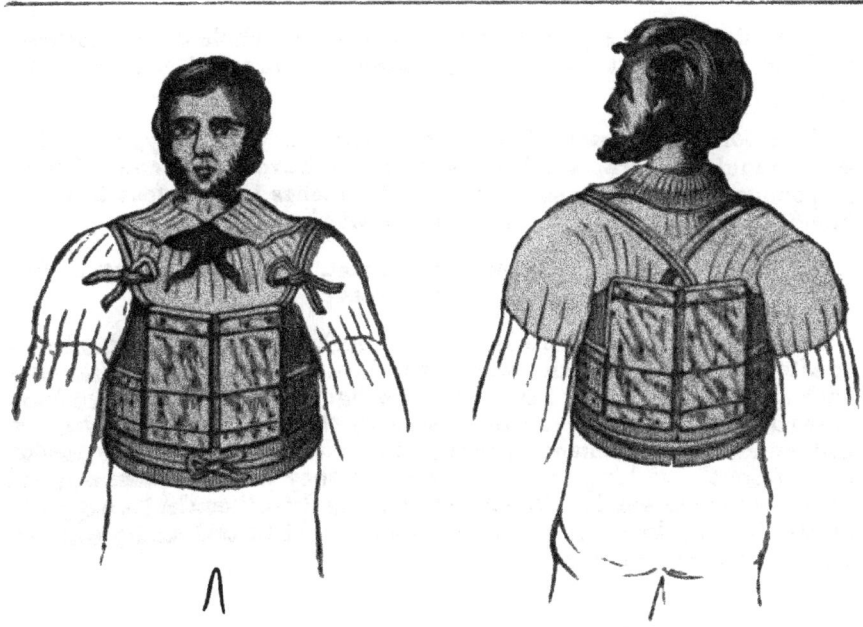

Fig. 3.   Fig. 4.

CORK MATTRESSES.

It is necessary that the Mattress should, when in use as a Life Belt, be as narrow as possible, so that the centre of buoyancy shall be as high as practicable for the man; they have a hinge in the middle, and on one side there is a pocket extending over the whole of the mattress, for hair. The mattress is ribbed across to prevent the cork from moving.

The Cork Mattresses, at present on trial in the navy, are stuffed with granulated cork, weigh 13 pounds, and have a buoyancy of about 60 pounds. Their dimensions:—5 feet 6 inches long, 1 foot 10 inches broad, and 3 inches deep. They will float three men.

Extracts from ADMIRAL RYDER's pamphlet, on "*Life-saving at Sea, by Cork Mattresses.*"

"If two hammocks are lashed or toggled together, either before or after the men are in the water, the latter can float between them with an arm over each hammock (see fig. 5); and it will be found that the two hammocks can be steered by the hands in the water over and outside the hammocks, while the raft is propelled by the feet away from the sinking ship. Six men, if they are self-possessed, and have been exercised in "hammock floating drill," could be supported by two hammocks; but of course there would be ordinarily only two men to each couple of hammocks."

Sec. VIII. CORK MATTRESSES.

In a heavy breaking sea, the best way for the men to secure themselves from being washed away from the hammocks, should be the subject of experiment; also how to protect themselves from the blows of the breaking sea.

A single man will probably best support himself by the aid of his hammock, if he secures the ends together (see fig. 6), and places himself in the middle (see fig. 7). Although the buoyancy is, as has already been said, enough to support three men if necessary, yet this will only be the case if they preserve their presence of mind, and *do not attempt to raise themselves out of the water sufficiently to immerse the hammock.* If the lashing has eight turns, one man should place himself between the second and third turns outside, another between the fourth and fifth inside, and the third between the sixth and seventh

Fig 6.

outside, so as to separate them as much as possible; the worst swimmer, or a wounded man, might be placed in the inside berth.

It will probably be found preferable to have a toggle and a becket always secured at each end of the hammocks, which will very much facilitate and expedite the securing the ends together, as some time must necessarily elapse before the clews and laniards can be disentangled from the lashings, during which interval the man, fatigued, perhaps wounded, and too probably an imperfect swimmer, may succumb and sink.

The hammock-ends should be so secured that the lashing is outside, as the drawing the hammock-ends together will then tighten the lashing; whereas if the lashing is inside it will be loosened, and the hammock consequently get adrift.

Note 1.—This refers to that portion of the lashing which connects the adjoining round turns.

Note 2.—Two pair of eyelet holes, one pair at each end of the hammock for the lashing to reeve through should always be fitted before the hammock leaves the dockyard.

When a man inside the circle of his hammock, after having secured the ends together, sees no immediate prospect of assistance from friend or enemy, he will begin, no doubt, to think how he can continue to support himself by aid of his hammock, with the least possible fatigue. He would find it difficult to unsling his hammock, but he could easily, *if the ends of the hammock are toggled together*, cut the clews close to the canvass, and if he knotted the nettles of the two clews together, he could make a long meshed net (with two laniards), in which, if the ends of the laniards were secured to each side of his hammock raft, he could sit with comfort and relieve the strain on his arms—remembering that the seat had better be sufficiently low under the water not to raise more than his chest out of it. In the position described he could not sink, even if he fainted or fell asleep.

Hints issued by the Royal Humane Society.
TO BATHERS.

AVOID bathing within TWO hours after a meal. Avoid bathing when exhausted by fatigue or any other cause. Avoid bathing when the body is cooling after perspiration; but, bathe when the body is warm, provided no time is lost in getting into the water. Avoid chilling the body by sitting or standing NAKED on the banks or in boats, after having been in the water.

Avoid remaining too long in the water—leave the water immediately there is the slightest feeling of chilliness.

Avoid bathing altogether in the open air, if, after having been a short time in the water, there is a sense of chilliness with numbness of the hands and feet.

The vigorous and strong may bathe early in the morning, on an empty stomach. The young and those who are weak, had better bathe three hours after a meal—the best time for such, is from two to three hours after breakfast.

ALL that is necessary to keep a person from drowning in deep water, is to keep the water out of the lungs. Suppose yourself a bottle—your nose is the nozzle of the bottle, and must be kept out of the water; if it goes under, don't breathe at all till it comes out; then, to prevent its going down again, keep every other part under—head, legs, arms, all under but your nose; do that, and you cannot sink in any depth of water. All you need do to secure this, is to clasp your hands behind your back, and point your nose upwards towards the heavens, and keep perfectly still. Your nose will never go under water, unless you raise your chin, hand, knee, or foot higher than it.

INSTRUCTIONS FOR SAVING DROWNING PERSONS, BY SWIMMING TO THEIR RELIEF.

1st.—When you approach a person drowning in the water, assure him with a loud and firm voice, that he is safe.

2nd.—Before jumping in to save him, divest yourself, as far and as quickly as possible, of all clothes ; tear them off if necessary, but if there is no time, loose, at all events, the foot of your drawers if they are tied, as if you do not do so, they fill with water and drag you.

3rd.—On swimming to a person in the sea, if he be struggling, do not seize him then, but keep off for a few seconds till he gets quiet, for it is sheer madness to take hold of a man when he is struggling in the water ; if you do, you run a great risk.

4th.—Then get close to him and take fast hold of the hair of his head, turn him as quickly as possible on to his back, give him a sudden pull and this will cause him to float, then throw yourself on your back also and swim for the shore, both hands having hold of his hair, you on your back and he also on his, and of course his back to your stomach. In this way you will get sooner and safer ashore than by any other means, and you can easily thus swim with two or three persons ; the writer has even, as an experiment, done it with four, and gone with them forty or fifty yards in the sea. One great advantage of this method is that it enables you to keep your head up, and also to hold the person's head up you are trying to save. It is of primary importance that you take fast hold of the hair, and throw both the person and yourself on your backs. After many experiments it is usually found preferable to all other methods. You can, in this manner, float nearly as long as you please, or until a boat or other help can be obtained.

5th.—It is believed there is no such thing as a death-*grasp*, at least it is very unusual to witness it. As soon as a drowning man begins *to get feeble* and to lose his recollection, he gradually slackens his hold *until he quits* it altogether. No apprehension need therefore be felt *on that* head when attempting to rescue a drowning person.

SAVING DROWNING PERSONS.

6th.—After a person has sunk to the bottom, if the water be smooth, the exact position where the body lies may be known by the air-bubbles, which will occasionally rise to the surface, allowance being of course made for the motion of the water, if in a tide-way or stream, which will have carried the bubbles out of a perpendicular course in rising to the surface. A body may be often regained from the bottom before too late for recovery, by diving for it in the direction indicated by these bubbles.

7th.—On rescuing a person by diving to the bottom, the hair of the head should be seized by one hand only, and the other used in conjunction with the feet in raising yourself and the drowning person to the surface.

8th.—If in the sea, it may sometimes be a great error to try and get to land. If there be a strong "outsetting" tide, and you are swimming either by yourself, or having hold of a person who cannot swim, then get on your back and float till help comes.

9th.—These instructions apply alike to all circumstances, whether the roughest sea or smooth water.

TREATMENT OF THE APPARENTLY DROWNED.

Issued by the Royal National Life-Boat Institution.

The leading principles of the following Directions for the Restoration of the Apparently Dead from drowning are founded on those of the late DR. MARSHALL HALL, combined with those of DR. H. R. SILVESTER, and are the result of extensive inquiries which were made by the Institution in 1863-4 amongst medical men, medical bodies, and coroners throughout the United Kingdom. These directions have been extensively circulated by the Institution throughout the United Kingdom and in the Colonies. They are also in use in Her Majesty's Fleet, in the Coast-guard Service, and at all the Stations of the British Army at home and abroad.

I.

Send immediately for medical assistance, blankets, and dry clothing, but proceed to treat the Patient *instantly* on the spot, in the open air, with the face downwards, whether on shore or afloat; exposing the face, neck, and chest to the wind, except in severe weather, and removing all tight clothing from the neck and chest, especially the braces.

The points to be aimed at are—first and *immediately*, the RESTORATION OF BREATHING; and secondly, after breathing is restored, the PROMOTION OF WARMTH AND CIRCULATION.

The efforts to *restore breathing* must be commenced immediately and energetically, and persevered in for one or two hours, or until a medical man has pronounced that life is extinct. Efforts to promote *warmth and circulation*, beyond removing the wet clothes and drying the skin, must not be made until the first appearance of natural breathing; for if circulation of the blood be induced before breathing has re-commenced, the restoration to life will be endangered.

II.—To Restore Breathing.

To Clear the Throat.—Place the patient on the floor or ground with the face downwards, and one of the arms under the forehead, in which position all fluids will more readily escape by the mouth, and the tongue itself will fall forward, leaving the entrance into the windpipe free. Assist this operation by wiping and cleansing the mouth.

If satisfactory breathing commences, use the treatment described below to promote warmth. If there be only slight breathing—or no breathing—or if the breathing fails, then :—

To Excite Breathing.—Turn the patient well and instantly on the side, supporting the head, and excite the nostrils with snuff, hartshorn, and smelling salts, or tickle the throat with a feather, &c., if they are at hand. Rub the chest and face warm, and dash cold water, or cold and hot water alternately on them. If there be no success, lose not a moment, but instantly :—

To Imitate Breathing.—Replace the patient on the face, raising and supporting the chest well on a folded coat or other article of dress.

Turn the body very gently on the side, and a little beyond, and then briskly on the face, back again, repeating these measures cautiously, efficiently, perseveringly, about fifteen times in the minute, or once every four or five seconds, occasionally varying the side.

[*By placing the patient on the chest, the weight of the body forces the air out ; when turned on the side, this pressure is removed, and air enters the chest.*]

2.—EXPIRATION.

The foregoing two Illustrations show the position of the Body during the employment of Dr. Marshall Hall's Method of Inducing Respiration.

On each occasion that the body is replaced on the face, make uniform but efficient pressure with brisk movement, on the back between and below the shoulder-blades or bones on each side, removing the pressure immediately before turning the body on the side.

During the whole of the operations let one person attend solely to the movements of the head and of the arm placed under it.

[*The first measure increases the expiration—the second commences Inspiration.*]

∴ The Result is *Respiration* or *Natural Breathing*;—and if not too late, *Life*.

Sec. VIII. RESTORING THE APPARENTLY DROWNED.

Whilst the above operations are being proceeded with, dry the hands and feet, and as soon as dry clothing or blankets can be procured, strip the body, and cover or gradually re-clothe it, taking care not to interfere with the efforts to restore breathing.

III.

Should these efforts not prove successful in the course of from two to five minutes, proceed to imitate breathing by Dr. Silvester's method as follows :—

Place the patient on his back on a flat surface, inclined a little upwards from the feet; raise and support the head and shoulders on a small firm cushion or folded articles of dress placed under the shoulder-blades.

Draw forward the patient's tongue, and keep it projecting beyond the lips; an elastic band over the tongue and under the chin will answer this purpose, or a piece of string or tape may be tied round them, or by raising the lower jaw, the teeth may be made to retain

I.—INSPIRATION.

the tongue in that position. Remove all tight clothing from about the neck and chest, especially the braces.

To Imitate the Movements of Breathing.—Standing at the patient's head, grasp the arms just above the elbows; draw the arms gently and steadily upwards above the head, and keep them *stretched* upwards for two seconds. *(By this means air is drawn into the lungs.)* Then turn down the patient's arms, and press them gently and firmly for two seconds against the sides of the chest. *(By this means air is pressed out of the lungs.)*

Repeat these measures alternately, deliberately, and perseveringly, about fifteen times in a minute, until a spontaneous effort to respire is perceived, immediately upon which, cease to imitate the movements of breathing, and proceed to INDUCE CIRCULATION AND WARMTH.

2.—EXPIRATION.

The foregoing two Illustrations show the position of the Body during th employment of Dr. Silvester's Method of inducing Respiration.

IV.—Treatment after Natural Breathing has been Restored.

To Promote Warmth and Circulation.—Commence rubbing the limbs upwards, with firm grasping pressure and energy, using handkerchiefs, flannels, &c. *(By this measure the blood is propelled along the veins towards the heart.)*

The friction must be continued under the blanket or over the dry clothing.

Promote the warmth of the body by the application of hot flannels, bottles, or bladders of hot water, heated bricks, &c., to the pit of the stomach, the arm-pits, between the thighs, and to the soles of the feet.

If the patient has been carried to a house after respiration has been restored, be careful to let the air play freely about the room.

On the restoration of life, a teaspoonful of warm water should be given; and then, if the power of swallowing have returned, small quantities of wine, warm brandy and water, or coffee should be administered. The patient should be kept in bed, and disposition to sleep encouraged.

General Observations.

The above treatment should be persevered in for some hours, as it is an erroneous opinion that persons are irrecoverable because life does not soon make its appearance, persons having been restored after persevering for many hours.*

Appearances which generally accompany Death.

Breathing and the heart's action cease entirely; the eyelids are generally half closed; the pupils dilated; the tongue approaches to the under edges of the lips, and these, as well as the nostrils, are covered with a frothy mucus. Coldness and pallor of surface increase.

---

* There is an authentic account of recovery after *submersion* for 20 minutes, and alleged ones after longer periods.

Cautions.

Prevent unnecessary crowding of persons round the body, especially if in an apartment.

Avoid rough usage, and do not allow the body to remain on the back unless the tongue is secured.

Under no circumstances hold the body up by the feet.

On no account place the body in a warm bath unless under medical direction, and even then it should only be employed as a momentary excitant.

RESTORING THE APPARENTLY DROWNED.
(Dr. Howard's Method.)

From the "LIFE BOAT," or Journal of the National Life-Boat Institution, February, 1873.

This Method is recommended by an American Doctor named HOWARD, and is by him called the "DIRECT METHOD."

The Direct Method.

Rule I. *Arouse the Patient.*—Unless in danger of freezing, do not move the patient an inch; but instantly expose the face to a current of fresh air; wipe dry the mouth and nostrils, rip the clothing so as to expose the chest and waist, and give two or three quick smarting slaps on the stomach and chest with the open hand.

If the patient does not revive, then proceed thus :—

Rule II. *To draw off water, &c., from the stomach and chest.*— Turn the patient on his face, a large bundle of tightly-rolled clothing being placed beneath his stomach, and press heavily over it for half a minute, or so long as fluids flow freely from the mouth.

Rule III. *To produce breathing.*—Place the patient on his back, the roll of clothing being so placed beneath it as to raise the pit of the stomach above the level of any other part of the body. If there be another person present, let him, with a piece of dry cloth, hold the tip of the tongue out of one corner of the mouth, and with the other hand grasp both wrists and keep the arms forcibly stretched back above the head. (This position prevents the tongue from falling back and choking the entrance to the windpipe, and increasing the prominence of the ribs tends to enlarge the chest; it is not, however, essential to success.)

Kneel beside, or astride, the patient's hips, and with the balls of the thumbs resting on either side the pit of the stomach, let the fingers fall into the grooves between the short ribs, so as to afford the best grasp of the waist. Now, using your knees as a pivot throw all your weight forward on your hands, and at the same time squeeze the waist between them, as if you wished to force everything in the chest upwards out of the mouth; deepen the pressure while you can count slowly one, two, three, then *suddenly* let go with a final push, which springs you back to your first kneeling position. Remain erect on your knees while you can count one, two; then repeat the same motions as before, at a rate gradually increased from four or five to fifteen times in a minute, and continue thus this bellows movement with the same regularity that is observable in the natural motions of breathing which you are imitating.

Continue thus far from one to two hours, or until the patient breathes; for awhile after, carefully deepen the first short gasps into full breaths, and continue the drying and rubbing, which should have been unceasingly practised from the beginning.

Rule IV. *After treatment. Externally.*—As soon as the breathing has become established, strip the patient, wrap him in blankets only, put him in a bed comfortably warm, but with a free circulation of fresh air, and leave him to perfect rest.

Internally.—Give a little hot brandy and water, or other stimulant at hand, for every ten or fifteen minutes for the first hour, and as often thereafter as may seem expedient.

Rules for the course to be followed by the Bystanders in case of Accidents, where Surgical Assistance cannot be at once obtained.

THE DANGERS TO BE FEARED ARE:—

Shock or collapse, loss of blood, and unnecessary suffering in the moving of the patient.

1.—In *shock* the injured person lies pale, faint, cold, and sometimes insensible, almost pulseless, scarcely breathing, in fact he is all but dead. Apply external warmth, by wrapping up in blankets or extra clothes. Bottles of hot water or hot bricks may also be wrapped in cloths, and placed along the sides, and between the legs and feet.

If the patient has not been drinking, give a small quantity of brandy, or whisky. This, however, should be done cautiously for fear of excessive reaction. Rather depend on external means to bring him round. Food should be given now and then (strong soup is the best).

2.—*Loss of blood.*—To check the flow of blood from a wound, the principle to be acted upon is to arrest the flow of blood to the part. Bleeding may be from an artery; it then comes in jets, or spirts, and is of a bright red colour, in such a case pressure must be made on the main artery of the limb above the wound, that is to say, on the side of the wound nearest the heart. Again, the blood may be from a wounded vein, it then flows in a stream (no jets), and is of dark colour; pressure, in this case, should be made below the wound, with the finger for a short time, when a small pad of lint and bandage over the wound will suffice. To check the current of blood in the main artery of the arms or legs, feel for the artery on the inner side of the limbs, when it will be recognised by its pulsation; lay a firm and even compress across the site or course of the artery, tie a handkerchief around the limb and compress, pass a piece of stick under the handkerchief, on the outer side of the limb, and twist it round until sufficient tightness is produced to arrest the flow of blood; then secure the stick by a piece of tape or twine to prevent its untwisting. In cases of emergency a handful of earth has been placed in the wound to check bleeding, it acts as an absorbant, entangles the blood, and allows coagulation or clot to take place, which acts as a mechanical plug.

Memo.—The arteries, in arms and legs, will be found coursing towards the inner side of both limbs. The pressure should be *above* the wound to arrest *arterial* bleeding, *below* the wound in *venous*. In applying the tourniquet or handkerchief, the pad or compress should be on the inner side, the screw of the former or knot of the latter should be on the outer side of limb—elevation of the wounded limb will contribute much to check the bleeding. A compress, or pad, may be made of a cork, piece of wood, or round stone wrapped up in a piece of rag or cloth. Care should be taken to examine the limb from time to time, and to lessen the constriction if it becomes very cold or purple, tighten up the handkerchief if the bleeding begins afresh.

If any of the limbs be fractured, they should be temporarily splinted with two or four pieces of bark or light wood tied at each end with handkerchiefs passed twice round like the old fashioned twice round necktie.

If a broken limb be not splinted the ends may be forced through the skin, if the patient be carried awkwardly.

To transport a wounded person safely, let the patient be laid on a door, shutter, settee, or some firm support, properly covered. Have sufficient force to lift steadily, and the bearers should *not* keep step.

Dislocations.

Whenever a bone is dislocated, there is a deformity at joint and loss of motion.

The sufferer usually becomes faint. While this condition exists, an attempt should be made to reduce the bone into position. It is surprising how easily this may be done if it be tried at once by extension and counter-extension by jack towels or sheets.

If, for example, the dislocation takes place at the shoulder joint, a clove hitch by towel should be applied above the elbow joint, and steady traction made in the direction of the axis of the bone, by a strong man, while two others make counter extension by another towel or sheet, in the arm-pit, crossed towards the sound shoulder. This may be held by two men, or made fast to a ring bolt; or the heel may be placed in the arm-pit, while a towel is clove-hitched above the elbow joint. The limb should be steadily pulled downwards, while the heel assists to force the head of the dislocated bone outwards and upwards into its natural position. A loud snap is always heard when the bone returns into its socket.

Dislocation at Hip.—Clove hitch above knee, second sheet on inside of thigh for counter-extension, and made fast to bolt on deck or ship's side.

Bruises.—Use hot fomentations at first. After inflammation has subsided, use stimulating applications, as vinegar and water, alcohol, &c.

Sprains.—Elevate the limb, keep the joint perfectly quiet, apply luke-warm lotions or fomentations. When inflammation has ceased, apply stimulating linaments, as soap or camphor linaments, and bandages; shower the part with cold water, alternating with warm water, or the hot and cold douches.

Burns or Scalds.

Do not cut the bladder. The readiest application may be flour from the dredging box, laid on thick as possible. Lime water and oil in equal parts is excellent, covered with cotton wadding.

The object is to exclude the air and prevent suppuration: do not change the dressing for four or five days unless there be profuse discharge or bad odour. Resinous ointment spread on lint sprinkled with turpentine is another excellent application. If the scald is extensive and on the body, *cold* applications are *not proper*. Keep the air from the wound; this can be done by the dressing already suggested for burns.

Sun Stroke.

Take the patient immediately into the shade, place in a semi-recumbent position; head raised, loosen the clothes about neck and chest; apply immediately ice or cold wet cloths to the head and nape of the neck, changing them frequently. The douche over head, spine, and chest, from a height of about three feet. Patient to be fanned to produce a cool current of air. Mustard to limbs and sides—stimulants.

Poisons.

In all cases of poisoning, the first step is to give the antidote if you know it and then evacuate the stomach. The last should be effected by a mustard emetic, a tablespoonful of mustard in a cup of water, or a table-spoonful or two of common salt in a tumbler of water. When vomiting has already taken place, copious draughts of warm water, or warm mucilaginous drinks, soap and water, or, oil should be given, to keep up the effect till the stomach has been thoroughly cleared.

ANTIDOTES.—*For any of the strong acids*:—Common chalk, oil or soap suds.

For Arsenic:—Magnesia, powdered charcoal, oil, and lime water.

For Prussic Acid:—Cold affusion—ammonia.

For Opium:—Keep patient walking, strong coffee, slap with flat ruler, sting with nettles, mustard emetics.

Asphyxia.

Asphyxia arises from carbonic acid, from charcoal fumes, and other gases interfering with the respiration. The face becomes turgid and livid, owing to the accumulation of impure blood. The patient in this case should be placed with the head high, so as to facilitate the flow of blood from the brain, which is congested; the clothes should be taken off and he should be dashed with cold water. Ammonia should be applied to the nose, &c.; the face and body should be sponged with brandy, or vinegar and water, friction all over, and artificial respiration if recovery is not evident.

Poisonous Fish.

At certain seasons, some fish, when eaten, produce poisonous effects. If vomiting has already occurred, it should be encouraged, by giving lukewarm water to drink, or it may be rendered more decided by the administration of the mustard emetic; after the stomach has been well cleared out and has had a little rest, a tablespoonful of castor oil, or half a drachm of rhubarb may be given with advantage.

Stings from Wasps or Bees.

The wound should be examined and the sting extracted, if left in the wound. The barrel of a watch key, pressed over it, will cause it to be dislodged, when it can be easily withdrawn. A little spirits of hartshorn, or Eau de Cologne, may be then applied to the part.

Rules for treatment of Cholera.

Cholera is almost invariably preceded by a painless diarrhœa, and this should, in all cases, be promptly treated.

When diarrhœa is present, go to bed and maintain the horizontal position, use abundance of blankets. Stay in bed until you are well, do not consider yourself well until you have had a natural movement from the bowels. Let no solid food be used, everything to be fluid as milk, soup, a soft boiled thin egg or arrowroot. Apply mustard plasters to the abdomen. In the absence of a physician, an adult can take ten drops of laudanum, and ten drops of spirits of camphor. A child of ten may take five drops of each. A child of five years may take three drops of each. Never chill the surface of the body by getting out of bed.

* Opium is dangerous to old people, as well as very young; and should not be repeatedly administered, without medical supervision.

Fever and Ague.

Fever and ague is always preceded by an ague fit: it has three stages, the cold, hot, and sweating stage.

1st.—The cold, when teeth chatter.

2nd.—The hot, with high fever.

3rd.—The sweating, when moisture appears and feeling of health returns.

In the event of there being no physician; in cold stage give *hot drinks*, hot foot bath, hot bottles to sides and limbs.

In hot stage, give cooling drinks, half teaspoonful of sweet spirits of nitre in water every two hours.

During sweating stage, rub with dry towels. In *intermission* give quinine in from two to ten grain doses every three hours, for a few doses: afterwards give ten drops of tincture of iron three times a day for a week. Avoid the hot sun, and damp evening and morning air. If singing in the ears should come on while taking quinine, the dose should be either lessened or suspended altogether.

To restore persons affected by cold.

For Frost-bite or Numbness.—Return warmth *gradually*, by warm water.

For a Frozen Limb.—Rub with snow, and place in cold water for a time. When sensation returns, place again in cold water; add heat *very gradually* by warm water.

If Apparently Dead or Insensible.—Strip entirely of clothes, and cover body, except mouth and nostrils, with snow or ice, or place in cold water. When body is thawed, dry it, place it in a cold bed; rub with warm hands under the cover,—continue this for hours. If life appears, give small injections of camphor and water; put a drop of spirits of camphor on tongue; then rub body with spirits and water, finally with spirit; then give tea, coffee, or brandy and water.

Fainting.

When a person suddenly grows pale and faint, he should be immediately placed, full length, on the floor, the head being kept low. The face may be dashed with cold water. In this position he will quickly recover—this is owing to the head having been placed low so as to facilitate the flow of blood to the brain.

Drunkenness.

Drunkenness, in a severe form, may cause death like apoplexy; it is poisoning by alcohol. The individual should be placed in a semi-recumbent position, with head on one side to favour vomiting, all the clothing about the neck being freely opened. A douche, from a height, of cold water over face, head, and neck, will probably rouse him. is difficult to get an emetic to act in this state, yet a table-spoonful of mustard, in half-a-tumbler of water, had better be given, to excite vomiting. If the respiration becomes embarrassed, artificial respiration should be used, as directed in case of drowning. The preparation called acetate water of ammonia, (to be had at any apothecary's) taken in ounce doses, every half-hour, is said to have a most magical effect in restoring drunken men to sobriety; about three doses ought to suffice.

Epilipsy or Fits.

If a man fall in a fit, insensible or struggling, it is probably epilepsy. The clothing about chest and throat should be thrown open; the head should be raised, and the patient prevented from hurting himself during his struggles. A piece of wood, or a cork, may be placed between the teeth, to prevent the tongue from being bitten.

Bite of a Mad Dog.

In the absence of a surgeon to excise the part, which is the proper treatment, the wound should be quickly washed, sucked, and salt freely applied to the bottoms of the teeth punctures. If not to be at hand a hot iron wire may be used as an actual cautery, or powder may be placed in the wounds and ignited. This has been done, by some English sportsmen, in Albania, when bitten by a dog.

Marsh Poison.

When men are employed on detached service in boats, or are otherwise exposed in a swampy region, they should be supplied with quinine to guard them against the marsh poison. Four grains should be administered before starting in the morning, and four on their return; but if they should be exposed for twelve hours, or if the exposure be over-night, the quinine should be continued until they *return* on board, and for fourteen days afterwards. The quinine *might* affect the head, by producing buzzing and noise in the ears.

When this happens, the dose should be reduced in quantity, or perhaps discontinued.

Strength for one dose :—

Quinine .... 4 grains.
Dilute sulphuric acid ....10 drops.
Water .... 2 ounces.
Rum .... 1 ounce.

By simple multiplication, the above formulæ may be mixed, at once, for any number of men daily.

DISINFECTANTS.

These substances are of various kinds and act in different ways. Air and water are nature's great purifiers, but when these cannot advantageously operate, it is necessary to apply one or other of the many powerful chemical agents which are available. The choice of such agent must depend somewhat upon the circumstances of the case, but for general use Carbolic Acid and its preparations, Condy's Fluid, Chloride of Lime, and Sir W. BURNETT's solution appear to be the most convenient and efficacious.

Carbolic Acid.—This is the disinfectant which is now generally supplied to H.M. Ships. In adding it to bilge water or other liquids, it should first be mixed with about 100 parts of water to completely dissolve it; for, if the acid is not properly mixed it will merely sink to the bottom, softening any bituminous cement with which the bottom of the ship may be coated. To purify the air of a cabin, a little of the acid, unmixed with water, should be placed in a saucer. McDOUGALL and CALVERT's disinfecting powders are preparations of Carbolic Acid.

Burnett's Solution.—Since the introduction of Carbolic Acid, this fluid is not so extensively used. One pint of the solution should be added to a gallon of water.

Condy's Fluid and Chloride of Lime are very efficient disinfectants. The latter deteriorates considerably by keeping, unless well preserved from the action of the air.

To Fumigate a Ship.—Chlorine is perhaps the most powerful agent for this purpose. It may be evolved from Chloride of Lime by moistening it with water, or dilute Sulphuric Acid. It may also be obtained by pouring four parts of Hydrochloric Acid on one part of Binoxide of Manganese placed in a shallow vessel, and gently heated. The vessel should be placed high up as the gas descends, and all exits should be closed up.

In many cases the burning of a small piece of sulphur will be found a very simple and efficacious means of disinfecting the air of a cabin or room. The room, however, must be made air-tight to retain the sulphurous acid gas, which is an irrespirable one.

Clothing, after infectious or contagious diseases, may be disinfected, at once, by being plunged into boiling water. Linen and washing apparel should be immersed in water $212°$; as uniforms or cloth clothing may be injured by water, it should be baked in an oven at a temperature from $210°$ to $250°$, but as ovens are not to be obtained at sea, it is worth remembering, in an emergency, that boiling water will act efficiently.

Cloths dipped in a solution of chloride of lime, one pound to the gallon, may be hung up in an inhabited room to disinfect the air.

Section 9.

MONEY, WEIGHTS AND MEASURES OF ALL NATIONS;

USEFUL RULES IN MENSURATION;

RULES FOR CALCULATING TONNAGE OF VESSELS,

FLOATING POWERS OF SPARS,

NUMBER OF SQUARE YARDS IN SAILS, &c.

MONEY.

America, U.S.

				English value.
		1 Cent	=	½d.
100 Cents	=	1 Dollar	=	4s. 2d.
5 Dollars	=	1 Half Eagle	=	20s. 10d.
10 Dollars	=	1 Eagle	=	£2 1s. 8d.
20 Dollars	=	1 Double Eagle	=	£4 3s. 4d.

£1 Sterling = 4 Dollars, 80 Cents.

Argentine Republic, or La Plata.

100 Centesimos	=	1 Dollar or Patacon	=	4s. 2d.
17 Patacons	=	1 Doubloon	=	about 64s. 8d.

£1 Sterling = 4 Patacons, 80 Cents.

Austria.

100 Kreutzers	=	1 Florin	=	1s. 11½d.

£1 Sterling = 10 Florins, 28 Kreutzers.

Paper money is the chief medium of payment.

Belgium.

Decimal system (same as France).

Brazil.

(Same as Portugal.)

The value of a paper Milreis varies from 2s. to 2s. 4d. sterling.

British India.

12 Pies	=	1 Anna	=	1½d.
16 Annas	=	1 Rupee	=	2s. 0d.
15 Rupees	=	1 Mohur	=	£1 10s. 1d.

£1 Sterling = 10 Rupees, 4 Annas.

				English value
Lac of Rupees	=	100,000	=	£10,000
Crore of Rupees	=	10,000,000	=	£1,000,000

The Independent States have each a separate coinage, all different.

Burmah.

4 Great Rweh (rees)	=	1 Bais	=	1½d.
4 Bais	=	1 Math	=	6d.
4 Maths	=	1 Tical	=	2s. 0d.

£1 sterling = 10 Ticals or Kyats.

Canada.

| | | 1 Cent | = | ½d. |
| 100 Cents | = | 1 Dollar | (currency 5s.) | 4s. 2d. |

Sometimes accounts are kept in £ s. d. currency.

| 12 Pence (currency) | = | 1 Shilling (currency) | | 10d. |
| 20 Shillings ,, | = | 1 Pound ,, | | 16s. 8d. |

In Nova Scotia the Pound currency = 16s. 0d.

Cape of Good Hope.
(Same as Great Britain.)

Chili.

| 100 Centavos | = | 1 Dollar or Peso current | = | 3s. 9d. |

China.

10 Cash	=	1 Candareen (Fun)	=	$\frac{7}{10}$d.
10 Candareens	=	1 Mace (Tsien)	=	7d.
10 Mace	=	1 Tael (Lëang)	=	5s. 10d.

Established Rate among Foreigners—Dollar = 4s. 2d.

Average rate—Dollar = 970 Cash.

Denmark.

| | | 1 Skilling | = a little more than ½d. |
| 96 Skilling | = | 1 Rixdaler or Daler | 2s. 2½d. |

£1 Sterling = 9 Rixdalers, 11 Skillings.

East Coast of Africa (Mozambique.)
(Same as Portugal.)

Egypt.
(Same as Turkey, but coins of various nations are in circulation.)
Falkland Islands.
(Same as Great Britain.)
France.

				English value.
100 Centimes	=	1 Franc	=	9¼d.
20 Francs	=	1 Napoleon	=	15s. 10¼d.

The par of exchange with London is about 25 Francs for £1 sterling.

Germany.

In December, 1871, one uniform system was introduced for all Germany; it is as follows :— 10 Pfennings=1 Groschen=1·175d.; 10 Groschen, or 100 Pfennings=1 Mark=11¾d. GOLD COINS—10 and 20 Mark pieces=respectively, 9s. 9½d. and 19s. 7d. Compared with the Old System, the 10 Mark piece is equal to 3 Thalers 10 Silber Groschen of North German Currency, or 5 Florins 50 Kreutzers of South German Currency, or 8 Marks 5½ Schillings of Lubec or Hamburg customary Currency, or 3 1·93rd Thalers of Bremen Gold reckoning. SILVER COINS.—During the period of transition the old silver coins of North Germany, down to the 5 Groschen piece, continue to be a legal tender, namely—the Thaler=3 Marks=2s. 11¼d.; the Double Thaler=6 Marks=5s. 10½d.; the 10 Groschen piece=1 Mark =11¾d.; and the 5 Groschen piece=5¾d.

Gibraltar.

16 Quartos	=	1 Real	=	4½d.
12 Reals	=	1 Dollar	=	4s. 2d.
100 Cents	=	1 Dollar	=	4s. 2d.

The Garrison accounts are kept in £ s. d.

Since May, 1872, the standard of value has been the Spanish Doblon d'Isabel, and the moneys of account are as follows :—

10 Decimas de Real Vellon	=	1 Real Vellon =	................2½d.
20 Reals de Vellon	=	1 Dollar =	.... 4s. 2d.
100 Reals de Vellon	=	1 Doblon =	.......£1 0s. 5d.

GOLD COINS.—Doblon ; 2 and 1 Dollars. SILVER COINS.—Dollar; Half-Dollar ; 2, 1, and ½ Reals of Plate. The Real of Plate=6d. sterling. BASE SILVER COINS.—Peseta of 4 Reals Vellon=9¼d. ; 2 and 1 Reals Vellon. BRONZE COINS.—Half Real Vellon ; Cuartillo= ¾d. ; and 2, 1, and ½ Decimas de Real Vellon.

Greece.
Decimal system (same as France).
100 Lepta   =   1 Drachma   =   9¼d.

Holland.
              English value.
100 Cents = 1 Guilder or Florin   = 1s. 8d.
   £1 Sterling = 12 Guilders or Florins.

Hong-Kong.
     1 Cent   = ¼d.
100 Cents = 1 Dollar   = 4s. 3d.

India.
(See British India.)

Italy.
Decimal system (same as France.)
100 Centimes = 1 Lira   = 9¼d.

Japan.
In 1871 a new Monetary System, based upon a gold standard, was introduced in Japan. The Yen, weighing 25·72 Troy grains of gold, $\frac{9}{10}$ fine, was constituted the fundamental unit of the system. The Yen is divided into 100 Sen, and the Sen into 10 Rin, as follows :—
  10 Rin..................= 1 Sen ....= ½d.
  100 Sen..................= 1 Yen.............. = 4s. 2d.
The Itsiboo was the coin in most common use, its intrinsic value being 1s. 4½d.

Mauritius
Government accounts are kept in £ s. d. Merchants use Dollars and Cents. Dollar=3s. 10d.

Mexico.

			English value.
100 Cents	= 1 Dollar	=	4s. 2d.

New South Wales.
(Same as Great Britain.)

Norway.

24 Skillingen	= 1 Mark or Ort	=	10¼d.
5 Mark	= Species-Daler	=	4s. 5¼d.

£1 Sterling = 4½ Species-Dalers.

Portugal.

The unit of account is the Rei, worth $\frac{1}{45}$d. sterling.
A Cruzado = 400 Reis. A Cruzado Novo or Pinto = 480 Reis.
£1 sterling = 4,500 Reis.

Russia.

100 Copecks	= 1 Silver Rouble	=	3s. 2d.
10 Roubles	= 1 Imperial	=	£1 11s. 8d.

£1 Sterling = 7 Roubles, 31½ Copecks.
Paper money is the chief medium of payment. The paper Rouble is worth about 2s. 6d. sterling.

Siam.

200 to 450 Cowries or Bier	= 1 P'hainung	=	1d. nearly
4 P'hainungs	= 1 Fuang	=	3¾d.
2 Fuangs	= 1 Salung or Miam	=	7½d.
4 Salungs or Miam	= 1 Tical	=	2s. 6d.
4 Ticals	= 1 Tamlung	=	10s. 6d.

Singapore.

100 Cents	= 1 Dollar	=	4s. 3d.
16 Annas	= 1 Rupee	=	1s. 11¾d.

Government accounts are kept in £ s. d. sterling.

Spain.
Decimal System (Same as France).

			English Value.
100 Centimos	= 1 Centimo =		$\frac{19}{200}$d.
	= 1 Peseta =		9¼d.

£1 Sterling = 25 Pesetas.

St. Helena.
(Same as Great Britain.)

Sweden.
100 Öre = 1 Riksdaler = 1s. 1¼d.

£1 Sterling = 18 Riksdalers.

Turkey.
40 Paras = 1 Piastre = 2$\frac{4}{11}$d.
100 Piastres = 1 Medjidie, or Lira Turca = 18s.

West Coast of Africa.
(Same as Great Britain.)

WEIGHTS AND MEASURES.

The United Kingdom of Great Britain and Ireland.

I.—IMPERIAL MEASURES OF LENGTH.

12	Inches	= 1	Foot
3	Feet	= 1	Yard
5½	Yards	= 1	Pole, Rod, or Perch
4	Poles, or 100 Links	= 1	Chain
40	Perches, or 10 Chains	= 1	Furlong
8	Furlongs, or 1760 Yards	= 1	Mile
3	Miles	= 1	League

The Fathom=6 Feet; A Cable's Length=The Tenth of a Sea Mile, or about 200 Yards; the mean Nautical Mile=1·151 Statute Miles=1853 Metres.

II.—IMPERIAL MEASURES OF SURFACE.

144 Sq. Inches	=	1 Sq. Foot
9 Sq. Feet	=	1 Sq. Yard
30¼ Sq. Yards, or 272¼ Sq. Feet	=	1 Sq. Rod, Pole, or Perch
40 Poles	=	1 Rood
4 Roods, 4840 Sq. Yards. 10 Sq. Chains, or 100,000 Sq. Links	=	1 Acre
100 Acres	=	1 Hide of Land
640 Acres	=	1 Sq. Mile

III.—IMPERIAL MEASURES OF CUBIC CAPACITY.

1728 Cubic Inches	=	1 Cubic Foot
27 Cubic Feet	=	1 Cubic Yard

IV.—IMPERIAL MEASURES OF CAPACITY FOR LIQUIDS.

8·665 Cubic Inches	=	1 Gill
4 Gills	=	1 Pint
2 Pints	=	1 Quart
4 Quarts	=	1 Gallon

The Imperial gallon is the standard of capacity for both wet and dry goods, and has a cubic content of 277·274 cubic inches, and contains 10lbs. Av. of distilled water at a temperature of 62° Fahrenheit.

V.—IMPERIAL MEASURES OF CAPACITY FOR DRY GOODS.

2 Pints	=	1 Quart
4 Quarts	=	1 Gallon
2 Gallons	=	1 Peck
4 Pecks (8 gallons)	=	1 Bushel
8 Bushels	=	1 Quarter

VI.—MEDICAL SUBDIVISIONS OF THE IMPERIAL PINT.

60 Minims	= 1	Fluid Drachm (f. ℨ.)
8 Fluid Drachms	= 1	Fluid Ounce (f. ℥.)
20 Fluid Ounces	= 1	Imperial Pint (O.)
8 Pints	= 1	Gallon

VII.—AVOIRDUPOIS WEIGHT.

27 11/32 Troy Grains	= 1	Dram (dr.)
16 Drams	= 1	Ounce (oz.)
16 Ounces (or 7000 Troy Grains)	= 1	Pound (lb.)
14 Pounds	= 1	Stone
28 Pounds	= 1	Quarter (qr.)
4 Quarters (or 112 lbs.)	= 1	Hundredweight (cwt.)
20 Hundredweight	= 1	Ton

This weight is used in almost all commercial transactions.

TROY WEIGHT.

Metric Equivalent. Grammes.

24 Grains	= 1	Pennyweight (dwt.)	= 1·555176
20 Pennyweights	= 1	Ounce (oz.)	= 31·10352
12 Ounces	= 1	Pound (lb.)	= 373·24224

This weight is used for gold, silver, and precious stones, and in philosophical experiments.

OLD APOTHECARIES' WEIGHT.

20 Grains	= 1	Scruple	= 20 Grains Troy
3 Scruples	= 1	Drachm	= 60 ,, ,,
8 Drachms	= 1	Ounce	= 480 ,, ,,
12 Ounces	= 1	Pound	= 5760 ,, ,,

Apothecaries compounded by this weight, but bought and sold their drugs by avoirdupois.

NEW APOTHECARIES' WEIGHT.
Introduced in 1864.

Ounce	=	437½ Grains.
Pound, 16 Ounces	=	7,000 ,,

Same as Avoirdupois.

Sec. IX.   WEIGHTS AND MEASURES.   321

Water and Provisions.

Ton	210 Gallons	Weight,	2100 lbs.
Butt	110 ,, ...	,,	1100 ,,
Puncheon	72 ,, ...	,,	720 ,,
Barrel	36 ,, ...	,,	360 ,,
Kilderkin	18 ,, ...	,,	180 ,,

Tanks.

	No.	Galls.	Length. ft. in.	Breadth. ft. in.	Depth. ft. in.
Whole	1	600	4 1	4 1	6 1
,, ...	2	500	4 1	4 1	5 1
,, ...	3	400	4 1	4 1	4 1
Half	4	200	4 1	2 1	4 1
,, ...	5	200	4 1	4 1	2 1
,, corner off	6	193	4 1	4 1	2 1
,, ...	7	200	3 3	3 3	3 3
Quarter	8	100	3 3	1 8	3 3
,, ...	9	100	3 3	3 3	1 9
Bilge	10	375	4 1	4 1	4 1
,, ...	11	264	4 1	3 7	4 1
,, ...	12	110	3 3	2 7	2 10

Provisions.

Biscuit	Bags	112 lbs.	
Rum	Puncheon	72	gallons.
	Hogshead	54	,,
	Barrel	36	,,
	Half-hogshead	25	,,
	Kilderkin	18	,,
	Small cask	12	,,
Salt Beef	Tierce	38	8-lb. pieces.
	Barrel	26	,, ,,
Salt Pork	Tierce	80	4-lb. pieces.
	Barrel	52	,, ,,

x

WEIGHTS AND MEASURES.

Flour	{ Barrel ....336 lbs. { Half-hogshead ....250 ,,	
Suet	{ Half-hogshead ....168 ,, { Small cask .... 70 ,, { Ditto .... 56 ,,	
Raisins	{ Barrel ....336 ,, { Half-hogshead ....224 ,, { Small cask (Firkin) .....112 ,,	
Pease	{ Barrel .... 5 bushels=40 gallo { Half-hogshead .... 3½ bushels=28 ,,	
Oatmeal	{ Barrel .... 7½ ,, { Half-hogshead .... 5½ ,, { Small cask .... 2¼ ,, { Ditto .... 2 ,,	
Sugar	{ Barrel ....392 lbs. { Half-hogshead ....280 ,, { Small cask ....112 ,,	
Chocolate	{ Half-hogshead ....108 ,, { Small cask .... 55 ,,	
Tea	{ Chest .... 83 ,, { Half-chest .... 36 ,,	
Vinegar	{ Puncheon .... 72 gallons. { Hogshead .... 54 ,, { Barrel .... 36 ,, { Half-hogshead .... 25 ,, { Kilderkin .... 18 ,, { Small cask .... 12 ,,	
Tobacco	{ Hogshead ....242 lbs. { Barrel ....160 ,, { Half-hogshead ....126 ,,	
Soap	{ Half-hogshead ....113 ,, { Barrel ....244 ,,	

India (Bengal.)

The Indian weights and measures vary in every district, but English equivalents are coming into use in large towns.

Guz (average) = 36 English Inches.
Bám or Danda (Fathom) = 2 English Yards.
Coss = 2·46 miles.

WEIGHTS.

Tola, the unit of postage = 180 Grains Troy. 5 Tolas = 1 Chitták. 16 Chittáks = 1 Seer = $2\frac{2}{35}$lbs. av. 40 Seers = 1 Imperial or Indian Maund = $82\frac{2}{7}$lbs. av.

For weighing gold and silver the Tola is divided into 12 Machas. Pure gold or silver is said to be 12 Machas fine.

LIQUID MEASURE.

Maund = 8 Palli = 9·81 British Imperial Gallons.

Madras.

The English foot and yard are now used by almost all native workmen. The Native Guz = 33 English Inches. The Baum or Fathom = $6\frac{1}{2}$ English Feet. A Kádàm = 10 English Miles.

WEIGHTS.

The Viss = 3·09lbs., the Maund 25lbs., and the Candy 500 lbs. English av.

Bombay.

LENGTH.

Guz = 27 English Inches.

The weights given under the head of Bengal are being introduced.

CAPACITY (GRAIN.)

Muda of 25 Parahs is = 14 British Quarters; the Candy of 8 Parahs = 4·48 British Imperial Bushels.

LIQUIDS.

Maund of 50 Seers = 7·71 British Imperial Gallons.

Canada.

The Metric System has been legalized as permissive.
(Same as Great Britian.)

Mauritius.
(Same as Great Britain.)

West Indies (British.)
(Same as Great Britain.)

Gibraltar.
(Weights and Measures chiefly those of Great Britain.)

Cape of Good Hope.

Weights and Measures in use are partly those of Great Britain, and partly the old Dutch Measures.

Rheinland Fuss = 1·03 feet, The Aam of 2 Ankers, of 2 Steeknen, of 8 Stoopen, of 2 Mengeln, of 2 Pintjes, of 4 Maajes = 34·16 British Imperial Gallons.

The old Amsterdam Pfund = 1·32 Troy lbs. or 1·09lbs. av. English. 92 old Amsterdam Pfund = 100lbs. av. English.

France.

The *Metre* is the fundamental unit of the measures of length as well as of all weights and measures. It is the ten millionth part of the distance from the Pole to the Equator.

The French weights and measures are given in full as representative of the Metric System.

Measures of Length.

(Of which the *Metre* is the element.)

			English value. inches.
1000th part of a Metre	= 1 Millimetre	=	·039
100th ditto	= 1 Centimetre	=	·394
10th ditto	= 1 Decimetre	=	3·937

WEIGHTS AND MEASURES.

			English value.	
	1 Metre	=	39·371 inches =	3·281 feet
10 Metres =	1 Decametre	=	32·809 feet =	10·936 yards
100 ,, =	1 Hectometre	=	328·09 feet =	109·363 yards
1000 ,, =	1 Kilometre	=	1093·63 yards =	0·621 miles
10000 ,, =	1 Myriametre	=	10936·33 yards =	6·214 miles

The Kilometre is sometimes termed a Metrical Mile. The mean Nautic Mile = 1853 Metres = 1·1507 Statute Miles. The Marine League = $\frac{1}{20}$ of a Degree = 5560 Metres.

MEASURES OF SURFACE.
(Of which the *Are* is the element.)

English value.

				Square Inches.
100 Sq. Millimetres	=	1 Sq. Centimetre	=	·155
100 Sq. Centimetres	=	1 Sq. Decimetre	=	15·500
				Square Feet.
100 Sq. Décimetres	=	1 Centiare or Sq. Metre	=	10·764
				Square Yards.
100 Sq. Metres	=	1 Are or sq. Decametre	=	119·603
				Acres.
10 Sq. Ares	=	1 Decare	=	0·247
100 Sq. Decametres	=	1 Hectare	=	2·471
100 Sq. Hectares	=	1 Sq. Kilometre	=	247·113
100 Sq. Kilometres	=	1 Sq. Myriametre	=	24711·340

CUBIC, OR SOLID MEASURES.
(Of which the *Stere* or Cubic Metre is the element.)

English value.
Cubic Feet.

10th of a Stere	=	1 Decistere	=	3·532
10 Decisteres	=	1 Stere	=	35·317
10 Steres	=	1 Decastere	=	353·166

MEASURES OF CAPACITY.
(Of which the *Litre* is the element.)

English value.
Minims.

1000th of a cubic Decimetre	=	1 Millilitre	=	16·903

				Fluid Drachms.
10 Millilitres	=	1 Centilitre	=	2·817
				Imperial Pints.
10 Centilitres	=	1 Decilitre	=	·176
10 Decilitres	=	1 Litre	=	1·761
				Imperial Gallons.
10 Litres	=	1 Decalitre	=	2·201
				Imperial Bushels.
10 Decalitres	=	1 Hectolitre	=	2·751
10 Hectolitres	=	1 Kilolitre	=	27·512

TABLE OF WEIGHTS.
(Of which the *Gramme* is the element.)

English value.
Grains (Troy).

1000th of a gramme	=	1 Milligramme	=	·015
10 Milligrammes	=	1 Centigramme	=	·154
10 Centigrammes	=	1 Decigramme	=	1·543
10 Decigrammes	=	1 Gramme	=	15·432
10 Grammes	=	1 Decagramme	=	154·323

lbs. oz. drs. av.

10 Decagrammes	=	1 Hectogramme	=	0 3 8·438
10 Hectogrammes	=	1 Kilogramme	=	2 3 4·383

lbs. av.

10 Kilogrammes	=	1 Myriagramme	=	22·047
10 Myriagrammes	=	1 Quintal Metrique	=	220·466
10 Quintals	=	1 Ton or Millier	=	2204·66

Spain.
The weights and measures are the same as those of France.

Portugal.
(Same as France.)

Belgium.
(Same as France.)

Italy.
(Same as France.)

Brazil.
(Same as France.)

Holland.
(Same as France.)

Elle = Metre. Palm = Decimètre. Duim = Centimetre.
Streep = Millimetre. Roede = Decametre. Mijle = Kilometre.
Bunder = Hectare. Kan and Kop = Litre. Maajte = Decilitre.
Vingerhoed = Centilitre. Scheppel = Decalitre. Vat or Ton, and Mud or Zak=Hectolitre. Last=Kilogramme. Ons=Hectogramme.
Lood=Decagramme. Wigtje=Gramme. Korrel=Decigramme.

Greece.
(Same as France.)

Pecheus = Metre. Palame = Decimetre. Daktylus = Centimetre.
Gramme=Millimetre. Stadion=Kilometre. Skoinis=Myriametre.
Stremma = Are. Koilon = Hectolitre. Litra = Litre.
Kotyle = Decilitre. Mystron = Centilitre. Kybos = Millilitre.
Mnă=1½ Killogramme. Drachme=Gramme. Obolos=Decigramme.
Kokkos=Centigramme. Tonos=29·5 English cwt.

Argentine Republic.
(Same as France.)

United States of North America.

In 1866 the Metric System was legalized as permissive.

The measures of *Length, Distance, Surface,* and *Solidity,* are the same as those of Great Britain. The weights also are the same; but articles formerly sold by the cwt. are now generally sold by the Centner or Quintal, of 100 lb.

For Dry Goods, the Gallon= ·969 Imperial Gallon; and the Bushel= ·96 Imperial Bushel. For Liquids, the Gallon= ·833 Imperial Gallon.

Germany (North).
Prussia.

The Metric System became compulsory for the whole German Empire from January 1st, 1872. The old system was as follows:—

MEASURES OF LENGTH.

English value.

12 Linien	=	1 Zoll	=	1·029 Inches
12 Zoll	=	1 Fuss	=	12·356 ,,
2 Fuss	=	1 Elle	=	2·06 Feet
12 Fuss	=	1 Ruthe	=	4·12 Yards
2000 Ruthen	=	1 Meile	=	4·68 Miles

MEASURE OF CAPACITY (DRY GOODS).

16 Metzen	=	1 Scheffel	=	1·51 Bushels

MEASURE OF CAPACITY (LIQUIDS).

2 Ossel	=	1 Quart	=	1 Quart
60 Ossel	=	1 Anker	=	7·56 Gallons
2 Ankers	=	1 Eimer	=	15·12 ,,
2 Eimers	=	1 Ohm	=	30·24 ,,

WEIGHTS.

10 Corn	=	1 Cent	=	·094 Drams Avoir.
10 Cents	=	1 Quentche	=	·941 ,,
10 Quentchen	=	1 Loth	=	9·406 ,,
30 Loth	=	1 Zollpfund	=	1·102 lbs. Avoir.
100 Zollpfund	=	1 Centner	=	110.232 ,,

Russia.
LENGTH.

The English foot is the standard. Sashine=7 feet. Verst=0.66 mile. The Marine Sashine=the English Fathom.

CAPACITY (LIQUIDS).

Tscharkey=·86 of a Gill. Vedro=2·7 Gallons. Anker=8·11 Gallons.

CAPACITY (DRY GOODS).

Pajak = 1·44 Bushels.

WEIGHTS.

Lana = 1¼ oz. av.   Funt (Pound) = 14½ oz. av.
Pud (or Pood) = 1-qr. 8-lbs. av.   Berkovitz = 3-cwt. 0-qrs. 25-lbs.

Austria.

LENGTH.

Zoll = 1·04 Inches.   Fuss = 1·04 Feet.   Elle = 2·07½ Feet.
Klafter = 6·22 Feet.   Meile = 4·71 Miles.

CAPACITY (DRY GOODS).

Metze = 1·69 Bushels.

CAPACITY (LIQUIDS).

Kanne = 1·25 Pints.   Mass = 1·25 Quarts.
Viertel = 3·11 Gallons.   Eimer = 12·46 Gallons.

WEIGHTS.

Quentchen = 2·47 Drachm.   Unze = 1-oz. 3·7-dr.
Pfund = 1·23-lbs. av.   Centner = 123·47-lbs. av.

Sweden.

The Decimal System has been introduced.

LENGTH.

Fot = 11·66 Inches.   Famn = 5·843 Feet.
Mil = 6·62 Miles.

WEIGHTS.

Skälpund = ·938 of a lb. av.   Centner = 83·72-lbs.

Norway.

LENGTH.

The Decimal System has been introduced.
Fod = 12·36 Inches.   Favn = 6·175 Feet.

Denmark.

LENGTH.

Fod = 12·36 Inches.   Favn = 6·175 Feet.   Miill = 4·68 Miles.

WEIGHTS.
Pund = 1·102 lbs. av.   Centner = 110·23 lbs. av.

Turkey.
LENGTH.
Pike or Drâ = 27 Inches.   Berri = 1·04 Miles.
WEIGHTS.
Rottolo=1·24 lbs. av.   Cantar=124·7 lbs. av.   Oke=2·83 lbs. av.

Egypt.
LENGTH.
Pike Draâ = 27 Inches.   Gasab = 8·5 Feet.
WEIGHT.
Cantar of 100 Rottolos = 98·05 lbs. av.

China.
LENGTH.
Chih (Canton) = 14·1 Inches.   Chang = 11·75 Feet.
Li—average length is about ⅓ Mile.
WEIGHTS.
¾ Tael = 1 oz. av.   12 Taels = 1 lb. av.   Catty = 1·33 lbs. av.
FOR WEIGHING GOLD, SILVER, AND MEDICINES.
10 Cash=1 Fau.   10 Fau=1 Tsien.   10 Tsien=1 Tael=579·8-grs. Troy.

Japan.
LENGTH.
Shaiku=the English Foot.   Ken=the English Fathom.
Ri=2·45 Miles.
WEIGHTS.
Tael (ichi nomme)=2·13 drs. av.   Catty (ikkin)=1·33 lbs. av.
Picul (hiak-kin)=100 Catties=133·33 lbs. av.

NOTE.—For a detailed explanation of the Money, Weights, and Measures of all countries, see *The Merchant's Handbook*, by W. A. Browne, L.L.D. Stanford, Charing Cross.

USEFUL RULES IN MENSURATION.

I.—*To find the circumference of a circle, the diameter being given, and vice versa.*
    (1.) Multiply the diameter by 3·1416, the result will be the circumference.
    (2.) Divide the circumference by 3·1416, and the result will be the diameter.

II.—*To find the area of a circle, the diameter being given, and vice versa.*
    (1.) Multiply the square of the diameter by ·7854. Result—area.
    (2.) Divide the area by ·7854, and take the square root of the quotient, the result will be the diameter.

III.—*To find the surface of a cylinder, the diameter and length being given.*
    Multiply the diameter by 3·1416, to get circumference of the section and again multiply this product by the length.

IV.—*To find the volume of a cylinder, the diameter and length being given.*
    Multiply the square of the diameter by ·7854, to get the area of the section, and then multiply this area by the length.

V.—*To find the surface of a sphere.*
    Multiply the diameter of the sphere by its circumference, and its product will be the surface.

VI.—*To find the volume of a sphere.*
    Multiply the cube of the diameter by ·5236, and the product will be the volume required.

VII. *To find the volume of a cone.*
    Proceed as in IV, and divide the result by three.

VIII. To find the area of an ellipse.

Multiply the greatest and least diameters together, and that product by ·7854, the result will be the area.

IX. To find the circumference of an ellipse.

Multiply half the sum of the two diameters by 3·1416, and the result will give the circumference sufficiently accurate for practice.

X. Area of a triangle = ½ base, multiplied by the perpendicular.

XI. Square inches divided by ·7854 = circular inches. Circular inches multiplied by ·7854 = square inches.

To find the Weight of a known substance of given dimensions.

Having found the volume, if the result be in cubic feet, multiply by the weight of a cubic foot; but if the volume be in cubic inches, divide the product so obtained by 1728.

Area of a square = square of one side.
Side of a square = square root of area.

Area of a parallelogram = length × perpendicular breadth.

Square inches	×	·007	=	square feet.
Cubic inches	×	·00058	=	cubic feet.
,, ,,	×	·003607	=	Imperial gallon (nearly.)
,, feet	×	6·232	=	,, ,,
Avoirdupois lbs.	×	·00893	=	cwts.
,, ,,	×	·00045	=	tons.

Scale of Measurements adopted for ascertaining amount of Coal at Depôts, and Capacity of Stowage Room.

1 ton of Welsh coal = 40 cubic feet.
1 ton of North Country coal = 43 ,, ,,

TO CALCULATE THE FLOATING POWER OF SPARS.

The weight they will sustain is the difference between their own weight and that of the water they displace.

To ascertain the weight, multiply the square of the mean diameter by ·7854 to find the area; multiply the area by the length to find the cubic contents; and the product by the weight of a cubic foot of the material.

EXAMPLE.

Topmast—length = 64 feet; mean diameter = 21 inches.

$$64 \text{ feet} = 768 \text{ inches; and } \frac{21^2 \times \cdot 7854 \times 768}{1728} = 154 \text{ cubic feet}$$

154 × 64·18 (the weight in lbs. of 1 cubic foot of salt water) = 9883·7 lbs.

154 × 36·3 (the weight in lbs. of 1 cubic foot of Norway spars) = 5590·2 lbs.

Floating power of spar   4293·5 lbs. = 38 cwt.

To find the content of square or four-sided timber, multiply the mean breadth by the mean thickness, and the product by the length.

To find the solid content of a tree, multiply the square of ¼ of the mean girth by twice the length.

In making use of tanks, casks, &c., for floating purposes, the internal capacity is immediately obtained by multiplying the gallons it holds by 10, and subtracting the weight of the tank or cask. This gives the floating power.

In constructing a raft, it should be borne in mind that all the weight of human beings is to be placed *on* it, and that a great quantity of provisions and water may be safely carried *under* it.

For instance: a cask of beef slung beneath would be 116 lbs; above, 300 lbs.

IRON TANKS AND CASKS.

Capacity in Gallons.	Weight when Empty.			Measurement in Cubic Feet.		Capacity in Gallons.	Weight when Empty.			Measurement in Cubic Feet.	
	cwt.	qrs.	lbs.	ft.	in.		cwt.	qrs.	lbs.	ft.	in.
600	10	1	14	98	3	375	6	2	25	61	8
500	8	2	16	82	0	264	5	2	2	43	9
400	6	3	25	65	8	110 CASK.	3	0	6	18	9
200	4	2	25	33	2	27	0	2	19		
193	5	1	20	32	4	13	0	1	12		
100	2	3	20	16	6	7	0	1	8		

To calculate the capacity of a cask, multiply half the sum of the areas of the two interior circles, viz: at the head and bung, by the interior length, for the contents in cubic inches; which, divided by 277·27 (the number of cubic inches in a gallon), reduces the result to that measure.

In calculating the floating powers of casks, the weight of the casks themselves need not be taken into consideration, as when they are submerged, it is inappreciable.

SPECIFIC GRAVITIES.

Barometer 30-in., Fahrenheit's thermometer 60°, distilled water is represented by 1000.

Material.	Specific Gravity.	Weight of One Cubic Foot in lbs.
Ash	690 to 845	43·12 to 52·81
Beech	854 to 690	53·37 to 43·12
Birch	792	49·50
Box	960	60·00
Brass	8399	524·94
Cedar (fresh)	909	56·81
,, (seasoned)	753	47·06
Coal	1232 to 1657	77·0 to 103·56
Copper	8607	537·93
Elm (seasoned)	588	36·75
Fir (New England)	553	34·56
,, (Riga)	753	47·06
Gold	19258	1203·62
Iron	7700	481·25
Lead	11446	715·38
Lignum Vitæ	1220	76·25
Mahogany	800	50·00
Oak (English)	934	58·37
,, (African)	972	60·75
Pine	660	41·25
,,	461	28·81
Silver	10312	644·50
Steel	7780	486·25
Teak	657	41·06
Tin	7291	455·69
Water (sea)	1027	64·18
,, (distilled)	1000	62.50
Walnut	671	41·94
Zinc	7028	439·25

To find the specific gravity of a substance — W = weight of body in air. X = weight in water. G = specific gravity. $G = \dfrac{W}{W-X}$

If the substance be lighter than water, sink it by means of a heavier substance, and deduct additional weight.

TIMBER.

Classification of Timber at Lloyd's.

FIRST CLASS.

English Oak.   African Oak.

The Live Oak of America.

The Morra and Greenheart of British Guiana.

The Teak and Saul of India.

The Iron Bark of Australia.

Timber is bought and sold by solid measure according to the number of cubic feet.

In measuring standing timber, the length is taken as high as the tree will measure 24 inches in circumference; less than that is not considered timber.

At half this height the measurement for mean girth is taken. One-quarter this girth is assumed to be the side of the equivalent square area.

An allowance is deducted for bark; in oak, $\frac{1}{10}$ to $\frac{1}{12}$ of circumference; less for other woods.

RULE FOR FINDING THE TONNAGE OF VESSELS.

(New Measurement.)

The length of the upper deck is divided into six parts, a, a'' and the depth taken at the foremost, aftermost, and midship points of division. These three perpendicular measurements are each intersected at four intermediate points, $b\ e''$, which divides them into five equal parts.

Sec. IX. FINDING TONNAGE OF VESSELS.

The length for tonnage, fg, is the length of the ship from after side of stem to fore side of sternpost, at a height half-way up the midship line of section a'.

The depths to be employed are—twice the depth of the midship line of section at a' added to once the depth of the foremost and aftermost sections, at a and a''.

The breadths used are—at the foremost section a, the (internal) breadth of the ship at b, added to the breadth at e. In the midship section a', three times the breadth at c' are added to once the breadth at e'. In the after section a'', once the breadth b'' is added to twice the breadth at e''.

Then, the product of the sum of the depths into the sum of the breadths, into the length for tonnage, divided by 3500, gives the number of tons for register.

If the vessel has a poop or half-deck, or a break in the upper deck, measure the inside mean length, breadth, and height of such part thereof as may be included within the bulkhead; multiply these three measurements together, and divide the product by 92·4; the quotient will be the number of tons to be added to the result as above ascertained.

For open vessels.—The depths are to be taken from the upper edge of the upper strake.

For steam vessels.—The tonnage due to the engine room is deducted from the total tonnage computed by the above rule. To determine this, measure the inside of the engine-room from the foremost to the aftermost bulkhead; then multiply this length by the amidship depth of the vessel, and the product by the inside amidship breadth at ·4 of the depth from the deck, and divide the final product by 92·4.

TONNAGE OF VESSELS.

(Old Builder's Measurement).

L = Length of keel between perpendiculars in feet.
B = Breadth of vessel in feet.

$$\text{Tonnage} = \frac{L \times B \times \tfrac{1}{2} B}{94}$$

The fore perpendicular is taken at the fore part of the stem, at the height of the upper deck.

The aft perpendicular is taken at the back of the stern posts, at the height of the upper deck.

The breadth is taken as the extreme breadth at the height of the wales, subtracting the difference between the thickness of the wales and the bottom plank.

To Find the Number of Square Yards in a Square Sail.

Head × Depth × Depth × ½ difference of Head and Foot.

EXAMPLE.

Head = 43¼ cloths = 29 yards.
Foot = 46 cloths   = 30¼ yards.
Depth = 12⅔ yards.

$29 \times 12\cdot 66 = 367\cdot 14$
$12\cdot 66 \times \cdot 75 = 9\cdot 50$
$\overline{376\cdot 64}$ square yards.

Sec. IX. TO FIND THE NUMBER OF SQ. YDS. IN SAILS.

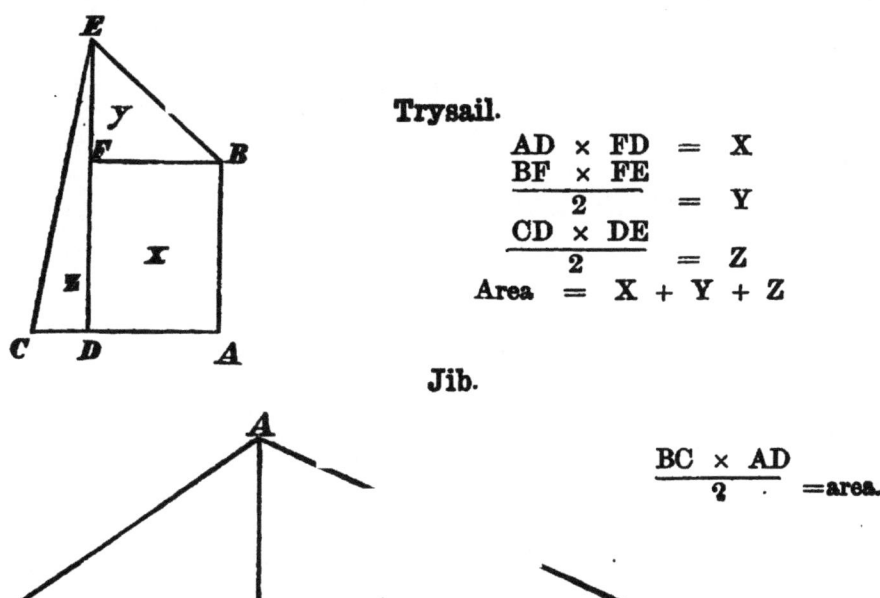

Trysail.

$$\begin{aligned} AD \times FD &= X \\ \frac{BF \times FE}{2} &= Y \\ \frac{CD \times DE}{2} &= Z \\ \text{Area} &= X + Y + Z \end{aligned}$$

Jib.

$$\frac{BC \times AD}{2} = \text{area}.$$

Section 10.

MISCELLANEOUS.

PARTICULARS OF DOCKS ABROAD.

Name of Port.	Dock.	Length over all.	Breadth of Entrance.	Depth over sill at high water, ordinary springs	Rise of Tide. Springs.	Rise of Tide. Neaps.	Remarks.
		feet	feet	feet	feet	feet	
Adelaide (Australia)	Patent Slip No. 1.	..	..	..	8	5	For vessels of 600 tons
	,, No. 2.	..	..	..	..	..	,, 1000 tons
Alexandria	Government.	235	58½	19	..	..	
Algiers	No. 1.	426	105	32	4	..	{ Repairs can be executed with facility.
,,	No. 2.	278	65	25	..	..	
Amoy	Amoy Dock	286	60	18	18½	14½	{ Amoy dock has 18 feet on blocks, from Aug. to Dec. In April the tide may not allow more than 16 feet
,,	Bellamy Dock	180	40	14	..	..	
Barcelona	Gridiron Slip	..	..	..	..	..	Adapted for small vessels
Bangkok	Dock	300	100	15	9 to 10	..	
Batavia	See Onrust.						
Bermuda	Iron Floating Dock	333	83'9	..	..	..	{ These dimensions are for the *inside*
Bombay	Lower }	256	51'9	16	12 to 17	..	{ These docks can be used as one.
,,	Middle }	183	..	..		..	
,,	Upper	200	47'7	..	..	..	
,,	Lower Duncan }	246	63½	16½	..	..	{ These docks can be used as one.
,,	Upper Duncan }	286	..	..	..	..	
,,	Gridiron.	..	..	..	..	..	For small vessels.
,,	Hydraulic Lift.	350	88	30½	..	..	{ Constructed to lift flat bottomed vessels, like the Indian troop ships of 8000 tons.
,,	Mazagon Old Dock	177	44	16 to 18	..	..	

Name of Port.	Dock.	Length over all	Breadth of Entrance.	Depth over sill at high water, ordinary springs.	Rise of Tide. Springs.	Rise of Tide. Neaps.	Remarks.
		feet	feet	feet	feet	feet	
Bombay	Mazagon Old Dock	234	59½	..	..	..	The two docks at Mazagon can be used as one.
,,	,, small Dock	150	34	9 to 11	..	..	
,,	,, new Dock	395	65·4	17½ to 19½	..	..	
Cadiz	No. 1.	344	..	24½	9½	..	In Carracas dockyard.
,,	Two	..	..	..	..	..	Fit for frigates.
,,	Government Slip	..	..	..	..	..	For small vessels.
,,	Patent Slip	..	..	..	..	..	Fit for vessels of 600 tons. Messrs. Retoriks.
Calcutta	Upper Union	350	72	19	12½	..	The highest tide, 18 feet, occurs in springs of August and September; the least water is in January and February.
,,	Lower Union	368	56	20½	..	..	
,,	East India	200	42	17	..	..	
,,	Calcutta Company	365	50	20½	..	..	
,,	Six other Private Docks	..	..	..	..	..	Capable of docking 1000-ton ships
,,	Kiddapore, Old Dock	226½	77	18·8	..	..	
,,	Steam Dock	211	72½	14·8	..	..	
,,	New Dock	207	83	16·8	..	..	
Callao	Floating Dock	300	100	..	..	..	Capable of docking a vessel of 2000 tons, drawing 21 feet.
Cape of Good Hope	Table Bay Dock	380	65	..	..	5	
,,	Simon's Bay	..	..	..	..	..	Patent slip fit for vessels of 900 tons.
Carthagena	Floating Dock	325	105	..	..	..	Can receive a vessel drawing 27 feet.
Constantinople	Two Docks. Nos. 1 and 2	220	60	25	..	..	
,,	One Dock. No. 3.	400	70	..	..	..	
Demerara	Sproston Dock	230	45	11	9	6	

Sec. X PARTICULARS OF DOCKS ABROAD.

Name of Port.	Dock	Length over all.	Breadth of Entrance.	Depth over sill at high water, ordinary springs.	Rise of Tide Springs.	Rise of Tide Neaps.	Remarks.
		feet	feet	feet	feet	feet	
Fu-chau	Dock.	300	60	18	17	14½	
Genoa	Government	285	75	19¼	..	..	Capable of receiving a vessel of 1200 tons
,,	Slip	620	60	23	..	..	
Havana	Floating Dock	302	85	25	..	..	Will take a ship of 3000 tons, and 20ft. draught
Hong-Kong	Aberdeen Dock	335	60	18	4¾	..	
,,	Hope Dock	410	84	24½	..	..	
,,	Union Dock, Kowloon	300	80	21	..	..	
,,	Slip at East Point	..	..	..	..	..	Capable of taking vessels of 300 tons
Hobart Twn.	Floating Dock	250	30	12	4½	3½	
,,	Four Slips	..	..	..	..	..	One capable of taking vessels of 1000 tons One ditto - 500 tons Two ditto - 300 tons
Leghorn	Dock	344	69	..	..	..	
Malta	Government outer	281	82	22	..	..	Can be used as one
,,	,, inner	253	90	25	..	..	
,,	Somerset Dock	468	80	33'6	..	..	
,,	Hydraulic Lift	326	62½	19	..	..	Constructed to lift a vessel of 3000 tons
Marseilles	No. 1 Dock	433	83'6	23	..	..	Repairing basin, large enough for steam-ships of 4000 tons to swing in.
,,	No. 2 ,,	346	83'6	19½	..	..	
,,	No. 3 ,,	280	62'6	19½	..	..	
,,	No. 4 ,,	280	62'6	19½	..	..	
Martinique	Government Dock	420	86	28	4½	..	Very good dock
Mauritius	Stevenson Dock	378	60	18½	3	2½	There are 4 docks at Mauritius
Melbourne	Williamstown Alfred Dock	421	80	27	2¾	..	
,,	Government Patent Slip	720	75	18	..	..	Will take a ship of 2000 tons
,,	Wright's Slip	600	..	..	..	..	Ditto 500 ,,
,,	Hobson's Bay Dock	153	32	14	3	2½	Ditto 600 ,,
,,	Duke's Slip	300	30	9	..	..	Ditto 50 ,,
Monte Video	Dock	..	..	..	..	..	Dry dock for small vessels
,,	Colonia patent slip	300	45	..	..	..	Will take a ship of 450 tons, and 9ft. draught

Name of Port.	Dock.	Length over all	Breadth of Entrance.	Depth over sill at high water, ordinary springs.	Rise of Tide. Springs	Rise of Tide. Neaps	Remarks.
		feet	feet	feet	feet	feet	
Nagasaki	Dock in construction	..	..	..	9	7½	When finished will take largest-sized vessels.
,,	Patent Slip	..	..	..	..	..	Will take a vessel of 1500 tons.
Naples	Government Dock	239	63	22	..	..	Also a temporary slip.
,,	Graving Dock	300	..	22	..	..	
Nicholson Port N. Z.	Patent Slip	..	..	..	..	..	Can take vessels of 1000 to 1250 tons.
Onrust (Batavia)	Wooden Floating Dock	164	45	20	2	..	Will take a vessel 144 ft. long, drawing 13½ ft.
,,	Iron Floating Dock	..	..	..	..	..	
Otago, N.Z.	Dock	330	50	19½	21	17½	A ship of 1500 tons, and drawing 21½-ft., has been docked.
Pola	Balance Dock	468	83	28	3	..	
,,	Dock	452	80	27	..	..	
Rangoon	Dalla Dock	..	52	13	21	14	
,,	Floating Dock	300	50	..	..	..	Vessels drawing 15-ft. can be taken at any time of tide.
,,	Gridiron	210	.	..	..	..	
Rio de Janeiro	Cobras Island, No. 1.	301	70	25	4	3	A slip capable of taking a vessel of 2000 tons, and 420-ft. long.
,,	,, No 2.	400	74	25	..	..	
,,	On Coal Island	380	45	18	..	..	
Saigon	Gunboat Dock	..	..	..	12½	..	No particulars.
,,	Floating Dock	..	..	..	..	..	
San Francisco	Hunter's Point Dock	450	93	22	4	..	
,,	Floating Dock	..	..	..	..	..	Will take a vessel of 1500 tons.
,,	Government Floating Dock, Mare Island	..	..	..	..	..	Sectional dock, with a trifling power of 4000 tons.
Savannah, U.S.	Dry Dock	345	..	18	..	..	
Shanghai	Dare Dock	360	72	20	10	7	Both old and new docks are fitted with steam pumps.
,,	Two Mud Docks	300 about	..	..	..	..	
,,	Old or Point Dock	432	62	18	..	..	
,,	New or Lower Dock	390	48	14	..	..	
,,	Pootung Dock	380	75	21	..	..	

Sec. X.   PARTICULARS OF DOCKS ABROAD.   347

Name of Port.	Dock.	Length over all	Breadth of Entrance.	Depth over all at high water, ordinary springs	Rise of Tide. Springs.	Rise of Tide. Neaps.	Remarks.
		feet	feet	feet	feet	feet	
Singapore	New Harbour	420	43	15	10	7½	Can admit two vessel at a time
,,	Tanjong Pagar	450	65	20	..	..	
Sourabaya	Iron Floating Dock	202½	47	17	6 to 8	..	
,,	Wood ditto	144	45	17	..	..	Vessels 126-ft. long, 26-ft. beam, and draught 10¾-ft. can enter.
Swinemund	Iron Floating Dock	..	..	..	..	..	For vessels of 20-feet draught not iron plated
Spezzia	No. 1 Dock	393	85	27¾	..	..	No information received as to the completion of these docks.
,,	No. 2 ,,	328	82	26	..	..	
,,	No. 3 ,,	328	82	26	..	..	
St. Thomas' Island West Indies	Patent Slip	..	..	..	..	..	Can take a vessel of 900 tons
Suez	Dock	416	78	29½	6	4	Two miles S.E. of town
Sydney	Fitzroy Dock. Cockatoo Island	350	58	20½	4¾	4	
,,	Morto—Waterview Bay	345	75	19	..	..	
Trieste	Dock	437	74	23·8	3	..	
,,	Patent Slip	..	..	..	..	..	Has capacity for two vessel of 1100 tons each
Valparaiso	Floating Dock	..	..	..	..	..	
Whampoa	Cooper's Dock	450	60	18½	7 to 8	..	
,,	Lamont	320	60	18½	..	..	
,,	Granite	550	80	..	..	..	
,,	Wood Dock, No. 1	240	..	..	..	..	Can take 14-ft. at Springs ,, 11-ft. ,,
,,	No. 2	160	..	..	..	..	
,,	Mud Dock	164	..	12½	..	..	
,,	Ditto	120	..	10½	..	..	
Yokohama	Yokosuka Dock	407	72	21	6½	4½	Very convenient dock

Dimensions of a few Ships classed according to length.

H.P. nominal.	Ship's Name.	Tonnage, Old Measurement	Length.	Breadth	Draught of Water.
			Ft. In.	Ft. In.	Ft. In.
2600	*Great Eastern* -	22,500	680·0	83·0	28· 0
1350	*Northumberland*	6621	400·0	59·3½	26· 2
1250	*Warrior* - -	6039	380·0	58·0	26· 6
1000	*Inconstant* - -	4066	337·0	50·3½	24· 6
600	*Royal Yacht* -	2345	300·0	40·3	16· 0
800	*Ariadne* - -	3214	280·0	50·0	22· 4
600	*Defence* - -	3668	280·0	54·0	25· 0
800	*Royal Oak* - -	4045	273·0	58·5	26· 7
120	*Flirt*, gun vessel	464	155·0	25·0	8·11

To lighten the *Warrior* one inch it would be necessary to remove 42 tons of weight, or about 500 tons to lighten the ship one foot.

The quickest way, probably, of doing this would be to take out the chain cables, say 128 tons, anchors 40 tons, water 90 tons, shot 113 tons, powder 23 tons, leaving more than 100 tons to be made up by the removal of about half the guns.

To Find the Weight that will give a Ship one inch Immersion (approximately).

Cut off the two right hand figures of the *tonnage* of the Ship, and the remainder will give the weight in Tons that will immerse the Ship one inch when at the load line.

Thus *Minotaur* is 6621 Tonnage; therefore, the weight of 66 tons would immerse her one inch.

FOREIGN SEA TERMS AND PHRASES.

Sails.

ENGLISH.	FRENCH.	ITALIAN.	SPANISH.
Sail	Voile	Vela	Vela
Mainsail	Grande voile	Vela di Maestra	Vela mayor
Main Topsail	Grand hunier	Gabbia	Gavia
Main top-gallant sail	Grand perroquet	Gran velaccio	Juanete mayor
Main royal	Grand cacatois	Contro velaccio	Sobre mayor
Foresail	Misaine	Trinchetto	Sobre trinquete
Fore-top sail	Petit hunier	Parrochetto	Sobre velacho
Fore top-gallant sail	Petit perroquet	Piccolo velaccio o velaccino	Juanete de proa
Fore royal	Petit cacatois	Contro velaccino	Sobre de proa
Mizen	Artimon	Mezzana	Cangreja mesana
Mizen top-sail	Hunier d'artimon	Contra mezzana	Sobre mesana
Mizen top-gallant sail	Perruche	Belvedere	Perico
Mizen royal	Cacatois d'artimon	Contro Belvedere	Sobre perico
Fore topmast staysail	Voile d'étaie du petit hunier	Trinchettina	Vela de estay de velacho
Main-stay sail	Grand voile d'etaie	Carbonari	Vela de estay mayor
Gaff top sail	Voile à corne	Vela di pico	Cangrejo
Jib	Foc	Fiocco	Foque
Flying jib	Clinfoc	Controfiocco	Petifoque
Trysail	Brigantine	Brigantina	Vela de estay
Spanker	Voile à gui	Randa	Cangreja

Foreign Sea Terms and Phrases.—*(continued.)*

ENGLISH.	FRENCH.	ITALIAN.	SPANISH.
Lower studding sail	Bonnette basse	Scopa mare	Rastrera
Topmast studding sail	Bonnette de hunier	Coltellaccio di gabbia	Alas de gavia
Top-gallant studding sail	Bonnette de perroquet	Coltellaccio di velaccio.	Alas de juanete

Masts, Yards, &c.

ENGLISH.	FRENCH.	ITALIAN.	SPANISH.
Mast	Mât	Albero	Palo
Yard	Vergue	Pennone	Vergo
Boom	Baume	Boma	Botavara
Mainmast	Grand mât	Albero d'Maestra	Palo mayor
Foremast	Mât de misaine	Albero di trinchetto	Palo de trinquete
Mizenmast	Mât d'artimon	Albero di mezzana	Palo de mesana
Bowsprit	Beaupré	Bompresso	Bauprés
Topmast	Mât de hunier	Albero di gabbia	Mastelero
Top-gallant mast	Mât de perroquet	Albero di velaccio	Mastelero de juanete
Lower yard	Vergue basse	Pennone maggisre	Verga mayor; Verga de trinquete
Topsail yard	Vergue de hunier	Pennone di gabbia	Verga de gabia
Top-gallant yard	Vergue de perroquet	Pennona di velaccio	Verga de juanete
Royal yard	Vergue de cacatois	Pennone di contro	Verga de sobre juanete
Top	Hune	Coffa	Cofa

Foreign Sea Terms and Phrases.—*(continued.)*

ENGLISH.	FRENCH.	ITALIAN.	SPANISH.
Cross-tree	Barres traversières	Crocette	Crucetas
Cap	Chouquet	Testa di moro	Tamborete
Truck	Pomme du Mât	Galetta	Perilla ; Laletta ; Bola de la tope
Jib-boom	Baton de foc	Asta di fiocco	Botalon de foque
Studding-sail boom	Bout de bonnette	Asta di coltellaccio	Botalon de ala

Miscellaneous.

Ship	Vaisseau, bâtiment	Nave, baptimento	Buque
Deck	Pont	Coperta	Cubierta
Hold	Cale	Stiva	Bodega
Cabin	Chambre	Camera	Cámara
Windlass	Vireveau	Mulinello	Molinete
Capstan	Cabestan	Argano	Cabrestante
Bitts	Bittes	Bitte	Bitas
Compass	Boussole, Compas.	Bussola	Aguja, compás.
Helm	Barre	Timone	Timon
Anchor	Ancre	Ancora	Ancla
Cable	Cable	Gomena	Cable ; Cadena
Forward	Par avant	A prora	A proa
Aft	Par derrière	A poppa	A popa
Fore and aft	Allonges	Poppa a prora	Abarloado ; Prolongado
Athwart	A'travers	Al traverso	Por el través
Hawser	Grelin	Gherlino	Guindaleza
Starboard	Tribord	Dritta	Estribor
Larboard (port)	Babord	Sinistra	Babor

Foreign Sea Terms and Phrases.—*(continued.)*

ENGLISH.	FRENCH.	ITALIAN.	SPANISH.
Below	En bas	Abasso	Abajo
Aloft	En haut	Ariva	Arriba
Avast	Bosse	Basta	Forte !
Shrouds	Haubans	Sartie	Obenques
Stay	Etaie	Straglio	Estay
Backstays	Galhaubans	Patarazzi	Brandales ; Volantes
Bobstay	Sous barbe	Briglia	Barbiquejo
Lanyards	Rides	Ride	Acolladores
Tackle	Palan	Paranco	Aparejo
Braces	Bras	Bracci	Brazas
Tack	Amure	Mura	Amura
Sheet	Ecoute	Scotta	Escota ; Escotin
Haulyards	Drisse	Drizza	Driza
Downhaul	Carguebas	Caricabasso	Cargadera
Clewlines	Cargues points	Controscotte	Chafaldete
Buntlines	Cargues fonds	Mezzi	Brioles
Brails	Cargues de voile à gin	Serrapennoni	Candalizas
Reeftackle	Palanquin de ris	Paranchini di terzarnolo	Amante de rizos
Bowline	Bouline	Boline	Bolina
Gasket	Raban de ferlage	Matafioni	Tomador
Crew	Equipage	Equipaggio	Tripulacion
Boatswain	Maitre d'equipage	Nostromo	Contramaestre
Sailmaker	Voilier	Veliere	Velero
Carpenter	Charpentier	Falegname	Carpintero
Steward	Depensier	Dispensiere	Despensero
Cook	Cuisinier	Cuoco	Cocinero de equipage

Foreign Sea Terms and Phrases.—(*continued.*)

ENGLISH.	FRENCH.	ITALIAN.	SPANISH.
Seaman	Matelot	Marinaro	Marinero
Boy	Mousse	Mozzo	Jóven
Belay	Amarrez	Date volta	Amarra !
Let go	Dèmarrez	Larga	Larga !
Hoist away	Hissez	Issa	Iza !
Lower away	Amenez	Amaina	Arria !
Haul	Halez	Ala	Hala !
Handsomely	En Garante	Poco a poco	Poco á poco; Sobre vuelta
Hold on	Tenez	Aggnanta	Aquanta !
Heave away	Virez	Virate	Vira !
Slack	Filez	Filate	Arria !
Bear a hand	Vite	Presto	Pronto ! Volado ! Echa una mano
Boat	Bâteau	Batello	Bote
Oar	Rame	Remo	Remo
Bailer	Escop	Sassola	Vertedor
Clear	Clair	Chiaro	Clara
Foul	Embrouille	Impegnato	Enredado
Windward	Au vent	Sopravento	Barlovento
Leeward	Sous le vent	Sottovento	Sotavento
Catch hold	Saissez	Arresta	Agarra !
Look out	A'vigie	Guardate	Vigia,—Tope,—Serviola
All right	Tout droit	Va bene	Beuno ; No hay novedad ; Esta bien, &c.

Phrases.

Heave away the windlass	Virez au cable	Vira l'argano	Vira cabrestante
Sheet home the topsails	Bordez les ecoutes des huniers	Alate le scotte delle gabbie	Caza á beser los escoti nes de las gavias

z

Foreign Sea Terms and Phrases.—(continued.)

ENGLISH.	FRENCH.	ITALIAN.	SPANISH.
Hoist the topsails	Hissez les huniers	Issate le gabbie	Iza gavias
Hoist the topgallant sails	Hissez les perroquets	Issate i velacci	Iza juanetes
The anchor is aweigh	L'ancre est deplantéc	L'ancora ha lasciato	El ancla ha largado
Hoist the jib	Hissez le foc	Issate il fioco	Iza foque
Fill the head sails	Brassez pleine les voiles de l'avant	Fate portar le vele di proa	Braza en viento las velas de proa
Set the mainsail	Mettez la grande voile	Spiegate la vela di maestra	Amura mayor
Ready about	Prét pour virer de bord	Pronto a virare	Alista á virar
Helm's a-lee	Adieu va	Sotto la barra	Orzo á la banda
Raise tacks and sheets	Levez les lofs et debordez les ecoutes	Smura	Levanta punos sobre bolinas
Mainsail haul	Brassez d'arriere	Tira-molla a poppa	Cambia al medio
Let go and haul	Brassez d'avant	Tira-molla a prora	Descarga á proa
Brace sharp up	Halez ferme les bras de dessous le vent	Bracciate bene a sottovento	Braza bien á cenir
Square the yards	Brassez quarre les vergues	Bracciate in croce	Braza en cruz
Hard up the helm	La barre au vent	Barra al vento	Andar todo,—Todo de arribada
Steady the helm	Appuyez la barre	Barra in mezzo	Cana á la via
Up mainsail	Carguez la grande voile	Imbrogliate la maestra	Carga mayor

Foreign Sea Terms and Phrases.—*(continued.)*

ENGLISH.	FRENCH.	ITALIAN.	SPANISH.
Round in the weather braces	Tenez les bras du vent	Alate i bracci di sopravento	Afirma á barlovento
Haul down the jib	Amenez le foc	Alabasso il fioco	Arria foque
Up foresail	Carguez la misaine	Imbrogliate il trinchetto	Carga trinquete
Back the main-topsail	Brassez a culer les voiles d'arriere	Bracciate in faccia la gabbia	Poner la gavia por delante; Poner en facha la gavia
Let go the anchor	Mouillez l'ancre	Date fondo all'ancora	Fondo al ancla
Clew up the topsails	Carguez les huniers	Imbrogliate le gabbie	Carga gavias
Brail the spanker up	Cargues l'artimon	Imbrogliate la randa	Carga cangreja
Pay out the cable	Filez le cable	Filate la gomena	Fila cadena; Fila cable; Arria cadena
Furl the sails	Ferlez les voiles	Serrate le vele	Aferra el velámen
Hoist the ensign	Hissez le pavillon	Alzate la bandiera	Iza la bandera
Coil the ropes up	Rouez la manœuvre	Raccogliere la manovra	Aduja los cablos
Right the helm	Droit la barre	Il timone alla via	Cana á la' via; Cana al medio; A la via, &c.
Ship ahoy!	Bâtiment ho!	Olá bastimento	Ah del buque
What ship is that?	Quel est le nom du navire?	Come si chiama il bastimento?	Como se llama el buque?
Where are you from?	D'ou venz vous?	D'onde venite?	De dónde viene?
Where are you *bound to?*	Ou allez vous?	Dove andate?	Adónde vá?

Foreign Sea Terms and Phrases.—*(continued.)*

ENGLISH.	FRENCH.	ITALIAN.	SPANISH.
How many days out?	Combien de jours en route?	Quanto tempo in viaggio	Cuántos dias de navegacion?
Please to report me	Faites rapport de moi	Annunziatemi	Diga qué noticias trae
Ship's papers	Papiers du bord	Documenti del bastimento	Los papelas del buque
Register	Contrat de construction	Lettere di costruzione	Registro de construccion
Bills of lading	Connaissements	Conoscimentí	Conocimientos
Manifest	Manifest	Manifesto	Manifiesto
Bill of health	Le certificat de santè	Certificato di sauità	Patente de sanidad
Log book	Livre de loc	Quadarmo di chiasnola	Cuaderno de bitácora

CONVERSION TABLES
OF
EQUIVALENT WORDS, LETTERS, AND FIGURES,

Arranged for Telegraphing, for Reduction, and for Comparisons; and for Reporting Meteorological Observations.

Equivalents for French Words Descriptive of Weather.

Beau	Fine	b*
Belle	Smooth, still	—
Brouillard	Fog	f
Brumeux-se	Foggy	g
Ciel	Sky	—
Clair-e	Clear	b

* Beaufort Letters, with additions.

Sec. X. CONVERSION TABLES. 357

Equivalents for French Words.—*(continued.)*

French	English	Symbol
Couvert-e	Overcast	o *
Coup (de vent)	Heavy squall	q q or q̲
Eclair	Lightning	l
Eclaireux-se	Lightning around	ll or l̲
Faible	Light, slight	—
Fort-e	Much, strong	(— or · or repetition)
Grains	Squally	q or q̲
Grand-e	Great, much	(— or · or repeated letter)
Humide-ité	Damp, humidity	—
Intense-ité	Intense-ity	(— or · or a repetition)
Legèr-e-ment	Light-ly	—
Mauvais-e	Bad, threatening	u
Nebeleux-se	Misty, hazy, obscure	m
Neige-ant	Snow-ing	s
Nuage-s-eux-se	Cloud-s-y	c
Orage-ux	Storm-y	q̲w̲w̲
Pluie-s	Rain-s	r̲r̲
Pluvieux-se	Rainy	r
Presque	Almost, slight	—
Rafales	Sudden squalls	q
Serein-e	Serene, settled	b c
Sombre	Gloomy, dark	g or gg
Tempête	Tempest	w̲ w̲ q̲ q̲
Tempestueux-se	Tempestuous	—
Tonnerre	Thunder	t
Tonnant-e	Thundery	t t
Très	Very, excessively	(— or · or a repetition)

* Beaufort Letters, with additions.

Table of French and English Words for the State of the Sea.

État de la Mer.	State of Sea.
Calme	Dead calm
Assez calme	Calm
Très belle	Very smooth
Belle	Smooth
Tranquille	Still
Faible houle	Slight swell
Petite houle	Do.
Un peu houleuse	Some swell
Un peu de mer	Rather rough
Risée	Squally
Agitée	Disturbed irregularly
Houleuse	Considerable swell
Très houleuse	Much swell
Moutonneuse	Crested waves
Creusée	Cross sea
Grosse houle	Great swell
Haute mer	High sea
Trés-grosse mer	Very large and high sea

TERMS APPLICABLE TO THE WINDS ABROAD.

North.—*Tramontana* and *Gli Secchi*, or dry winds, by the Italians.

N.E.—The *Gregale* of the Italians and Maltese; the *Bora* of the Adriatic.

East.—*Solano* and *Levanter* of the Straits; *Levante*; *Bentu de Sole*; and when light, *Chocolatero* by the Italians.

S.E.—*Scirocco*, the hot debilitating wind of South Italy and Africa; *Maledetto, Levante, Molezzo*, and in the Adriatic, when strong, *Furiante*.

South.—*Mezzo Giorno*; from S. to S.W. *Simoom, Shume* or *Siume*, on the African Coast.

S.W.—*Vendavales*; also *Lebeches*, and *Virazones*, by the Spanish. *Libeccio*; when gusty, *Labeschades*; and when very stormy, accompanied by lightning, rain, &c., it is called *Ouragans* by the Italians. *Labbetch*, in Algeria; and *Siffanto* in the Adriatic.

West.—It is called in the Straits of Gibraltar the *Liberator*. *Ponente* (strong) by the Italians.

N.W.—The *Mistral, Mistrasu*, the *Bize*, and *Grippe*, also the *Vent de cers* of France. *Maestro, Maestrale*, of the Italians; and when light it is called *Mamatele* by the Sicilians.

N.N.W.—*Provenzale* by the Italians of Livorno.

A Sea Breeze.—*Imbattu*.

A Land Breeze.—*Vento di Terra* or *Rampinu*.

Land Squalls.—*Raggiature*, by the Italians.

Mountain Storms.—*Burrasche*, South Italy, and *Raffiche* in Corsica.

Golfada, a hard gale. *Bonaceia*, calms between land and sea breezes, in Italy.

The meeting of opposing winds is called *Contrastes* by the Spaniards.

ROPE MAKING.

ROPE.	WHAT MADE FROM.	HOW LAID UP.
Yarns	Hemp	Right-handed
Strands	Yarns	Left or right handed
Hawser-laid rope	3 strands	Opposite way to the strands
Shroud-laid rope	4 strands and a heart	Right-handed
Cable-laid rope	3 hawser-laid ropes	Left-handed
Spunyarn	3 to 9 yarns	Right-handed
Sennit	Yarns	Plaited
Nettle stuff	2 or 3 left-handed yarns	Right-handed
Foxes	Short yarns laid up by hand	Left-handed

Rope is measured by its circumference, and is manufactured in lengths of 113 fathoms, and in different sizes, up to 28 inches; it is made up in coils up to 5-inch; above that it is sent on board in the length.

Running rigging is hawser-laid, right-handed.

A four-stranded rope is about one-fifth weaker than a three-stranded one.

Wire rope usually consists of 6 strands round a hempen core; each strand consists of 6 wires round a smaller hempen core, 36 wires in all.

Coir rope is equal in strength to hempen rope of the same size, and is but two-thirds the weight.

*Two men can worm and serve seven fathoms of 3½ inch rope in an hour, or worm, parcel, and serve 3 fathoms of 7-inch in an hour.

Three men can worm, parcel, and serve 2 fathoms of 12-inch in an hour.

---
*Boyd.

One man can make 9 feet of 9 yarn sennit in an hour.

Six men can make an Elliott's eye in a 24-inch cable in one day.

The strength of a yarn may be called in round numbers 100lbs.

Proportionate Strength of Rope of different sizes.

To find what number of parts of a small rope are equal to a large rope:—

> Divide the square of the circumference of the larger rope by the square of the circumference of the smaller, and the result will be the number of parts of the smaller equal to the larger.

Practical rule for ascertaining the strength of rope (hawser laid):—

> Square the circumference, and divide by 3 for the breaking strain, in tons; by 4 for the proof strain; by 6 for the working strain.

To find what weight a rope will lift when rove as a tackle:—

> Multiply the weight the rope is capable of suspending by the number of parts at the moveable block, and subtract one-fourth of this for resistance.

To determine the relative strength of chain and rope:—

> Consider the proportional strength to be 10 to 1, using the diameter of the chain and the circumference of the rope. Half-inch chain may therefore replace 5-inch rope.

Rule for calculating the weight of rope:—

> Three strand, hawser laid, 25 thread yarn, tarred. Multiply the square of the circumference by the length in fathoms, and divide by 4·24 for the weight in lbs.

The divisor for hempen cables is 4·79.

Table of Size of Rope Stropping.

Blocks.	Rope.	Blocks.	Rope.	Blocks.	Rope.	Blocks.	Rope.
Inches.	Inches.	Inches.	Inches.	Inches.	Inches.	Inches.	Inches.
5	1	10	3	15	5	20	7
6	1½	11	3½	16	5	21	8
7	2	12	4	17	6		
8	2¼	13	4½	18	6		
9	2½	14	5	19	7		

Table showing the Sized Chain or Wire Rope which is used as a Substitute for Hempen Rope.

Hemp.	Chain.	Wire.	Hemp.	Chain.	Wire.
Inches.	Inches.	Inches.	Inches.	Inches.	Inches.
3	$\frac{5}{16}$	1½	8	$\frac{7}{8}$	3½
4	$\frac{3}{8}$	1¾	9	1	4
5	½	2	10	1⅛	4½
6	⅝	2½	11	1¼	5
7	¾	3			

Table No. 1.—CHAIN CABLES.

SCALE OF PROOFS SHOWING THE TENSILE STRAIN TO WHICH CHAIN CABLES ARE SUBJECTED BEFORE BEING RECEIVED FOR THE USE OF HER MAJESTY'S NAVAL SERVICE.

Diameter of Iron of Common Links.	Common Links. Mean Length six diameters of the Iron; not to be over more than 1/10 of a diameter.	Common Links. Mean Width 3·6 diameters of the Iron; not be over or under more than 1/10 of a diameter.	Stay Pins weight of each not to exceed	Weight of 100 fathoms of Cable, in eight lengths, including four swivels and eight joining shackles, not to be exceeded by more than 1/20th part.* Cwt.	qrs.	lb.	Weight by which to be proved equal to 630 lbs. per circular 1/8 inch.	Strain to be withstood by three links in each fifteen fathoms, before the fifteen fathoms are proved.
In.	In.	In.	Ozs.	Cwt.	qrs.	lb.	Tons.	Tons.
2¾	16½	9·9	72	363	0	0	136⅛	190·5
2½	15	9·0	54·69	300	0	0	112½	157·5
2⅜	14¼	8·55	47·5	270	3	0	101½	142
2¼	13½	8·1	40	243	0	0	91⅛	127·5
2⅛	12¾	7·65	33·584	216	3	0	81¼	113·75
2	12	7·2	28	192	0	0	72	100·8
1⅞	11¼	6·75	23	168	3	0	63¼	88·5
1¾	10½	6·3	18·76	147	0	0	55⅛	77
1⅝	9¾	5·85	15	126	3	0	47½	66·5
1½	9	5·4	11·81	108	0	0	40½	60·75
1⅜	8¼	4·95	9	90	3	0	34	51
1¼	7½	4·5	6·839	75	0	0	28⅛	49
1⅛	6¾	4·05	4·983	60	3	0	22¾	34·12
1	6	3·6	3·5	48	0	0	18	27
⅞	5¼	3·15	2·344	36	3	0	13¾	20·6
¾	4½	2·7	1·473	27	0	0	10⅛	15
11/16	4⅛	2·475	1·137	22	2	21	8½	12·75
⅝	3¾	2·25	·854	18	3	0	7	10·5
9/16	3⅜	2·025	·622	15	0	21	5⅝	8·25
½	3	1·8	·437	12	0	0	4½	6·75
7/16	2⅝	1·575	·293	9	0	21	3½	5·25

* The tensile strain is supplied to each of the eight lengths separately, and not to the whole length of 100 fathoms at one time; and each length is to be provided with a shackle and shackle bolt, to be tested as part of the chain.

Table No. 2.—ANCHORS.

SCALE OF PROOFS SHOWING THE TENSILE STRAIN TO WHICH ANCHORS ARE SUBJECTED BEFORE BEING RECEIVED FOR THE USE OF HER MAJESTY'S NAVAL SERVICE.

TEST OF ANCHORS IN TONS, PROPORTIONED TO THEIR WEIGHT IN CWTS.

Weight	Test	Weight	Test	Weight	Test	Weight	Test
Cwt.	Tons.	Cwt.	Tons.	Cwt.	Tons.	Cwt.	Tons.
100	67¼	75	56¼	50	42⅜	25	24⅜
99	67⅛	74	55¾	49	41¾	24	23⅝
98	66⅞	73	55¼	48	41⅛	23	23⅛
97	66½	72	54¾	47	40½	22	22⅜
96	65½	71	54¼	46	39⅞	21	21⅝
95	65⅜	70	53¾	45	39¼	20	20¾
94	65	69	53¼	44	38⅝	19	19⅞
93	64½	68	52⅝	43	37⅞	18	19
92	64	67	52⅛	42	37¼	17	18½
91	63⅜	66	51½	41	36⅝	16	17⅝
90	63¼	65	51	40	35¾	15	16⅝
89	62½	64	50½	39	35⅛	14	15⅝
88	62⅛	63	50	38	34½	13	14¾
87	61⅝	62	49½	37	33¾	12	13⅜
86	61⅛	61	48⅞	36	33⅛	11	12⅝
85	61	60	48⅜	35	32⅜	10	12
84	60½	59	47¾	34	31⅝	9	11⅛
83	60	58	47¼	33	30⅞	8	10⅛
82	59½	57	46⅝	32	30⅛	7	9¼
81	59	56	46	31	29⅜	6	8¼
80	58½	55	45⅜	30	28⅝	5	7⅜
79	58⅛	54	44¾	29	27⅞	4	6⅜
78	57⅝	53	44⅛	28	27⅛	3	5⅜
77	57¼	52	43⅝	27	26¼	2	4⅜
76	56¾	51	43	26	25⅜	1	3⅜

NOTE.—The strain is applied on the arm, or on the palm, at a spot which, measured from the extremity of the bill, is one-third of the distance between it and the centre of the crown.

PAINTING SHIP.

White Lead is the principal ingredient in all ordinary colours used in painting, the quality is therefore of the greatest importance; it is most difficult to get it free of adulteration. The cheap kinds are adulterated by "byrates," which renders it of a less compact body, and causes it to be much more easily acted upon by the atmosphere. White lead improves by keeping. In mixing, the oil and turpentine should be thoroughly incorporated with the white lead. If patent dryers are used, the proportion is about half-an-ounce to a pound of colour.

Zinc White is more durable than white lead, it is extremely pure, but possesses little body.

Vegetable Black.—This is the cheapest and best black for all ordinary work. In a dry state it resembles soot, and being free from grit does not require grinding. It should be mixed with boiled oil.

Vermillion—in a state of powder may be tested by placing a dust of it on a piece of clean white paper, and crushing it with the thumb-nail. If pure, it will not change its colour by any amount of rubbing; but if adulterated, it will become a deep chrome yellow, or assume the appearance of red lead, with which article it is mixed in order to cheapen it. This accounts for the unstable quality of the inferior kinds of vermillion. When using vermillion, if necessary to give two coats, *both* should be of the best quality.

Blue.—The most serviceable blue for the painter is French ultramarine. It is a permanent, kindly working colour, and affords a variety of clear tints when mixed with white. It is a brilliant blue, preserves its purity when reduced in tone by the addition of white.

It may be deepened by Prussian blue or indigo, or by a trifling addition of vegetable black.

Green—Like black, should be mixed with boiled oil, or boiled oil and varnish; and *not with* linseed oil and turpentine.

Mixing Colours.

Cream Colour.—Chrome yellow, the best Venetian red, and white lead.

Fawn Colour.—Burnt sienna, ground very fine, mixed with white lead.

Drab.—Raw or burnt umber and white lead, with a little Venetian red.

Purple.—White lead, Prussian blue, and vermillion.

Violet.—White lead, French ultramarine, vermillion, and a small portion of black.

French Gray.—White lead and Prussian blue, tinged with vermillion.

Salmon Colour.—White lead, tinged with the best Ventian red.

Imitation of Gold.—Mix white lead, chrome yellow, and burnt sienna, until the proper shade is obtained.

Proportions for Mixing Black Paint.

	Black ....25 lbs.	
Linseed	Oil, raw .... $\frac{3}{4}$ gallon.	
	Oil, boiled .... $\frac{3}{4}$ gallon.	
	Litharge... ....$1\frac{1}{2}$ lbs., or patent dryers.	

Total weight, $43\frac{1}{2}$ lbs. mixed colour.

1 lb. of black paint will cover 5 square yards.

Proportions for Mixing White for 'Tween Decks.

White lead ....25 lbs.
Patent dryers ....$1\frac{1}{2}$ lbs., or liquid—$\frac{1}{2}$ pint.
Linseed oil, raw .... $\frac{3}{4}$ gallon.
Turpentine .. .. $\frac{1}{4}$,,

Total weight, $38\frac{1}{2}$ lbs. mixed colour.

Sec. X.   MIXING COLOURS.   367

White for out-door work, as above, omitting one-fourth of the turpentine, and substituting oil. 1-lb. of white paint will cover 4 square yards. On new work one-third less is covered.

Stone Colour.—Same proportions as the white, stained with burnt umber, and spruce ochre ground in oil to the tint required.

Mast Colour.—Same proportions as out-door white, stained with spruce yellow ochre, and a little Venetian red to the tint required.

Copper Colour Paint.
(A very good recipe.)

Six parts of Spruce Ochre; One part of Venetian red; and One part of Black.

Bronze Paint.

Chrome green	2	lbs.
Ivory black	1	oz.
Chrome yellow	1	oz.
Good japan	1	gill.

Grind all together, and mix with linseed oil.

Removing Old Paint.

Nothing is so efficacious as heat, applied by a small brazier with a handle.

One part of pearl ash mixed with three parts of quick stone lime, (by slaking the lime in water and then adding pearl ash,) laid over paint work and allowed to stand 14 or 16 hours, will soften it, so that it can be easily scraped off.

Gilding.

Books of gold leaf contain twenty-five leaves. Gilders estimate their work by the number of "hundreds," it will take (meaning one hundred leaves) instead of the number of books.

Gold leaf should fall freely from the book on the leaves being opened, without any particle sticking to the paper.

The simplest way to use gold leaf is as follows:—Procure a clean sheet of silver or tissue paper, of not too great a density, and rub it over lightly on one side only with a piece of white wax; beeswax or wax candle will do. The paper should be placed on something flat, so that the wax is spread evenly.

After waxing a sheet of paper it should be cut into squares a little larger than the leaves of the book of gold, which should be opened, and the waxed side of the tissue paper gently pressed on the gold leaf. On removing the paper the gold leaf will be found attached to it, and is ready for use. All that the gilder has to do, is to cut it into convenient strips, and pressing it on to the sized surface, the gold will readily leave the paper. The work should be finished by gently dabbing it with a pad of cotton wool.

Gilt work exposed to the weather lasts much longer if it receives a coat of clear varnish when finished.

Size.—Oil gold size gives the best results, but requires time— twelve hours or so—before it gets "*tacky*" enough. Japanner's gold size on the other hand, dries very quickly, and is useful when pressed for time.

In estimating amount of gold leaf required for gilding ordinary grooved mouldings round boats, &c.—one leaf covers about nine inches.

USEFUL PRACTICAL RECIPES.

Paint for Tarpaulins.

Add twelve ounces of beeswax to one gallon of linseed oil; boil it two hours; prime the cloth with this mixture, and use it in place of boiled oil for mixing the paint.

To kill Knots.

One pint of vegetable naptha, one teaspoonful of red lead, quarter of a pint of japanner's gold size, seven ounces of orange shellac. Add them together, set in a warm place to dissolve, and let the whole be frequently shaken. Coat the knots with it.

Another.

Mix red or white lead powder in strong glue size, and apply it warm. Or, cover with lime for twenty-four hours, then coat with red lead and glue.

Distemper for between decks may be used instead of, and looks better, than whitewash.

Putty.

Well dried and sifted whiting (112 lbs.), and two gallons linseed oil well mixed, left for three or four days, and then worked up again before using.

To 56 lbs. of whiting add three pounds of glue boiled in half a gallon of water. No more water should be used than is necessary to soak the whiting.

Jelly or parchment size may be used instead of glue. A little alum added has a good effect in hardening.

To make Size.

Parchment chippings make the best. Put them in an iron kettle and fill with water; let it stand twenty-four hours; then boil for five hours, occasionally taking off the scum. Strain through a coarse cloth. If the size is required to be kept any length of time, dissolve three or four ounces of alum in boiling water, and add to every bucketfull. Then boil again till it becomes very strong, and strain a second time.

Boiled Oil for quick Drying purposes.

With two gallons of linseed oil mix two pounds of litharge and one pound of red lead. Keep it on the fire, allowing the heat to increase gently; boil for three hours; then remove it from the fire, and let it stand twenty-four hours; take off the scum which has formed on the top, and it will be ready for use.

Boiled Oil.

To sixteen gallons of oil, when boiling, add one pound of red lead and one pound of powdered litharge. Time for boiling, six hours.

Another.

Four-and-a-half gallons of raw linseed oil, one pound of copperas, two pounds of litharge. Put the litharge and copperas in a cloth bag, and suspend in the middle of the kettle. Boil four hours and a half over a slow fire, then let it stand.

Baked Oil.

Into an iron pot put half-gallon of linseed oil, to which add half-pound of litharge; boil carefully; keep simmering for fourteen or fifteen hours, and when a little cooled will draw into a thread, the oil will be ready for use.

Black Varnish.

To asphaltum and spirits of turpentine, add a small portion of baked oil.

Another.

To baked oil dissolved in spirits of turpentine, add Frankfort black.

Vermillion Varnish.

Baked oil dissolved in spirits of turpentine, with Vermillion added.

Lac Varnish.

Five parts of lac, one of turpentine dissolved in five times its weight of alcohol; keep gently warm until fluid, then strain.

Spar Varnish.—Boiled oil and resin.

Varnish for Painting on Glass.

Take one ounce of clear resin: melt it in an iron vessel; when all is melted, let it cool a little, but not harden; then add turpentine, sufficient to keep it in a liquid state. When cold, use it with colours ground in oil.

White Varnish, for Range Tables, etc.

White resin and turpentine dissolved in a close bottle, placed in *hot* water on a stove. Or—Canada balsam, one ounce; spirits of *turpentine*, two ounces; mix them together. Before varnishing, the *work* should be sized.

French Polish.

Five ounces of naptha; one ounce of shellac; one dram of myrrh; ten grains of isinglass; and six drams of olive oil.

French Polishing.

Pour a little linseed oil into one cup, and some polish into another; take a piece of woollen rag, and having rolled it up into a ball, saturate it with polish, and cover with a piece of linen drawn tightly over it. Apply one drop of oil and one drop of polish to the surface of the pad, and it is ready for use. The work having been thoroughly smoothed, the polishing is commenced with free, circular strokes, applied with very slight pressure, and taking care that every part receives an equal amount of polish. Finish off with a little spirits of wine applied on a clean rubber.

Spirits of Wine.

In buying spirits of wine the most simple test is to pour a small quantity into a cup, set it on fire, and dip a finger into it; if it burns out quickly it is good; but if it is long in burning, and leaves any dampness remaining on the finger, it is mixed with inferior spirits.

Oil Polish.

Dissolve resin in turpentine to about the consistency of treacle, add to that two pints of linseed oil to one of resin and turpentine.

STAINING.

Black Stain.

Quarter pound logwood; six ounces green copperas; eight nut galls; half gallon vinegar; and lay on warm.

Another.

Boil half pound logwood in two quarts of water, add one ounce pearl-ash, half ounce verdegris, and half ounce copperas. Strain it off, and put in half pound rusty steel filings; and lay on warm.

Another.

Boil half pound logwood in two quarts of water, add one pint vinegar, and half pound rusty steel filings; and lay on warm.

Another.

Place in a breaker two pounds of copperas, two pounds nut galls; eight pounds alum pounded up, eight pounds old iron, and twenty five pints vinegar. When wanted for use, boil as much as is required, and lay on hot. The longer it is kept the better it becomes.

Rosewood Stain.

Cover strips of logwood with water, boil it, add a very small piece of pearl-ash (this must be used very cautiously) lay on hot, and then streak with tincture of steel.

To Dye Bright Red.

Take two pounds of real Brazil dust, and four gallons of water, put in veneers until well covered. boil for three hours, and let them cool; then throw in two ounces of alum, one ounce of nitric acid, and keep luke warm until the mixture has struck through.

To Dye Yellow.

Take four pounds of the roots of barberry, reduced to saw dust, and four gallons of water, add quarter pound tumeric. Put in as many veneers as the liquor will cover; boil for three hours, often turning them. When cool, add two ounces of nitric acid, to assist the dye in striking through.

To Colour Light Woods.

Curly veined birch and béech regularly brushed with aquafortis, and dried at the fire, look remarkably like mahogany.

Another method is to smear the surface with a strong solution of per-manganate of potash. About five minutes will generally suffice. When the action is ended, the wood is carefully washed, dried, and afterwards oiled, and polished in the ordinary way.

Black Gun Polish.

Four ounces resin, two ounces lamp black, three ounces beeswax, two ounces shellac, one quart linseed oil. Boil 50 minutes, then add a half-pint of turpentine. For bronzed guns omit the lamp black.

Black Polish for Iron.

Coal tar one pint, lamp black one ounce, heel-bore half ounce, spirits of turpentine quarter pint, beeswax one ounce. The beeswax and heel-bore to be dissolved in the turpentine. Then add the lamp black, and tar; mix, warm it well, and apply hot.

Another.

Beeswax ½ lb.
Heel-bore 4 ozs.   } Dissolved in turpentine.
Lamp black 4 ozs.

LACQUERING.

This is done in two ways, called cold lacquering and hot lacquering. By the former a little lacquer is laid carefully and evenly over the work with a camel hair varnish brush, it is then placed in an oven or hot stove for a minute or two to set the lacquer. By the second method the work is heated first, and the lacquer brushed quickly over it. The great difficulty is to know the exact degree of heat, and this knowledge cannot be attained except by experience.

To prepare Brass for Lacquering.

The surface of the work being cleaned as well as possible, place it in aquafortis and water, leave it there for some hours and then put it into hot sawdust, and shake about until thoroughly dry and clean.

Brass Lacquer.

(1) Seed-Lac, Dragon's blood, annatto, and gamboge, of each four ounces; saffron, one ounce; spirits of wine, ten pints.

(2) Methylated spirits of wine, one gallon; seed-lac bruised, ten ounces; red sanders, half ounce; dissolve and strain.

Iron Lacquer.

Three pounds asphaltum, half-pound shellac, one gallon turpentine.

Marine Glue.

One part India rubber, twelve parts mineral naptha or coal tar, heat gently, mix, and add twenty parts of powdered shellac. Pour out on a slab to cool; when used, to be heated to about 250° Fahr.

Glue to Resist Moisture.

One pound of glue melted in two quarts of skimmed milk. When strong glue is required, add powdered chalk to common glue.

Glue Cement-to Resist Moisture.

One pound glue, one pound black resin, quarter pound red ochre, mixed with the least possible quantity of water.

Cement for Cloth or Leather.

Sixteen pounds of gutta percha cut small, four pounds India rubber, two pounds pitch, one pound shellac, two pounds linseed oil, melted together and well mixed.

Mortar.

One part of lime to three, or three and a half, of sharp river sand.

Coarse Mortar.

One pârt of lime to four of course gravelly sand.

Concrete.

One part of lime to four of gravel and two of sand.

Dubbing.

Two pounds of black resin, one pound of tallow, one gallon train oil.

Waterproofing for Boots.

Linseed oil one quart, beeswax six ounces, spirit of turpentine four ounces, Burgundy pitch one ounce. Melt wax and oil together and dissolve the pitch in the turpentine; pour both into a jar, and place it in a saucepan with water, boil and stir till well mixed. Being very inflammable, beware of fire getting near it.

To Waterproof Cloth.

Make after the following manner two separate solutions:—1st· Dissolve one pound of sugar of lead in one gallon of water. 2nd, Dissolve one pound of alum in one gallon of water. Dip the cloth first in the solution of lead, and, when nearly dry, dip it in the solution of alum, then dry it in the air or before the fire. This process may be used for coats after being made up.

COOKERY.

Boiling meat entails a loss in weight of about 30 per cent. The water should never be higher than 160° Fahr.; if hotter the meat becomes hard and shrunken; the lower the temperature the better are the nutritive juices kept in. The larger the pieces of meat the better. Put the meat into boiling water, let it boil for five minutes, and then reduce the temperature of the water, either by pouring in cold water, or by reducing the fire until it is about 160° Fahr., that is, as hot as the finger can be put into without scalding. Allow a quarter of an hour for every pound the meat weighs.

Roasting—The loss is a little less than in boiling. The meat should be exposed at first to a great heat for the purpose of keeping in the juice. Allow a quarter of an hour a pound.

Receipts for Cooking—Meat Soup. 16½ lbs. meat, 1 lb. onions, 1 lb. flour, 5 ozs. salt, ¼ oz. pepper, 5 ozs. sugar, small faggot of herbs, and 3½ galls. water. Separate the *large* bone from the meat, also the gristle, cut the meat into pieces of about 4 ozs., take 8 ozs.

of the fat and chop it up, slice the onions, put the fat in the boiler; when melted, add the onions, stir them well, so that they do not get brown, in five minutes add the meat, which keep stirring and turning over for five minutes longer; the meat ought to be warm through; then add the boiling water by degrees, let it simmer gently for an hour, mix the flour with cold water very smooth, add it to the soup, with the salt, pepper, sugar and herbs; simmer gently for thirty minutes, keep stirring it to prevent the flour from settling at the bottom. The great error commonly committed in making soup, is doing it too rapidly, which renders the meat hard and tasteless. Bones and scraps of meat should be collected after every meal and put down to simmer for next days' soup.

Irish Stew.—16½ lbs. meat, 16 lbs. potatoes, 4 lbs. onions, 6 ozs. salt, 1 oz. pepper, ¼ lb. flour. Cut the meat away from the bone, and then into pieces of a ¼ lb. each, the loin and neck of mutton into chops, disjoint the shoulder, and cut the blade bone into four pieces, if the leg cut into slices, ¾ inch thick, rub them with the salt, pepper, and flour, and place the meat in the boiler with some fat, brown it on both sides, then add the onions whole, and then the potatoes; stew gently for two hours; keep the fire down and well covered during the cooking.

How to Soak and Plain Boil the Rations of Salt Meat.—To each pound of meat allow ½ pint of water or a pint if handy; do not let the pieces weigh more than 3 or 4 lbs. each. Let them soak about eight hours, or all night if possible. Wash each piece with your hand to extract as much salt as possible; it is then ready for cooking. If less time is allowed, cut the pieces smaller, or parboil the meat for twenty minutes in the above quantity of water, which throw off and add more; simmer gently for three hours and serve. Vegetables or dumplings can be boiled with it.

Salt Meat, to prepare hurriedly.—Warm it slightly on both sides—this makes the salt draw to the outside—then rinse it well in a pannikin of water. This process is found to extract a great deal of salt, and to leave the meat in a fit state for cooking.

How to Stew Fresh Beef, Pork, Mutton, and Veal.—Cut or chop 2 lb. of fresh beef into ten or twelve pieces, put these into a saucepan with 1½ teaspoonful of salt, 1½ teaspoonful of sugar, ½ teaspoonful of pepper, 2 middle-sized onions sliced, ½ pint of water. Set on the fire for ten minutes until forming a thick gravy. Add a table-spoonful of flour, stir on the fire a few minutes; add a quart and a-half of water; let the whole simmer until the meat is tender. Beef will take from two hours and a-half or three hours; mutton and pork about two hours; veal one hour and a-quarter to one hour and a-half. If onions, sugar, and pepper, are not to be had, meat thus prepared will even then make a good dish; with ½ lb of sliced potatoes or 2 ozs. of preserved potatoes; ration vegetables may be added, also a small dumpling.

For a hurried dinner cut your rations into pieces about the size of a penny, but three or four times thicker; skewer them on a piece of iron wire or hard stick. A few minutes will cook them if hung before the fire. *Vegetables* must be carefully washed and cleaned from insects. Green vegetables must be boiled fast in plenty of water, and drained at once when done. They sink when sufficiently cooked. *Potatoes* take from twenty to thirty minutes boiling; they show signs of breaking when they are done, which can be ascertained by sticking a fork into them. *Carrots* and *parsnips* take from twenty to forty-five minutes boiling. Young nettles, sweet docks, turnip tops, or the young leaves of mangel wurtzel, make good green food. A little pepper and salt should be added to season them. Dandelion leaves, especially when young, make a most agreeable salad. *Dried* and *compressed vegetables* of all kinds should be soaked from four to six hours in pure water, and then boiled slowly; if there is any bad taste, from putrefaction having commenced, a little chloride of lime will remove it. The "*mixed compressed vegetables*" should be boiled in a little water for about half-an-hour; the *cabbage* to be boiled in sufficient water for half-an-hour; the *carrots* and *turnips* to be boiled for about fifteen minutes; *potatoes* to be boiled in sufficient water for half-an-hour. *Rice* should be washed and soaked and then boiled in plenty of water, *without salt*, for twenty or twenty-five minutes; then some salt should be thrown in, and the water drained off. Each grain will then be separate.

To make Tea.—If possible it should be made in a vessel solely used for that purpose; on service this is generally impossible, but it renders great care on the part of the cook all the more essential. Before the tea is made the kettle must be well washed, and heated with a little hot water and well rinsed. The water for the tea should then be put in, and boiled before the tea is put in. Care to be taken that the water is boiling fast when this is done. If possible the boiling water should be poured from one kettle into another containing the dry tea. The lid should then be put on, and the pot placed beside (but not on) the fire for four or five minutes before serving it out. Much depends on the softness of the water; if the water is hard, add, when possible, a small teaspoonful of soda to the camp kettle full (for five men each).

To make Coffee.—The same rules apply, as regards cleanliness and the description of water, as in making tea. Sometimes there is only time to prepare it by boiling; but, if possible, it is better to heat the coffee in the lid of the kettle; then put it in a kettle, and pour the boiling water on it, leaving it to stand near the fire for five minutes, when it will be fit for use. When there is time to do so, it should be strained through a cloth of some sort; when made, the dregs should be collected and well boiled. If this decoction is poured over fresh coffee the result of the second making will be found strong and aromatic. To clear coffee some cold water should be poured in from a height. The cold water sinks through the coffee, and carries down the suspended particles.

COOKERY FOR THE SICK.

Arrow Root.—Take a small tablespoonful of arrow root, mix or blend it smoothly in two tablespoonful of cold water. Then add half a pint of boiling water, and stir until it thickens; add sugar. A little wine will render it more agreeable.

Water Gruel.—Rub a tablespoonful of oaten meal smooth in a little cold water, add this to a pint of boiling water (on the fire), stir gently for a quarter of an hour. Let it settle; pour off from dregs, and add a little salt and butter.

Barley Water.—Wash a handful of barley, and simmer in 3 pints of water. A little lemon peel will flavour it.

Beef Tea.—Take one pound of rump steak, mince it fine, and mix it with one pint of cold water; place by the fire to heat *very* slowly. It may stand two or three hours before it is allowed to simmer, then boil for fifteen minutes, skim and serve. Good in fever and debility.

Egg Flip.—Beat up the raw yolk of 2 eggs, add half a pint of milk. Nourishing and useful in fever.

Lemonade.—Half a lemon squeezed into a pint of warm water; sweeten and allow to cool. Or, one ounce of lime juice instead of the lemon.

WATER.

In all localities where the quality of the water is suspicious, condensed water should, if possible, be used for drinking and cooking purposes. When this is not feasible the water should be carefully filtered and boiled.

Two barrels, one inside the other, having a space of 4 to 6 inches clear all round between them, filled with layers of sand, gravel, and charcoal, form an excellent filter. The inside one, without a bottom, rests on three stones placed in layers of sand, charcoal, and coarse gravel; the water, flowing or being poured into the space between the two barrels, and having thus to force its way through the substances into the inner barrel, becomes purified.

The water should be drawn off by means of a pipe, running through the outer into the inner barrel. Animal charcoal is the best. When, after a time, it ceases to act, it should be removed and well dried. It can then be used again with advantage. It is impossible to use too much of it.

The popular French plan of purifying turbid water *(alunage de l'eau)* simply consists in the addition of a small quantity of alum. It clears the water very rapidly, but merely converts the lime carbonate into sulphate which remains in solution.

HOISTING IN CATTLE.

Hoisting Bullocks in.—Hook on the yard and stay to a strop round the beasts' neck, putting his fore leg through it to prevent its choking him.

Cattle should be hoisted in by the horns only when the latter are very strong, and the animals themselves very small.

Hoisting Horses in.—If troublesome, cover his eyes, hobble him, and sling him with a broad mat or canvas slings under his belly, with a crupper from his haunches round the chest, to keep the slings from shifting. Lead the halter through a ring-bolt on deck, and take the slack through as he comes in. A strop round the nose, hove short with a short stick or toggle, will rapidly tame an unmanageable horse.

Horses should be kept slung in their stalls at sea, with their hoofs just resting on the deck.

To Find the Time to Fire the Daylight Gun.

Daybreak, or the commencing of morning twilight, occurs after the sun has crossed the crepus culum, or twilight circle, which is 18° below the Horizon.

RULE.

Under the latitude put the declination, marking them with their proper names N. or S.; if the names are alike take the difference; if unlike take the sum, and under this put the constant angle 108°; take the sum and difference. Add together the log. secants of the two first terms (rejecting tens in the index), and the half haversines of the two last. With the sum as a log. haversine, take out the corresponding angle in time at the top of the page, which will be the ending of evening twilight, and which, subtracted from twelve hours, will give the beginning of morning twilight, or daybreak, in *apparent time*.

```
          °   ′
Lat.    22.17 N.           Sec. 0·033708
Decl.   15· 3 N.           Sec. 0·015158
         7·14
       108·
       ─────
       115·14              ½ had. 4·926591
       100·46              ½ hav. 4·886675
       H.  M.              Hav.  9·862132
        7  49
       12   0
       ─────
        4  11  Morning twilight begins.
```

To Find the Time of Sunset.

See ordinary Nautical Tables; or Burdwood's Azimuth Tables between Lat. 30° and 60°; and Davis' Azimuth Tables between Lat. 30° and the Equator.

USEFUL NOTES ON THE MARINE STEAM ENGINE.

From the principle that Heat and Work are mutually convertible.

The construction of a perfect Engine, by which we mean one which can reconvert the whole of the heat developed by mechanical energy into work, is impossible.

It is estimated that a pound of ordinary coal is capable of producing nearly 10,000,000 foot pounds or units of work. Whereas, the best performance of a pumping engine on shore does not exceed 1,000,000 foot pounds or units of work per pound of coal. From careful experiment it has been shown that with an engine of good design and workmanship, and working with a low consumption of fuel, very little over *ten per cent.* of the total amount of heat parted with by the steam is converted into useful work; over 76 per cent. is imparted to the condensing water, and the remainder diffused by radiation, &c.

As a rule we may assume that the more distant the extremes of temperature between the boiler and the condenser the larger will be the proportion of heat turned into work. In the steam engine a certain quantity of power of doing work is expended, in order that a certain amount of work may be done, and the proportion which the useful work done, bears to that expended, is called the efficiency.

Causes of Waste Heat :—

1st.—The heat lost by the imperfect combustion of the fuel supplied to the furnace, loss by conduction and radiation, and by the high temperature at which the furnace gas escapes by the funnel. This latter loss is estimated at from from 700° to 800° Fahrenheit. *

* From 300° to 600° Fahr. preferable.

THE MARINE ENGINE.

A pound of carbon burnt into carbonic oxide gas evolves but about ⅔ths of the heat which is produced, if it is burnt into carbonic acid; therefore, for each pound of carbon allowed to escape in the form of carbonic oxide, ⅔ths of the heating effect which might have been obtained by its complete combustion in the furnace is wasted.

2nd.—The heat carried away by the steam when it leaves the cylinder.

3rd.—The heat power necessary to overcome the friction of the engine, &c., and that wasted in agitating the water in which the propeller works.

To, in any way, diminish this waste is the object of improvement in the economy of the marine steam engine.

The work done by the steam driving the piston may be called the indicated work, being registered by a self-acting instrument called the indicator.

All mechanical work is done by the exertion of a force through space, and is calculated and expressed as a quantity by multiplying the mean amount of the force into the space through which it acts.

$$\text{FORMULÆ FOR HORSE POWER} = \frac{P.V.}{33000}.$$

Where P equals the mean effective force of the steam on the area of the piston in square inches, the mean effective pressure being obtained from a diagram taken by the indicator; and V equals the velocity of the piston in feet per minute, sweeping through the cylinder; and 33000 the number adopted by James Watt to represent in foot pounds the work of one horse in a minute of time, *i.e.*, 33000 pounds raised through a space of one foot per minute equivalent to one horse-power.

The force acting upon the piston of a steam engine is the excess of the forward pressure above the back pressure exerted by the steam behind the piston as it comes through the regulating valves from the boiler. There are two stages in the action of the steam—the admission and expansion. During the admission, the steam is coming from the boiler into the cylinder, and it exerts a pressure less than that in the boiler, only by the amount to overcome the friction of the pipes, passages, and valve ports. The admission is terminated by the cut off, that is, by the closing of the valve which admits the steam into the cylinder. Then follows the expansion of the steam which is confined in the cylinder, as it drives the piston before it and exerting a continually diminishing pressure. For ordinary practical calculation we assume that the pressure varies inversely as the volume $P \propto \frac{1}{V}$. It is clear that work continues to be done by the steam driving the piston so long as the pressure behind the piston, or forward pressure, continues to be greater than the pressure in front, or the back pressure, exerted by the steam which has already done its work, and which the piston is expelling from the cylinder. Hence, the higher the forward pressure and the lower the back pressure, the greater is the efficiency of the steam in an engine. The greatest useful work is obtained by making the expansion cease when the forward pressure is just equal to the back pressure, added to the pressure equivalent to the friction to be overcome.

The pressure at which the steam should be used in the promotion of the economy of fuel is limited only by the strength of the boiler. Compound engines—by which is meant one in which the mechanical action of the steam commences in a smaller cylinder, and is completed in a larger cylinder before passing to the condenser, giving a wider range for expansion—are therefore more efficient for long voyages at a high speed, the efficiency of the steam being sometimes represented by an expansive force acting upon the piston from about 60lbs. per square inch above the atmospheric pressure, to from 7 to 8 lbs. below the atmospheric pressure, before being expelled to the condenser, effecting a large economy of fuel in proportion to the power used.

Heavier cylinders and a larger space are required for compound engines, but the forces acting upon the piston, and therefore throughout the working machinery, may be more equally distributed throughout the stroke, than with a simple cylinder engine, and at the same time to compensate for such increased weight in proportion to power, the boilers and surface condensers may be considerably reduced in size.

The average amount of coal burnt in the different kinds of engines under favourable circumstances would be for

Compound, or double cylinder, expansive engines } 2lbs. per indicated horse-power per hour.

Single cylinders, with surface condensers, superheaters and improvements } 3 to 3½ lbs.

Ordinary jet condenser   4½ lbs

High pressure, exhausting into the atmosphere } 6 lbs.

The increased pressure used with compound engines requires the boiler to be of a circular form to reduce the number of stays for strengthening within the shell and to render the interior accessible. Boilers are made of the circular form, with tubes returning over the furnace, dry uptakes, and working when new from 50 to 60lbs. pressure per square inch above that of the atmosphere.

General Proportions :—

	Ft.	In.
Diameter of shell of boiler	11	0
„ of furnace tubes	2	8
Thickness of shell	0	⅞
„ of furnace tubes	0	½
„ of brass tubes	0	⅛ full.
Heating surface, per Nom. H.P.	18 sq. ft.	
Fire grate surface „	·7 „	
Weight of water „	2¾ cwt.	
Boilers proved with water pressure at per square inch	140 lbs. / 230 lbs.	
Surface condensers, cooling surface in small tubes, per N.H.P.	15 sq. ft.	
Aggregate length for every 1000 N.H.P. about	from 10 to 14 miles.	

The weight of the boiler—about one ton for every 16 I.H.P.

The weight of the water—about one ton for every 40 I.H.P.

The ordinary square form of boiler, with the tubes returning over the furnaces, working (when new) at a pressure of about 30 lbs. per square inch. The thickness of plates would be about—

The shell of the boiler { Bottoms	$\frac{1}{2}$	an inch.
{ Other parts	$\frac{7}{16}$	,,
Tube plates	$\frac{5}{8}$	,,
Tops and sides of furnaces	$\frac{3}{8}$	,,
Heating surface, per nominal H.P.	19 to 20	sq. ft.
Surface of fire bars—grate surface	·7	,,
Superheating surface	3	,,
Weight of water, per nominal H.P.	2$\frac{3}{4}$	cwt.
Weight of boiler, per nominal H.P.	5	,,

Real power of the engine, from 6 to 7 times the nominal. Boilers proved at 60 lbs. by water pressure.

    BOILER TUBES—Diameter ... from 2$\frac{3}{4}$ to 3 inches.
                      Length ... ,, 6 to 6$\frac{1}{2}$ feet.
                      Thickness ... ,, $\frac{1}{8}$ inch.

IN CONDENSERS communicating with the boiler for the purpose of providing drinking water, on the average one ton of coals will generate enough steam to make a little more than 7 tons of water.

The Indian troop-ships while running allow one ton of coals for seven tons of water condensed.

When condensing only, the opportunity should be taken to burn up the very small coal, &c.

Working Boilers with Sea Water.

When sea water is used it is not advisable in practice to allow the water in the boiler to contain more than a double charge of salt, or 2 lbs. of salt to 32 lbs of water. To maintain this, then it would be necessary to expel one-half the water that entered the boiler,—say at a temperature of 250°—the remaining half of the water being converted into steam and utilised. Now, because the sensible temperature of the steam is the same as that of the water, it must not be supposed that one-half the value of the fuel has been lost by this process. It must be taken into account that the steam has absorbed as much heat as could, if a change of state had not taken place, have raised its temperature to 1212°; therefore, the loss incurred from priming the boilers and the prevention of deposit would be found only to be about 12 per cent. of the fuel used.

CENTRIFUGAL PUMP FOR A GUN-BOAT:—

No. of revolutions per minute	200
Diameter of wheel	12 inches
Width of discharging orifice	2 ,,
Average discharge per minute	6 tons

Centrifugal pumps, driven by engines of about 40 horse-power, will throw 60 tons of water per minute.

From Mr. Froude's Trials in "Greyhound:"—

"The power actually indicated throughout the towing trials on an average realized from 33 to 42 per cent. of the power expressed in indicated horse-power, at trial trip of the engines for the same speed of ship."

"A considerable portion of this loss of power is said to be due to the action of the propeller diminishing the water pressure against the after part of the vessel."

N.B. This action causes an increase of resistance when the propeller is allowed to revolve freely, as compared with fixing it up and down for sailing purposes.

USEFUL NOTES ON COALS.

Vide Main and Brown.

As much as possible avoid breaking the coal in its transit to the coal bunkers; weigh a basket or bag frequently to estimate total weight received.

Coal that has been Exposed to the Atmosphere.

The power for raising steam is reduced one half, after lying in the open air for six months, by disintegration and loss of heating properties. The destruction of coal by exposure depends on its ability to absorb oxygen, converting the hydro-carbons into water and carbonic acid.

Combustion of Coal.

The gas which escapes from the coal during the first process of combustion, will, if it be supplied freely with oxygen, and its temperature be sufficiently high, ignite in our fire places and furnaces. Coal containing a great quantity of hydrogen is called bituminous.

NEWCASTLE COAL is of this character.

CANNEL COAL is more particularly adapted for gas manufactories on this account.

WELSH COAL, or Anthracite, is coal containing but little hydrogen; its heating power is considerably greater than bituminous coal, because a great quantity of the heat in the former case is employed in converting a portion of the solid matter into a gaseous form, and consequently heat is absorbed by this process. For the same reason, Coke, from which the bituminous matter has been previously expelled, is preferable for furnaces where great local heat is required, besides which it is smokeless.

Welsh coal, when good, partaking of this character, is preferred in *our* war ships and fleets, the localized heat derived being more beneficial in our present form of boiler with its necessarily contracted combustion chamber, within which the gasses mix, but Welsh coal long in store or on board ship loses its cohesion, and becomes *small*, the waste is then very great, unless a more lively bituminous coal can be mixed with it; from this cause a mixed coal is better adapted as a general rule, and an attempt is made to properly supply the furnaces with oxygen to prevent the smoke from the bituminous portion of the coal. Usual mixture,—one-third North Country, two-thirds Welsh.

To find the Co-efficient of Speed.

Suppose *Agincourt* and *Northumberland* running equal distances, or equal time,—say 4 hours,—the *Agincourt* makes 5821 revolutions, and the *Northumberland* 5962; then, taking *Agincourt's* co-efficient as 1, the *Northumberland's* would be 1·024.

Therefore, for practice, if *Northumberland* is required to go the same speed as *Agincourt*—say, at 40 revolutions,—she would have to go $40 \times 1\cdot024 = 41$ (nearly) to keep station.

RULE.—Take the number of revolutions in a given time as a divisor into the number of revolutions for the same time in any other vessel—the quotient is the co-efficient.

Another approximate method is obtained by dividing the pitch of screw of any vessel by the pitch of the screw of flag-ship or leading vessel, the result being the co-efficient, which is used as before.

In paddle-wheel vessels the circumference of the wheel divided by 101· will give a constant, which, if multiplied by the number of revolutions per minute, will give the speed of wheel in knots per hour.

This 101 is obtained by dividing the feet per knot, 6080, by $60 = 101\cdot33$.

PROVISIONS ALLOWED FOR THE UNDERMENTIONED NUMBER OF MEN AND DAYS, SHEWING THE GROSS WEIGHT AND MEASUREMENT.

Article.	500 Men for 90 Days.							800 Men for 10 Days.					
	Quantity.	Packages.			Gross Weight.	Measurement.	Quantity.	Packages.			Gross Weight.	Measurement.
		Barrels.	Half Hgsheds.	Small Casks.	Cases.				Barrels.	Half Hgsheds.	Small Casks.	Cases.	
						lbs.	Cb. Ft						lbs.	Cb. Ft
Biscuit..........lbs.	56250	563 *bags*				57376	2812 1/2	10000	100 *bags*				10200	500
Rum............galls.	703	19	..	1	..	8214	195	125	3	..	1	..	1559	37
Sugar...........lbs.	5625	14	..	1	..	6668	145	1000	2	1	..	..	1239	27
Chocolate......lbs.	2812	..	..	..	29	3396	64 5/8	500	..	..	..	5	595	11 3/16
Tea............lbs.	703	..	..	..	7	847	35	125	..	..	..	2	187	8
Pork............lbs.	22500	75	..	..	..	39750	750	4000	12	2	..	..	7080	134
Peas............lbs.	7500	23	..	1	..	8794	235	1333	4	..	..	..	1496	40
Celery Seed.....lbs.	29	..	..	..	1	52	1	6	..	..	..	1	7	1/17
Salt Beef.......lbs.	11250	37	1	..	..	19970	377	2000	6	1	..	..	3540	67
Flour...........lbs.	6318	19	..	..	..	7258	190	1125	3	..	1	..	1278	34
Suet............lbs.	526	..	..	5	..	1275	25	94	..	..	1	..	181	4
Raisins.........lbs.	1053	..	..	..	10	1780	40	188	..	..	..	2	340	8
Oatmeal........lbs.	1212	..	..	8	..	1616	40	216	..	1	..	..	289	7
Mustard........lbs.	202	..	..	..	2	266	7 1/2	36	..	..	..	1	73	2
Pepper.........lbs.	101	..	..	..	2	146	4	18	..	..	..	1	36	1
Vinegar........galls.	202	..	8	..	..	2536	80	36	1	..	..	..	450	10
Boiled Beef.....lbs.	8438	..	..	..	234	12870	390	1500	..	..	..	42	2310	63
Preserved Potatoes lbs.	1406	..	..	..	14	1088	60 7/8	250	..	..	..	3	346	10 7/16
Rice............lbs.	1407	..	6	..	..	1854	42	250	..	1	..	..	309	7
Lemon Juice....lbs.	375	..	..	..	10	930	20	62	..	..	..	2	186	4
Candles.........lbs.	2700	..	..	..	54	3348	81	750	..	..	..	15	930	22 1/2
Tobacco........lbs.	1800	11	..	1	..	2650	125	300	2	..	..	..	458	20
Soap............lbs.	4050	..	..	..	41	5022	108 1/2	800	..	..	..	8	992	21 3/16
						188606	5828						34081	1039

Sec. X.   391

Proviross Weight and Measurement.

ARTIC	90 Days.			200 Men for 90 Days.						
					Packages.					
	Cases.	Total Weight.	Measurement.	Quantity.	Barrels.	Half-Hogsheads.	Small Casks.	Cases.	Total Weight.	Measurement.
		lbs.	Cubic feet						lbs.	Cubic feet
Biscuit	to e	11474	562 1/12	22500	{ according to stowage }				23000	1125
Rum	—	1776	42	281	—	11	—	—	3256	77
Sugar	—	1596	35	2250	—	8	—	—	2724	56
Chocolate	11	704	13 1/2	1125	—	—	—	13	1492	26 1/2
Tea	3	198	9	281	—	—	—	3	363	15
Pork	—	8122	158	9000	10	30	—	—	16100	310
Peas	—	1728	45	3000	—	13	—	—	3497	91
Celery Seed	1	7	1/2	12	—	—	—	1	14	1/2
Salt Beef	—	4162	81	4500	7	12	—	—	7310	154
Flour	—	1624	40	2531	—	11	—	—	3113	77
Suet	—	268	5 1/2	211	—	—	5	—	670	14 1/2
Raisins	2	340	8	422	—	—	—	4	680	16
Oatmeal	—	264	8	480	—	—	4	—	528	16
Mustard	1	73	2	80	—	—	—	2	146	4
Pepper	1	73	2	40	—	—	—	1	73	2
Vinegar	—	588	12 1/2	80	—	—	5	—	1048	22 1/2
Boiled Beef	47	2585	78 1/2	3376	—	—	—	94	5170	156 1/2
Preserved Potatoes	6	385	12	562	—	—	—	6	775	23 1/2
Rice	—	520	16	562	—	—	—	—	780	24
Lemon Juice	2	186	4	156	—	—	—	4	372	8
Candles	11	682	16 1/2	1080	—	—	—	22	1364	33
Tobacco	—	610	25	720	5	—	—	—	1145	50
Soap	—	992	22 1/2	1620	—	—	—	17	2046	44 1/2
Total		38957	1200						75666	2346 1/2

Sec. X.   CLOTHING REQUISITES.   393

FOR HOME STATION.

Average Quantity of Clothing for 100 Men for 90 Days.

Article.		Quantity.	Packages.		Weight.	Measurement.
			Bales.	Casks.		
					lbs.	Cubic Feet.
Blue Cloth, No. 1.	yds.	37	1	—	56	2
Duck	,,	300	1	—	200	4
Coaling Duck	,,	150	1	—	101	2
Flannel	,,	210	1	—	75	4
Serge	,,	320	2	—	182	8
Drill	,,	150	1	—	73	2
Jean	,,	30	1	—	13	$\tfrac{8}{12}$
Stockings	pairs	10	1	—	—	$\tfrac{6}{12}$
Shoes	,,	50	—	1	137	7
Half Boots	,,	50	—	1	168	10
Combs	No.	12	—	—	—	—
Towels	,,	12	1	—	—	--
Blankets	,,	12	1	—	62	4
Beds	,,	4	—	—	34	5
Bed Covers	,,	12 }	1	—	29	$1\tfrac{3}{12}$
Shirts	,,	25 }				
Handkerchiefs	,,	50	1	—	4	$\tfrac{6}{12}$
Sou' Westers	,,	6 }	1	—	— }	
Materials for Trowsers	,,	25 }				
Marks of Distinction	,,	—	—	—	— }	1
G. C. Badges	,,	— }	1	—	— }	
Marine Necessaries	,,	— }	—	—	—	

LOGARITHMS.

Sec. X.

Proportional Parts.

Nat. Nos.	0	1	2	3	4	5	6	7	8	9	1	2	3	4	5	6	7	8	9
10	0000	0043	0086	0128	0170	0212	0253	0294	0334	0374	4	8	12	17	21	25	29	33	37
11	0414	0453	0492	0531	0569	0607	0645	0682	0719	0755	4	8	11	15	19	23	26	30	34
12	0792	0828	0864	0899	0934	0969	1004	1038	1072	1106	3	7	10	14	17	21	24	28	31
13	1139	1173	1206	1239	1271	1303	1335	1367	1399	1430	3	6	10	13	16	19	23	26	29
14	1461	1492	1523	1553	1584	1614	1644	1673	1703	1732	3	6	9	12	15	18	21	24	27
15	1761	1790	1818	1847	1875	1903	1931	1959	1987	2014	3	6	8	11	14	17	20	22	25
16	2041	2068	2095	2122	2148	2175	2201	2227	2253	2279	3	5	8	11	13	16	18	21	24
17	2304	2330	2355	2380	2405	2430	2455	2480	2504	2529	2	5	7	10	12	15	17	20	22
18	2553	2577	2601	2625	2648	2672	2695	2718	2742	2765	2	5	7	9	12	14	16	19	21
19	2788	2810	2833	2856	2878	2900	2923	2945	2967	2989	2	4	7	9	11	13	16	18	20
20	3010	3032	3054	3075	3096	3118	3139	3160	3181	3201	2	4	6	8	11	13	15	17	19
21	3222	3243	3263	3284	3304	3324	3345	3365	3385	3404	2	4	6	8	10	12	14	16	18
22	3424	3444	3464	3483	3502	3522	3541	3560	3579	3598	2	4	6	8	10	12	14	15	17
23	3617	3636	3655	3674	3692	3711	3729	3747	3766	3784	2	4	6	7	9	11	13	15	17
24	3802	3820	3838	3856	3874	3892	3909	3927	3945	3962	2	4	5	7	9	11	12	14	16
25	3979	3997	4014	4031	4048	4065	4082	4099	4116	4133	2	3	5	7	9	10	12	14	15
26	4150	4166	4183	4200	4216	4232	4249	4265	4281	4298	2	3	5	7	8	10	11	13	15
27	4314	4330	4346	4362	4378	4393	4409	4425	4440	4456	2	3	5	6	8	9	11	13	14
28	4472	4487	4502	4518	4533	4548	4564	4579	4594	4609	2	3	5	6	8	9	11	12	14
29	4624	4639	4654	4669	4683	4698	4713	4728	4742	4757	1	3	4	6	7	9	10	12	13
30	4771	4786	4800	4814	4829	4843	4857	4871	4886	4900	1	3	4	6	7	9	10	11	13
31	4914	4928	4942	4955	4969	4983	4997	5011	5024	5038	1	3	4	6	7	8	10	11	12
32	5051	5065	5079	5092	5105	5119	5132	5145	5159	5172	1	3	4	5	7	8	9	11	12
33	5185	5198	5211	5224	5237	5250	5263	5276	5289	5302	1	3	4	5	6	8	9	10	12
34	5315	5328	5340	5353	5366	5378	5391	5403	5416	5428	1	3	4	5	6	8	9	10	11
35	5441	5453	5465	5478	5490	5502	5514	5527	5539	5551	1	2	4	5	6	7	9	10	11
36	5563	5575	5587	5599	5611	5623	5635	5647	5658	5670	1	2	4	5	6	7	8	10	11
37	5682	5694	5705	5717	5729	5740	5752	5763	5775	5786	1	2	3	5	6	7	8	9	10
38	5798	5809	5821	5832	5843	5855	5866	5877	5888	5899	1	2	3	5	6	7	8	9	10
39	5911	5922	5933	5944	5955	5966	5977	5988	5999	6010	1	2	3	4	5	7	8	9	10
40	6021	6031	6042	6053	6064	6075	6085	6096	6107	6117	1	2	3	4	5	6	8	9	10
41	6128	6138	6149	6160	6170	6180	6191	6201	6212	6222	1	2	3	4	5	6	7	8	9
42	6232	6243	6253	6263	6274	6284	6294	6304	6314	6325	1	2	3	4	5	6	7	8	9
43	6335	6345	6355	6365	6375	6385	6395	6405	6415	6425	1	2	3	4	5	6	7	8	9
44	6435	6444	6454	6464	6474	6484	6493	6503	6513	6522	1	2	3	4	5	6	7	8	9
45	6532	6542	6551	6561	6571	6580	6590	6599	6609	6618	1	2	3	4	5	6	7	8	9
46	6628	6637	6646	6656	6665	6675	6684	6693	6702	6712	1	2	3	4	5	6	7	7	8
47	6721	6730	6739	6749	6758	6767	6776	6785	6794	6803	1	2	3	4	5	5	6	7	8
48	6812	6821	6830	6839	6848	6857	6866	6875	6884	6893	1	2	3	4	4	5	6	7	8
49	6902	6911	6920	6928	6937	6946	6955	6964	6972	6981	1	2	3	4	4	5	6	7	8
50	6990	6998	7007	7016	7024	7033	7042	7050	7059	7067	1	2	3	3	4	5	6	7	8
51	7076	7084	7093	7101	7110	7118	7126	7135	7143	7152	1	2	3	3	4	5	6	7	7
52	7160	7168	7177	7185	7193	7202	7210	7218	7226	7235	1	2	2	3	4	5	6	7	7
53	7243	7251	7259	7267	7275	7284	7292	7300	7308	7316	1	2	2	3	4	5	6	6	7
54	7324	7332	7340	7348	7356	7364	7372	7380	7388	7396	1	2	2	3	4	5	6	6	7

Sec. X.   LOGARITHMS.   395
                                              Proportional Parts.

Nat. Nos.	0	1	2	3	4	5	6	7	8	9	1	2	3	4	5	6	7	8	9
55	7404	7412	7419	7427	7435	7443	7451	7459	7466	7474	1	2	2	3	4	5	5	6	7
56	7482	7490	7497	7505	7513	7520	7528	7536	7544	7551	1	2	2	3	4	5	5	6	7
57	7559	7566	7574	7582	7589	7597	7604	7612	7619	7627	1	2	2	3	4	5	5	6	7
58	7634	7642	7649	7657	7664	7672	7679	7686	7694	7701	1	1	2	3	4	4	5	6	7
59	7709	7716	7723	7731	7738	7745	7752	7760	7767	7774	1	1	2	3	4	4	5	6	7
60	7782	7789	7796	7803	7810	7818	7825	7832	7839	7846	1	1	2	3	4	4	5	6	6
61	7853	7860	7868	7875	7882	7889	7896	7903	7910	7917	1	1	2	3	4	4	5	6	6
62	7924	7931	7938	7945	7952	7959	7966	7973	7980	7987	1	1	2	3	3	4	5	6	6
63	7993	8000	8007	8014	8021	8028	8035	8041	8048	8055	1	1	2	3	3	4	5	5	6
64	8062	8069	8075	8082	8089	8096	8102	8109	8116	8122	1	1	2	3	3	4	5	5	6
65	8129	8136	8142	8149	8156	8162	8169	8176	8182	8189	1	1	2	3	3	4	5	5	6
66	8195	8202	8209	8215	8222	8228	8235	8241	8248	8254	1	1	2	3	3	4	5	5	6
67	8261	8267	8274	8280	8287	8293	8299	8306	8312	8319	1	1	2	3	3	4	5	5	6
68	8325	8331	8338	8344	8351	8357	8363	8370	8376	8382	1	1	2	3	3	4	4	5	6
69	8388	8395	8401	8407	8414	8420	8426	8432	8439	8445	1	1	2	2	3	4	4	5	6
70	8451	8457	8463	8470	8476	8482	8488	8494	8500	8506	1	1	2	2	3	4	4	5	6
71	8513	8519	8525	8531	8537	8543	8549	8555	8561	8567	1	1	2	2	3	4	4	5	5
72	8573	8579	8585	8591	8597	8603	8609	8614	8621	8627	1	1	2	2	3	4	4	5	5
73	8633	8639	8645	8651	8657	8663	8669	8675	8681	8686	1	1	2	2	3	4	4	5	5
74	8692	8698	8704	8710	8716	8722	8727	8733	8739	8745	1	1	2	2	3	4	4	5	5
75	8751	8756	8762	8768	8774	8779	8785	8791	8797	8802	1	1	2	2	3	3	4	5	5
76	8808	8814	8820	8825	8831	8837	8842	8848	8854	8859	1	1	2	2	3	3	4	5	5
77	8865	8871	8876	8882	8887	8893	8899	8904	8910	1915	1	1	2	2	3	3	4	4	5
78	8921	8927	8932	8938	8943	8949	8954	8960	8965	8971	1	1	2	2	3	3	4	4	5
79	8976	8982	8987	8993	8998	9004	9009	9015	9020	9025	1	1	2	2	3	3	4	4	5
80	9031	9036	9042	9047	9053	9058	9063	9069	9074	9079	1	1	2	2	3	3	4	4	5
81	9085	9090	9096	9101	9106	9112	9117	9122	9128	9133	1	1	2	2	3	3	4	4	5
82	9138	9143	9149	9154	9159	9165	9170	9175	9180	9186	1	1	2	2	3	3	4	4	5
83	9191	9196	9201	9206	9212	9217	9222	9227	9232	9238	1	1	2	2	3	3	4	4	5
84	9243	9248	9253	9258	9263	9269	9274	9279	9284	9289	1	1	2	2	3	3	4	4	5
85	9294	9299	9304	9309	9315	9320	9325	9330	9335	9340	1	1	2	2	3	3	4	4	5
86	9345	9350	9355	9360	9365	9370	9375	9380	9385	9390	1	1	2	2	3	3	4	4	5
87	9395	9400	9405	9410	9415	9420	9425	9430	9435	9440	0	1	1	2	2	3	3	4	4
88	9445	9450	9455	9460	9465	9469	9474	9479	9484	9489	0	1	1	2	2	3	3	4	4
89	9494	9499	9504	9509	9513	9518	9523	9528	9533	9538	0	1	1	2	2	3	3	4	4
90	9542	9547	9552	9557	9562	9566	9571	9576	9581	9586	0	1	1	2	2	3	3	4	4
91	9590	9595	9600	9605	9609	9614	9619	9624	9628	9633	0	1	1	2	2	3	3	4	4
92	9638	9643	9647	9652	9657	9661	9666	9671	9675	9680	0	1	1	2	2	3	3	4	4
93	9685	9689	9694	9699	9703	9708	9713	9717	9722	9727	0	1	1	2	2	3	3	4	4
94	9731	9736	9741	9745	9750	9754	9759	9763	9768	9773	0	1	1	2	2	3	3	4	4
95	9777	9782	9786	9791	9795	9800	9805	9809	9814	9818	0	1	1	2	2	3	3	4	4
96	9823	9827	9832	9836	9841	9845	9850	9854	9859	9863	0	1	1	2	2	3	3	4	4
97	9868	9872	9877	9881	9886	9890	9894	9899	9903	9908	0	1	1	2	2	3	3	4	4

NATURAL SINES, TANGENTS, &c. SEC. X.

Angle	Sine.	Diff. for 10'	Arc.	Tangent	Cotan·	Versin.	Cosine.	—
0°					infinite			90
1				·0175				89
2								88
3								87
4					14·301			86
5					11·430			85
6	·1045		·1047	·1051	9·514			84
7	·1219		·1222	·1228	8·144			83
8					7·115			82
9				·1571	·1584	6·314		81
10	·1736		·1745		·1763	5·671	·0152	·9848	80
11	·1908					5·145	·0184	·9816	79
12									78
13									77
14	·2419			·2493				76
15	·2588			·2679				75
16				·2867				74
17					3·271			73
18			·3142	·3249	3·078			72
19			·3316					71
20			·3491		2·747			70
21			·3665	·3839				69
22				·4040	2·475		·9272	68
23			·4014	·4245	2·356			67
24				·4452	2·246	·0865	35	66
25				·4663	2·145		63	65
26			·4538	·4877	2·050	·1012	88	64
27	·4540		·4712	·5095	1·963		10	63
28	·4695			·5317	1·881	·1171	29	62
29	·4848			·5543			46	61
30					1·732			60
31	·5150					·1428		59
32			·5585			·1520		58
33	·5446			·6494	·5	·1613		57
34	·5592			·6745	·4	·1710		56
35	·5736				·4	·1808		55
36	·5878				1·376	·1910		54
37	·6018				1·327			53
38	·6157				1·280	·2120		52
39	·6293				1·235	·2229	·7771	51
40	·6428		·6981	·8391	1·192	·2340		50
41	·6561		·7156	·8693	1·150	·2453	·7547	49
42	·6691		·7330		1·111		·7431	48
43	·6820		·7505	·9325	1·072		·7314	47
44			·7679				·7193	46
45	·7071		·7854			·2929	·7071	45
—		Diff.		Cotan	Tangent		Sine	Angle

INFORMATION AND INSTRUCTIONS FOR DIVERS.

Siebe and Gorman's Patent Apparatus.

The apparatus consists of a patent double acting air pump for supplying one or two Divers with air: two men can work under a ships bottom independent of each other from the same pump. By an ingenious arrangement with a lever, the supply of air can be thrown into one outlet from both cylinders so that the one Diver can work to a depth of 160 feet or 26 fathoms. When two Divers work from the pump they can with safety descend to 90 feet or 15 fathoms. The pressure gauges denote the depth at which the Divers are working.

The helmet is made of copper tinned, with a segmental neck ring to unscrew the head piece from the collar by one-eighth of a turn, lock pin to prevent it unscrewing, inlet and outlet valve, inflating arrangement so that the Diver can rise to the surface, and regulating cock, under control of diver to regulate his supply of air. Four lenses front, two side, and top glass: the India rubber dress is fastened to the helmet by means of brass bands in segments, two front, two back; with screws and wing nuts. A spanner is provided to fit the nuts for screwing tight the bands.

India rubber dress made of tanned twill with vulcanized collar and cuffs, the collar forming a water-tight joint on helmet collar, cuffs kept tight by flexible rings. Boots have lead soles, 16-lbs. weight each. Back and front lead weights fitted to helmet with clips, weight of each 40-lbs. There is a shackle used for protecting the stomach from pressure, when diving in deep water. Leather belt with metal pipe holder, knife in water-tight brass case, used for the Diver to clear away ropes. The air pipe is made of five ply India-rubber and canvas, with spiral wire imbedded with gun metal joints lashed in each, of 45-ft. and 30-ft. lengths, the pipe stands pressure to 100-lbs. square inch.

Care must be taken in dressing the Diver, but when once under water the apparatus is self acting, so that the man has only to think of the work he is going to perform. It is always advisable in a ship's crew to have the same men to attend the Diver, that is, the signal man and men to work the pump; the ship's diver should be according to Admiralty regulation,—down at least once a month.

Divers at the depth of 32 feet under water have upon the surface of their whole body a more than ordinary pressure of 20,000 lbs. weight, yet when we consider the uniformity of that pressure, and its equability, which causes no dislocation of the parts, all the external being equally affected with it, and being internally supported by the air and other elastic fluids, which constantly endeavour the more to expand themselves as they are compressed—if we also consider the firm texture of the membranes and other solid parts of the human body, and the incredible force they are able to bear, as has been demonstrated by experiments—we shall not much wonder that Divers complain of no sensible pain, though they be pressed with so great a weight of water, besides the ordinary pressure of the air, which our bodies are continually exposed to, and which is equal to a depth of 32 feet of water, or 20,000 lbs. so that the whole pressure to which a Diver is exposed at 32 feet of water is 40,000 lbs., and in ratio as he descends.

THE FOLLOWING TABLE REPRESENTS THE PRESSURE IN LBS. ON THE SQUARE INCH AT A GIVEN DEPTH OF WATER:—

20 feet	8½ lbs.	90 feet	39 lbs.
30 ,,	12¾ ,,	100 ,,	43½ ,,
40 ,,	17¼ ,,	110 ,,	47¾ ,,
50 ,,	21¾ ,,	120 ,,	52¼ ,,
60 ,,	26¼ ,,	130 ,,	56½ ,,
70 ,,	30½ ,,	140 ,,	60¾ ,,
80 ,,	34¾ ,,	150 ,,	65¼ ,,

As to the effect of compressed air upon the lungs, and the general construction of the human body, scientific men vary, and as we have no data from which we could form actual conclusions, we can only arrive at a result by the effects produced upon Divers that have come under our notice. The first time a man descends under water he is ordinarily suffering from inherent nervousness, occasioned by the fact of undertaking a thing that has hitherto been unknown to him, consequently there is an increased pulsation and peculiar gasping for breath; and it happens in some cases to be so strong that it would be unadvisable whilst in that condition to allow them to descend, but rather by making them acquainted with the working of the apparatus, and the example of teaching others to descend, it will gradually remove the nervous weakness which attends many constitutions. When the nervousness is overcome it would be advisable that they should descend slowly, swallowing their saliva, and not demanding too much air, resting at times to recover their equilibrium; and if the pressure should cause too great a pain in the head, by gently ascending a few feet it will gradually remove it, and the descent can be continued.

Remarking the effects of compressed air on the lungs, MR. BRUNEL, when inspecting an accident over the Thames Tunnel, the bell not being able to enter the hole, he took a rope from one of the attendants and plunged into it, remaining about two minutes, much above man's natural powers. The compressed air in bell being about 32 lbs. on the square inch, it is reasonable to suppose that the lungs act as air-sponges, and were saturated with this condensed vital element as he left the bell. Would it not, therefore, take more time than in ordinary circumstances to exhaust this double provision of breath?

The rule as to coming up depends very much upon the constitution of the Diver. A man at all sanguineous should ascend rather slowly; the brain being suddenly relieved from the pressure causes a sudden rush of blood to the head, and it may cause unpleasant and serious consequences. We should advise a Diver to ascend at a rate of not more than 2 feet every second—that is for a strong constituted man. Nor should a Diver, for at least two hours before commencing operations, take any food. If any kind of refreshments be required, a biscuit with a small quantity of drink, or anything that will not excite the digestion, may be taken during the operations.

Instructions for the Dressing of Diver and the Management of Apparatus.

Previous to the Diver being dressed, place the fly-wheel on crank shaft, and fix the handles at right angles, oil the pistons with olive or neatsfoot oil, and also the bearings and other working parts; let the pump be worked for a few minutes; also pour some water into the cistern to keep the cylinders cool, to prevent the air being heated and thus becoming rarefied. The piping should be laid on the deck or place from whence the descent is to be made, in a serpentine form, so that the pipe does not kink. Remove the nut on air-nozzle, and connect the air-pipe. The attendant should place his finger over the joint at the end of air-pipe, and let the pump work so that he can test the working of the air-pump's valves, and blow out any dust that may be in the pipes. If the water in the cistern becomes heated it should be renewed.

Put on the guernsey frock, a pair of drawers and stockings according to the temperature of the water; place helmet cushion on shoulders, and tie the tapes under the arm-pits. Then put on india-rubber dress, tie the outside collar piece round neck, and round each wrist place one or more vulcanised india-rubber rings. A piece of linen should be placed between the flesh and dress; the cuff-expanders should be used, so that the Diver can pass his hand; then put on helmet collar, place the vulcanized collar over the screws, put on the metal plates, and screw the dress between moderately tight with the wing nuts; be particular that the four nuts at joints are screwed up at the last. To keep the dress from chafing put on large overall stockings and canvas overall dress, then the boots with lead soles, and the leather belt with pipe-holder and knife. The attendant should blow through the outlet valve of the helmet; he can do so by placing his head in the interior, and placing his mouth to the hole where the air escapes; blow strongly; if in proper working order the valve will vibrate.

Connect air-pipe to inlet valve; previous to doing so pass it through the pipe-holder on belt, leading it under the "left arm" of

Diver, the signal line to be fastened round the body, and to pass up the front of the right shoulder; the head-piece (without the front-glass) can now be screwed on, which is done by one-eighth of a turn. Next attach the lead weights, one behind and one before; the lines of the back weight pass over the loops on head-piece; the small line should be fastened to the lower corner of weights and round the waist with a slip knot in front. The Diver now being dressed the air-pump must be set in motion; when all is ready for the Diver to descend, screw in the front glass.

Life Line.	1.	Pull	All right.
,,	2.	,, ...	According to Diver's Instructions.
,,	3.	,, ...	,, ,,
,,	4.	,, ...	Coming up.
Air Pipe.	1.	Pull	Sufficient air.
,,	2.	,, ...	More air (pump faster).
,,	3.	,, ...	According to Diver's instructions.
,,	4	,,	Haul up Diver.

To communicate with the Diver when underneath the water by word of mouth, we have made several experiments with very satisfactory results, and now have it in operation.

Divers' ladders are generally made of inch rope with ash rounds 22 inches long and weighted at the end. Some divers have the ladder only 20 feet long, to the last round a rope with a weight attached, which rests on the bottom; by that means they descend.

All now being ready let the Diver descend, and when he reaches the bottom, before he leaves the ladder, he must make fast a small leading line to the ladder. The line should be coiled in the hand with a loop round the wrist, and as he leaves the ladder he lets the rope gradually uncoil, so that if he is any distance off he can find his way back to the ladder if he wants to ascend; but if by accident he loses the line and is unable to find the ladder, he should make the signal to haul him up. In extreme cases the weights may be thrown off—that is, if he finds himself in any danger and he wishes to rise directly to the surface of the water; but this expedient should be seldom resorted to, as the signal for hauling up can always be given, and with presence of mind many difficulties can be overcome when hurry and excitement may cause the loss of the Diver's life.

The Diver should seldom go forward; he must generally go backwards; and if he meets with anything he must turn round and feel, particularly in the dark; but be careful to return the same way, otherwise he crosses the pipe and line; this precaution is very necessary. If entangled in the rigging make use of the knife at side to clear himself away.

As some men require more weight to sink them than others, we would recommend them to make a shot belt to buckle round the waist; it may be made any weight the Diver may think necessary.

On the deck or place from whence the Diver descends, two careful confidential persons must attend the signal line and air-pipe; they must attend them with the greatest vigilance, and keep them always moderately tight. If they should feel any irregular jerks which may be occasioned by falls or otherwise, they must haul him up immediately. The attendants on deck must from time to time give the signal that all is right, and if the Diver does not return the signal he must be immediately hauled up. Be particular that no conversation whatever is held with the attendants of the Diver when below, as it may take off their attention from the signal or any circumstance that may occur.

If the plungers of the air-pump or the other motions get slack, they must be screwed up with the spanners sent for that purpose, when the plungers will swell out a little; great care to be taken that they are put in the same way as they are taken out, and all other parts of engine put together according to the marks. Always use olive-oil for the air pumps; if not to be got, use well-cleaned neatsfoot oil. When done working, and the engine is to be put by some time or lifted about, unscrew the plug at the lower edge of the back of the box and draw out the water, to prevent it washing over into the box or cylinders, or splashing engine, or corroding the cistern if left standing. If the air-pump has been left standing by for any length of time, pour some warm water into the cistern, as it will warm the cylinders and soften the oil round the pistons, and the pump will work much easier.

To examine piston valves, withdraw piston from cylinder. When the valve can be unscrewed, to examine cylinder bottom valves, lean the chest back and unscrew iron plate on bottom of chest, and then unscrew the valve bonnets; this arrangement avoids the trouble of removing engine from chest.

Be careful that the leather washers are between the gun-metal joints, so that the air may not escape, also that the joints are screwed together moderately tight.

Should the waterproof dress, from constant use or accident, get leaky, it is easily repaired by laying two or three coats of varnish on each side of the seam, rubbing it with the finger as much as possible into the perforations made by the needle, allowing each coat to dry before the next is laid on; the sides of the seam may then be laid down, and two or three coats applied in the same manner to the channel of the seam, when the prepared strapping (which should have an extra coat laid on and dried) may be immediately applied, and well pressed down with the hand. Superfluous varnish may be removed with a piece of Indiarubber, but it is better to lay it on the proper width, so as not to require cleaning off, as too much friction sometimes does injury. Beginners should prove their seams by tapping them, when moistened externally with water.

India-rubber diving dresses should never be packed away in a wet or damp state; they must be thoroughly dried both in and outside before so doing, otherwise they will mildew, and become so rotten as to be of very little service afterwards.

In case the Diver urinates in the dress it should be turned inside out and washed with clean water, and then allowed to dry.

If the above directions are properly attended to, the diving dresses will last much longer.

The india-rubber pipes should also be thoroughly dried before packing away.

Should the dress and pipes be lying by for any length of time, and become hard, place them in a gentle heat, when they will become quite soft.

APPENDIX.—No. 1.

GENERAL REMARKS ON THE
WINDS AND WEATHER of the MALAY ARCHIPELAGO.

North of the Equator.—North-easterly winds prevail from December to March inclusive. This is the fine season, the winds blowing strong and steadily, except in the Sulu sea, where variables prevail.

Southerly and south-westerly winds prevail from May to September inclusive. This is the wet season, and the winds are variable in force and direction with bad weather. Sudden and violent squalls from the north-west occur in the Celebes and Sulu seas.

October and November are unsettled months; the N.E. monsoon not being fairly established before the middle of December.

South of the Equator.—West and north-west winds veering to north-east prevail from November to March. On coasts having a northern aspect, land and sea breezes, with unsettled weather, and rain will be found.

South-east and east winds prevail from May to September; generally fresh and steady, with fine weather, on coasts with a northern aspect, but bringing rain and bad weather to coasts open to the southward.*

New Guinea.—On the west coast of New Guinea are also two monsoons, the S.E. lasting from April to October, and the N.W. beginning with the end of October and terminating towards April.

---

* It has been remarked that between May and September, the season of the S.E. and S.W. monsoons, that in passing from one monsoon to the other, both in Gaspar strait and in the Molucca passage, the wind draws round gradually by the south.

In January, near this island, the wind sometimes varies from N.N.W. to N.E. In the spring the weather is often changeable, and in March, April, and May, it is squally. From June to September a great deal of rain falls; from October to May the weather is fine and calm, without either cloud or fogs.

Rainy Seasons.—In this archipelago, situated as it is in the vicinity of the equator, and within the regions of calms and doldrums caused by the meeting of the northern and southern wind systems, the wet and dry seasons are not strangely contrasted, as a great amount of rain falls more or less all the year round. The same monsoon is often stormy at sea, but fine near the land; as a rule, bad weather with rain is felt on coasts and islands that lie to windward, whilst leeward coasts enjoy fine weather.

Currents in the Eastern Passages to China.

The currents in the passages east of Java are very various, and have not yet been reduced to any fixed laws. The great irregularities they appear to be subject to is doubtless due to their geographical relations, lying as they do between the wind systems of the northern and southern hemispheres. But as their action is frequently of imimportance in endeavouring to make a passage against adverse winds, they require much attention. The notes which follow are given as a guide to their general character.

On the South Coast of Java, where the monsoons are liable to great deviations, there are some *remarkable reverse currents* experienced within a degree or two of the coast. During the easterly monsoon, April to November, a constant easterly current is encountered running against the monsoon, at times so strong as to ripple, but on an average 10 to 12 miles a day. The drift is frequently to S.E. two-thirds of a mile an hour. Captain M. H. JANSEN has, however, stated that in the east monsoon the current sets to the westward from full to change of the moon, and either to the eastward from the change, or that there was no current. It is certain that near the shore there is a considerable set to the westward in this monsoon.

In the west monsoon, December to April, the current sets sometimes to the S.S.E. and South, decreasing in force between 11° and 15° S. lat., and then ceases, and a strong westerly current is encountered increasing in velocity as Sunda strait is approached, amounting at times to 42 miles a day.

In **Baly Strait** the currents or tides run through the narrows with exceeding velocity, some say 6 knots, and cause great ripplings, eddies, and a boisterous sea, particularly near the shore of Baly during the S.E. monsoon, when the S.S.W. winds blow so strongly that it is often impossible to manœuvre a ship. The flood runs to the northward and the ebb to the southward, and it is high water, full and change, between 10h. and 1h. At neaps the tides are irrregular. They change first on the Java side of the strait, and about two hours later on the Baly side. During the east monsoon, May to November, the flood is often only found near the Java shore, and even then not to the northward of the strait, but during the west monsoon the northerly currents prevail. A tide often lasts for seven or eight hours.

Ombay Passage.—The currents are strong, with great ripplings, in the Ombay passage and other passages northward of Timor, generally setting to the N.E. during the west monsoon, and to the S.W. during the east monsoon; but in some places, close in shore, weak tides have been experienced. The strong current in the Ombay passage seems to cause a strong Easterly current along the north coast of Ombay during the east monsoon, May to November.

In June the S.W. current of Ombay passage attains its greatest strength, amounting to from 72 to 82 miles in 24 hours. Near the end of the east monsoon in August and September, there are strong easterly currents in Ombay passage, though in October they often run with great velocity to the south-westward.

Near the entrance of the straits of Allor and Pantar the current takes a northerly direction during the east monsoon, but during the west monsoon it sets out S.S.W.

Java to Amboina.—Ships from Java or Macassar bound to Amboina or Molucca channels during the east monsoon work along the north coasts Sumbawa, Flores, Ombay, &c., till they have reached the N.W. or North point of Wetta, or farther eastward if bound to Banda, and the voyage is often much accelerated by favourable currents.

Molucca Channels.—During the east monsoon, May to November, the current sets to the north-west along the western coast of New Guinea, and between the Ki and Arrou islands, and thence westward along the south coast of Ceram, at the rate of 1 or 1½ miles an hour, according to the strength of the wind, the velocity being greatest along the coast of New Guinea. At the same period an easterly current prevails on the north side of the islands extending from Timor to Timor Laut, so that a moderately fast vessel would experience no difficulty there in beating up against the monsoon. In the west monsoon the current in these seas usually sets with the wind, but its velocity is not so great as during the other season.

New Guinea.—Of the currents on the north coast of New Guinea we have few particulars, and these chiefly from D'URVILLE who sailed along it in August, 1827, where he found strong West and N.W. currents of more than a mile an hour. It is probable that this westerly drift is constant. Later information shows that it merges in the equatorial counter current.

APPENDIX No. 2.

CHINA.

Money, Weights, and Measures.

Money.—The only native coin in use in China is the *tsien*, called *cash* by the English, and *sapeque* by the French, who derive it from the Portuguese *sapeca*. It bears on one side the name of the province it is cast in, in Manchu letters, also the Chinese word "money"; and on the other side the name of the reigning Emperor, and above and below the words "current money" in Chinese characters.

Spanish, Mexican, and South American dollars, though not acknowledged by the government, are employed as a commercial medium throughout the maritime provinces and at the interior treaty ports. Lumps of stamped silver, called Sycee, pass current at a fixed standard of purity.

The nominal moneys of account are the *liang, tsien, fan,* and *li,* called by foreigners *tael, mace, candareen,* and *cash,* the proportion of which, one to the other, is decimal, but from various causes there is great diversity in the number of cash given in exchange for the tael. The terms *tael, mace, candareen,* and *cash,* are merely denominations of weight.

The circulating medium, in transaction with foreigners at the open ports, is chiefly in whole or broken dollars, clean or "chopped"* ; and the value of the dollar in relation to the tael is variable, the latter being approximately one third more.

Commercial Weights.—The unit of the table is the *liang* or *tael.*

    1 kernel of millet is equal to   1 *shu.*
    10 shu . . . . . = 1 *lui.*
    10 lui . . . . . = 1 *chu* or pearl.
    24 chu . . . . . = 1 *liang* or tael.
    1 tael = 1·333 oz. Avoirdupois = 37.796 grammes.

Also, 16 liang or taels =1 *kin* or catty = 1⅓ lbs. Avoirdupois.
    2 kin . . =1 *yin* . . = 2⅔ ,,   ,,
    30 kin . . =1 *kiun* . . = 40 ,,   ,,
    100 kiun . . =1 *tan* or picul = 133⅓ ,,   ,,
    120 kin . . =1 *shih* or stone = 160 ,,   ,,

The picul and catty are chiefly used in dealings with foreigners. The following equivalents will be found useful :—

    1 ton is equal to   16 piculs, 80 catties.
    1 cwt. . = 84 catties.

---

* "Chopped" dollars are those which are stamped all over and defaced with innumerable, private, commercial (or hong) marks. "Clean" dollars have no mark or stamp whatever.

```
1 lb. avoirdupois =   ¾ of a catty or 12 taels.
4 ozs.           =   3 taels.
1 picul          =   1·19047 cwt.
3,000 taels      =   302 lbs. troy.
```

Chinese weights and grain measures, also the linear long and land measures, all vary in different parts of the country, but as a general rule they are largest and longest in the southern provinces.

The difference in the values of the weights above a tael, as fixed by treaty, and those in common use in China, are as follows:—

```
        British Treaty.        French Treaty.        Common Weights.
Tael .  1·333 oz. avoir.    37·783 grammes.       1·328 oz. avoir.
Catty   1·333 lbs.  ,,      604·53     ,,         1·326 lbs.  ,,
Picul.  133·33     ,,  ,,   60·453 kilogr.        132.6      ,,   ,,
Stone   159·99     ,,  ,,   72·544     ,,         159·1      ,,   ,,
```

Measures.—The *li* is generally estimated by foreigners to be about one-third of a geographical mile.

```
Length.    1 grain is equal to   1 fun.
           10 fun           =    1 tsun or inch.
           10 tsun          =    1 chih or foot = 14·1 inches.
           10 chih          =    1 chang or pole = 11·75 feet.
           10 chang         =    1 yin.
Capacity.  1 grain of millet =   1 suh.
           6 suh            =    1 kwei.
           10 kwei          =    1 tsoh.
           10 tsoh          =    1 chau or handful.
           10 chau          =    1 choh or ladle.
           10 choh or 2 yoh =    1 koh or gill = 0·103 litre.
           10 koh           =    1 shing or pint = 1·031 litre.
           10 shing         =    1 tau or peck = 10·310 litre.
```

These are taken from the "Chinese Commercial Guide," by S. WELLS WILLIAMS,—an excellent work.

INDEX.

ABBREVIATIONS, hydrographical, 140.
Abbreviations, hydrographical, on foreign charts, 144.
Abstinence necessary in diving, 399.
Accidents, injuries, &c., rules for, where medical aid cannot be at once obtained, 302 to 310.
,, shock or collapse, 302.
,, loss of blood from a wound, 302.
,, arterial bleeding, 303.
,, to transport a wounded person, 303.
,, dislocations, 303, 304.
,, bruises and sprains, 304.
,, burns or scalds, 304.
,, sun stroke, 304.
,, poisons and antidotes, 305.
,, asphyxia, 305.
,, poisonous fish, 305.
,, stings from wasps, &c., 306.
,, treatment of cholera, 306.
,, caution as to using opium, 306.
,, fever and ague, 306.
,, frostbite, 307.
,, frozen limb, 307.
,, apparently dead from cold, 307.
,, fainting, 307.
,, drunkenness, 307.
,, epilepsy or fits, 308.
,, bite of mad dog, 308.
,, marsh poison, recipe for, 308, 309.
Aden, Gulf of, currents, 109.
Admiralty Charts, note on vast amount of information in, 143.
Admiralty Charts, tides on, 167.
Admiralty Knot, table for converting into statute miles, 201.
Admiralty Standard Compass, 37.
Africa, East Coast of, money of, 314.
,, West Coast of, money of, 318.
,, East Coast of, winds, 77.
Ague and fever, treatment of, 306.
Agulhas current, limits of, 110, 111.
,, ,, causes of, 111.
,, ,, recurving of, 111.
,, ,, temperature of, 111.
,, ,, warm and cold bands, 112.
Air pump for diving, management of, 400.
,, valve in working order, 400.
Aleutian current, 112.
Amazon River, stream from, 107.
American Coast, hurricanes, 78.
America (U.S.), money of, 313.
,, system of buoyage, 139.
Anchors and cables, boat cruising, 228.
Anchoring boats on rocky ground, 216.
Anchors, strength of, 364.
Aneroid barometer, description of, 93.
Angles, masthead, 9 to 16.
,, table of, 19.
Antidotes and poisons, 305.
Antarctic, cold, current, 111.
Apparently dead from cold, treatment of, 307.
Apparently drowned, treatment of, 293 to 301.
Apparent lights, 134, 135.
Apparatus for Diving, 397.
,, management of, 400.
Arabian sea currents, 109, 110.
,, ,, winds in, 77.
Arc of excess in sextants, 151.

ARC

Arc, length of, 206.
Area of sails, 33,
  ,,   ,,   deduction for reefs, 33.
Argand Burner, annual consumption of, 134.
Argentine Republic, money of, 313.
  ,,   ,,   weights and measures, 327.
Arrowroot, how to make, 378.
Arterial bleeding, *see Accidents.*
Artificial horizon, sextant, 194.
  ,,   ,,   pointing, 195.
Asphyxia, treatment of, 305,
Atlantic, correction for barometer in, 94.
Atlantic Ocean, average limit of trade winds, 74.
  ,,   ,,   rainy seasons, 74.
  ,,   ,,   cyclones, 74.
  ,,   ,,   variable belt, 74.
  ,,   ,,   currents, 103 to 108.
  ,,   ,,   Gulf Stream, 103 to 105.
  ,,   ,,   Guinea current, 105, 106.
  ,,   ,,   Equatorial, 106.
  ,,   ,,   stream from Congo, 106, *note.*
  ,,   ,,   Equatorial counter current, 107.
  ,,   ,,   Brazil and Guiana currents, 107, 108.
  ,,   ,,   stream from Amazon, 107.
  ,,   ,,   stream from R. Plata, 108.
  ,,   ,,   ice, 118.
  ,,   ,,   passage table, 121, 122.
Atlantic, tables of barometric pressure in, 95, 96.
Attack of a position, operations on shore, 252.
Australian current, 117
Australia, India and Australia passage tables, 123, 124.
Australia, India and China to England, passage table, 120.
Australia, tides in, 171.
Australia, west shore of, currents, 112.
Australian Coast, hurricanes on, 78.
Austria, money of, 313.

BEA

Austria, weights and measures, 329.
Azimuth Compass, 38.
Azimuth Tables, Burdwood's, 41, *note.*
Azimuth Tables, Davis', 381.

BAB-EL-MANDEB Strait, currents, 110.
Baked oil, 370.
Baly Strait, currents in, 406.
Bareca for beacon, 155.
Barley water, 378.
Barometer, 90 to 98.
  ,,   conclusions drawn from its movements must be checked for temperature, &c., 90.
  ,,   shifting of wind, 90, 91.
  ,,   sudden rise of, 90, 91.
  ,,   Admiral Fitz-Roy's scales for, 91.
  ,,   Buys—Ballot's law, 91.
  ,,   reading the, 92, 93.
  ,,   aneroid, description of, 93.
  ,,   average range of, 94.
  ,,   the fall of in hurricanes, 94.
  ,,   correction for, in Atlantic, 94.
  ,,   tables of pressure in Atlantic, 95, 96.
  ,,   table for converting millimètres into English inches, 97.
  ,,   table for converting English inches into millimètres, 98.
  ,,   measuring heights by, 200.
  ,,   revolving storms, 82.
  ,,   use of thermometer in conjunction with, 99.
Bars, crossing, in boat cruising, 229.
Bathers, hints to, issued by the Royal Humane Society, 291.
Battalion, complement of, operations on shore, 243.
Battalion, landing, &c., operations on shore, 243.
Battery, the, operations on shore, 244.
Beach Master's duties, 257, 264, 274.
Beaching or landing through a surf, 221.
  ,,   ,,   on a steep beach, 221.

INDEX. 413

BEA

Beaufort notation, weather, 85.
" " wind, 73.
" " French equivalent, 356.
Bearings on charts, how expressed, 141.
" sailing along a coast, 149.
Beef, how to stew, 376.
Beef tea, 379.
Belcher, Admiral, on resilvering sextant glasses, 152.
Belt, diving, 397.
Belgium, money of, 313.
" system of buoyage, 139.
" weights and measures, 326.
Bengal, Bay of, winds in, 77.
" " cyclones in, 78.
Bill of health, ship's papers, 226.
Bills of lading, ship's papers, 225.
Bill of sale, ship's papers. 225.
Bivouac, operations on shore, 248.
Black gun polish, to make, 373.
" polish for iron, 373.
" stain, to make, 371.
" varnish, to make, 370.
Black Stream, see *Kuro-Siwo*.
Blood, loss of, see *Accidents*.
Boarding a wreck or vessel under sail, 222, 223.
" merchant vessels, 223.
" officers, instructions to, in the capture of slavers, 230, 231.
" vessels, boat cruising, 228.
Board of Trade Catechism for the use of examiners in seamanship, 63 to 66.
" " Instructions, Rocket and Mortar apparatus, 281 to 283.
" " Merchant Shipping Acts, 67 to 69.
" " regulations for preventing collisions, 61.
Boat cruising (especially on East Coast of Africa), 226 *et seq*.
" " equipment. 226, 227.
" " rig, 227.
" " provisioning and cooking, 227, 228.
" " clothing and sleeping, 228.

BOI

Boat cruising, anchors and cables, 228.
" " boarding vessels, 228.
" " crossing bars, 229.
" " general remarks, 229.
" " careful steerage, 229.
Boat racing, 233 to 237.
" " rules for. 233, 235.
" " programmes for, 234, 237.
" " rowing regulations, 235.
" " sailing regulations, 236.
Boat's anchor, sounding, 155.
" details for, employed in disembarking troops, 259, 271.
" management of a large number of, disembarking troops, 273.
" particulars of steam, supplied to Navy, 209.
" instructions for working the engines of steam launches, 210 to 212.
" pulling, weights of, 213.
" practical hints for the guidance of officers having charge of. 214.
" rule of the road for, 214.
" conduct of men, 214.
" running dead before the wind, 215.
" if caught in a sudden squall, 215.
" passing to leeward of a vessel, 215.
" water ballast. 215.
" towing, 216.
" anchoring on rocky ground, 216.
" landing in a surf, 216.
" rules on management of open rowing boats in a surf, beaching them, &c., 216.
" in rowing to seaward, 217.
" on running before a broken sea, 218 to 221.
" beaching or landing through a surf, 221.
" do. do. on a steep beach, 221.
" boarding a wreck or vessel under sail, 222, 223.
" boarding merchant vessels, 223.
Boats' signals, disembarking troops, 261, 262, 267.
" steam, see *Steam Boats*.
Boiled oil, recipe for, 369.

414 INDEX.

BOI

Boiling, 375.
Boilers, 385.
Boiler tubes, 386.
Boilers, working with sea water, 386.
Boots, diving, 397.
  ,, to waterproof, 375.
Bombay, weights and measures of, 323.
Bottom, quality of, how expressed on foreign charts, 144.
Bourchier's life buoys, 284, 285.
Boxer's rocket, 281.
Braca or fathom, Portuguese, 142.
Brass lacquer, 373.
Brasse or fathom, French, 142.
Braza or fathom, Spanish, 142.
Brazil currents, 107, 108.
  ,, money of, 313.
  ,, weights and measures, 327.
Breeches buoy, 282.
Bright red, to dye, 372.
British India, money of, 313, 314.
  ,, ,, weights and measures, 323.
British West Indies, weights and measures, 324.
Brunel, Mr., experience in diving, 399.
Bruises, remedy for, 304.
Builder's contract, ship's papers, 225.
Buoying dangers, disembarking troops, 270.
Buoys, life, *see Life Buoys.*
Buoy systems, 136 to 139.
  ,, ,, of Corporation of Trinity House, 136.
  ,, ,, regulations for colouring buoys, 136, 137.
  ,, ,, Scotland, 137.
  ,, ,, France, 137.
  ,, ,, America, 139.
  ,, ,, Canada, 139.
  ,, ,, Holland, 139.
  ,, ,, Belgium, 139.
Burdwood's Azimuth tables, 41, *note.*
Burmah, money of, 314.
Burnett's solutions, disinfectant, 309.
Burns or scalds, cure for, 304.
Burt's nipper and buoy, 160.
Buys-Ballot's laws, barometer, 91.

CHA

CABLE'S length, 142.
  Cables, chain, strength of, &c., 315.
Carbolic acid, 309.
Carrington's spheroidal tables. 164 to 166.
Carrington's tables for converting mètres and decimètres into English feet and fathoms, 162, 163.
California, tides in, 171,
  ,, lower, hurricanes in, 78.
Camp equipment, operations on shore, 246.
Canada, money of, 314.
  ,, system of buoyage, 139.
  ,, weights and measures, 324.
Cape of Good Hope, money of, 314.
  ,, ,, weights and measures, 324.
Capture of Slavers, remarks on, 230 to 232.
Casks, floating power of, 334.
  ,, and tanks, capacity of, 334.
Catadioptric lights, 131.
Catechism for the use of examiners in seamanship issued by the Board of Trade, 63 to 66.
Cattle, hoisting in, 380.
Catoptric lights, 131, 132.
Cement, 374.
Centigrade thermometer, 99.
  ,, compared with Reaumur and Fahr., 101.
Centrifugal pump for gunboat, 387.
Certificate of registry, ship's papers, 224.
Ceylon, currents, 110.
Chain cables, strength of, 363.
Chain or wire rope equal to hempen rope, 362.
Charter party, ship's papers, 224.
Charts, signs and abbreviations on, 140, 141.
  ,, how constructed, 141.
  ,, soundings on, 141, 142.
  ,, underlined figures on, 141.
  ,, velocity of tides, how expressed on, 141.
  ,, tides, how measured, 141.
  ,, heights, how expressed, 141.
  ,, bearings, how expressed, 141.
  ,, natural scale for, 142.

INDEX. 415

CHA

Charts, meridians adopted in the construction of, 142.
   ,, note on vast amount of information in Admiralty, 143.
   ,, quality of bottom on, 144.
   ,, wind and current, note on, 76.
   ,, wind, 74 and 76.
   ,, current, 108.
Chili, money of, 314.
China, India and Australia to England passage table, 120.
China, India and Australia passage tables, 123, 124.
China, money of, 314.
   ,, weights and measures, 330.
China Seas, tides in, 170, 171.
   ,, typhoons, 78.
China Sea, winds, 77.
China, money, weights and measures, 407.
Chinese mariner's compass, 53.
Chinese currency, dollars, 408.
Chloride of lime, disinfectant, 309.
Cholera, treatment of, rules for, 306.
Chronometers, 188 to 193.
   ,, Admiral Shadwell on, 188.
   ,, mode of stowing, 188, 189.
   ,, relative position of, 189.
   ,, naturalization of, 189.
   ,, manner of winding, 190, 191.
   ,, comparing with standard, 191.
   ,, standard, 191, 192.
   ,, measuring meridian distances, 192.
   ,, meridian distances between Beirut and Sidon, 193.
Cirro-cumulus clouds, description of, 87
Cirro-stratus clouds, description of, 87..
Cirrus clouds, description of, 87.
Classification of timber (Lloyd's), 336.
Clearance, ship's papers, 226.
Cloth, to waterproof, 375.
Clouds, 86 to 88.
   ,, scale adapted for denoting amount of, 86.
   ,, what weather they indicate, 86.
   ,, description of, 87, 88.

CON

Clothing requisites for 100 men for 90 days, 393.
Coal, capacity of stowage room, 332.
   ,, average amount burnt in different kinds of engines, 385.
   ,, useful notes on, 388.
   ,, exposed to atmosphere, 388.
   ,. combustion of, 388.
   ,, different sorts of, 388.
Coarse mortar, 374.
Code of signals, 30, 31, 32.
Coffee, how to make, 378.
Cold Antarctic current, 111.
Cold, to restore persons affected by, 307.
Cold Wall of Gulf Stream, 103.
Collisions, how to avoid, 59.
   ,, aids to memory to avoid, 60.
   ,, extract from Board of Trade regulations for preventing, 61.
   ,, Merchant Shipping Acts, 67 to 69.
Colomb, Captain, on slave catching, 231.
Colomb's flashing signals, 23 to 28.
Colouring, 372.
Colours, mixing, 366.
Combustibles for lights, 133, 134.
Commercial weights, China, 408.
Compass, Admiralty Standard, 37.
Compass signals, 3, 4.
Compass, table for converting points of the, into degrees, &c., 49, 50.
   ,, to correct for variation, 51.
   ,, to correct compass courses, 51.
   ,, to convert true course into compass course, 51.
   ,, to correct the course at once for variation and deviation, 51.
   ,, of various nations, 52, 53.
   ,, standard, *see Standard Compass*.
   ,, variation of, notes on reporting, 148.
Compound Engines, 384.
   ,, ,, boilers, 385.
Concrete, 374.
Conduct of men in boats, 214.
Condy's fluid, disinfectant, 309.
Congo, stream from, 106.

INDEX.

CON
Conventional signs for charts, 141.
Cookery, recipes for, 375 to 379.
", for the sick, 378.
", boiling, 375.
", roasting, 375.
", meat soup, 375.
", Irish stew, 376.
", rations of salt meat, 376.
", salt meat to prepare hurriedly, 376.
", to stew fresh beef, pork, &c., 376.
", to make tea, 377.
", to make coffee, 378.
", arrowroot, 378.
", barley water, 378.
", beef tea, 379.
", lemonade, 379.
", water gruel, 378.
", egg flip, 379.
Cookery for sick, 378.
Cork mattresses, &c., 287 to 290.
", ", Admiral Ryder on, 288.
Correction for barometer, Atlantic, 94.
Courses and bearings, notes on reporting, 148.
Course table, Smith's, 46, 47.
Courses, to correct, 51.
Courses, ship's should only be directed by Standard Compass, 43.
Counter current, Pacific, 116.
Crossing bars, 229.
Cumulo-cirro-stratus or Nimbus clouds, description of, 88.
Cumulo-stratus clouds, description of, 88.
Cumulus clouds, description of, 87.
Currents, 102 to 118.
", how distinguished, 102.
", drift and stream, 102.
", Oceanic rivers, 102.
", changes of temperature of the surface water, 102.
", of the Atlantic Ocean, 103 to 108.
", Gulf Stream, 103.
", ", its colour, 103.
", ", cold wall, 103.
", ", warm water limits, 104.
", ", velocity of, 104.

CUR
Currents, Gulf Stream, average tempera- of, 104.
", ", climate where it prevails, 104.
", ", taking advantage of, 105.
", Guinea, limits of, 105.
", ", velocity and tempera- ture, 105.
", ", between C. Verde and Cape Palmas, 105.
", ", in the Harmattan season, 106.
", Equatorial limits of, 106.
", ", volocity of, 106.
", ", separation between it and Guinea cur- rent, 106.
", ", stream from Congo, 106 *note*.
", ", surface temperature, 107.
", Equatorial counter, 107.
", Brazil and Guiana Coast, 107, 108.
", River Amazon, 107.
", ", stream from, 107.
", Rio de La Plata, 108.
", Indian Ocean, 108, 109, 110.
", Arabian Sea, 109, 110.
", Sokotra, 109.
", Red Sea, 109.
", Gulf of Aden, 109.
", Bab-el-Mandeb Strait, 110.
", Ceylon, 110.
", Agulhas, limits of, 110, 111.
", ", causes of, 111.
", ", recurving of, 111.
", ", temperature of, 111.
", ", warm and cold bands, 112.
", cold Antarctic, 111.
", west shore of Australia, 112.
", Pacific, 112 to 118.
", Aleutian, 112.
", Kuro Siwo or Black or Japan Stream, 112 to 115.

CUR

Currents, Kuro Siwo, during the monsoons, 112.
," description and limits of, 113.
,, velocity, 113.
,, temperature of, 114.
,, hot and cold belts of, 114.
,, Kamchatka, 115.
,, Oya Siwo, 115, 116.
,, drift of N.E. Trade, 116.
,, Mexican, 116.
,, Pacific Counter, 116
,, Peruvian, 117.
,, ,, Bight of Panama, 117.
,, Equatorial, 117
,, Australian, 117.
,, S.E. drift, 117.
,, notes on reporting, 146.
,, in eastern passages to China, 405.
,, south coast of Java, 405.
,, chart, 108.
C. Verde, current between C. Palmas and, 105.
Cyclones, Atlantic, 74.
,, average limits of, 74.
,, hurricanes and typhoons, general notes on, 78 to 82.
,, Indian Ocean, 77.
,, Meldrum on, 82 *note*.
,, Pacific, 75.
Cyclonic storms, 74.

DANGER angle, 180.
Danish fathom or favn, 142.
Danish mariner's compass, 52.
Davis' tables, 381,
Daylight gun, 381.
Deep sea sounding line, 160.
,, ,, table, 158.
,, ,, time occupied in, 158.
Defence of posts, operations on shore, 254.
Denmark, money of, 314.
,, weights and measures, 329.
Determining positions, 177 to 180.
,, ,, by two objects, 177.
,, ,, by three objects, 178

DIS

Determining positions, projecting by circles 178.
,, ,, by straight line projection, 179.
,, ,, from a lighthouse, &c., when running along the shore, 179.
,, ,, the danger angle, 180.
Deviation in iron vessels affected by heeling of ship, 42.
,, ,, affected by funnel, 42.
Deviation of the compass, forms for registering, 48.
Deviation table, 44.
Diagram to illustrate tides, 176.
Dimensions of ships, 348.
Dinner, hurried, 377.
Dioptric lights, 131, 133.
Dip of the horizon, 198.
Dip table, 20.
Disembarking troops, &c., operations on shore, 255 to 277.
,, ,, before an enemy, 255.
,, ,, rendezvous, 256.
,, ,, beach master's duties, 257, 264, 274.
,, ,, details for boats employed in, 259.
,, ,, preparation and fitting boats, 260.
,, ,, boat's signals, 261, 262, 267.
,, ,, provisions, 262.
,, ,, officers of flotilla, 262.
,, ,, landing water, 265.
,, ,, landing horses, 266.
,, ,, boats to be prepared to re-embark troops, 268.
,, ,, transports, how to be distinguished, 269.

DIS

Disembarking troops, transports, scheme for anchoring, &c., 270.
" " precautions against attack at sea, 270.
" " buoying dangers, 270.
" " hospital ship, 270.
" " preparation of boats, &c., 271.
" " construction of jetties, &c., 272, 273.
" " management of large number of boats, 273.
" " preparation of land transport, 275.
" " re-embarking, 276.
Disinfectants, 309, 310.
Dislocations, *see Accidents*.
Distances at which objects can be seen at sea, 196.
" between principal ports of the globe, 120.
" measuring table of distances at which objects can be seen at sea, 196.
" measuring, table showing distance of horizon at different elevations, 197.
" dip of the horizon, 198.
" to find the distance of an object by two bearings, and distance run between them, 199.
" to convert knots into miles, 201.
" to convert miles into knots, 202.
" measuring, 181 to 184.
" at sea, 181.
" of a target, 181.
" Rapers' method, 182.
" log. of cosines of principal points of high land, 183.
" by sound, 183.
" measuring on shore, 186.
" extemporary measurements, 186.
" pacing, 187.
" step of track horses, 187.

DRO

Distances, walking horse, 187.
Distress signals, by day or night, 69.
Diurnal inequality, tides, 168, 169.
Divers, information and instruction for, 397
" signals to, 401.
Diving apparatus, 397.
" presence of mind in, 402.
" careful persons to attend, 402.
" care of dress, 403.
Docks abroad, particulars of, arranged alphabetically, 343.
Dollars, as Chinese currency, 408.
" chopped, 408.
Dove's law of storms, 82.
Dove's tables, 83, 84.
Dress of men, operations on shore, 244.
" Divers care of, 403.
" diving, 397.
Drift and stream currents, 102.
Drift, N.E. Trade, 116.
Drift, S.E., 117.
Drowned, apparently, treatment of the, 293 to 301.
" " medical assistance, 294.
" " to restore breathing, 294.
" " to excite breathing 294.
" " to imitate breathing, 295.
" " Dr. Marshall Hall's method, 296, 297.
" " to imitate the movements of breathing, 298.
" " Dr. Silvester's method, 297, 298.
" " treatment after breathing has been restored, 299.
" " general observations, 299.
" " appearances accompanying death, 299.

DRO

Drowned, apparently, cautions, 300.
" " Dr. Howard's direct method, 300.
" " Dr. Howard's after treatment, 301.
Drowning persons, instructions for saving, 292, 293.
Drunkenness, treatment of, 307.
Dubbing, 374.
Dutch elle, 143.
Dutch fathom or vaden, 142.
Dutch mariner's compass, 52.
Dyeing woods, &c., 372.

EAST Coast of Africa, money of, 314.
Egg flip, 374.
Egypt, money of, 315.
" weights and measures, 330.
" (slavers), 232.
Elle, Dutch, 143.
Engines, compound, 384.
" marine steam, useful notes, 382.
English Channel, tides in, 169.
English and French words for the state of the sea, 358.
English sea terms and phrases, 349 to 356.
Epilepsy or fits, treatment of, 308.
Equatorial counter current, 107 and 116.
" current, 106 and 117.
" " limits of, 106.
" " velocity of, 106.
" " separation between it and Guinea current, 106.
" " stream from Congo, 106, *note*.
" " surface temperature, 107.
Equipment, boat cruising, 226, 227.
Establishment of Port explained, 168.

FADEN or fathom, Prussian, 142.
Fahrenheit thermometer, rules for converting scale of, into Réaumur or Centigrade scales, 99, 100.

FOR

Fahrenheit thermometer, table showing comparisons of with Reaumur and Centigrade scales, 101.
Fainting, treatment of, 307.
Falkland Islands, money of, 315.
Famn or fathom, Swedish, 142.
Fathom or braca, Portuguese, 142.
" or brasse, French, 142.
Fathoms, table for converting into French mètres and decimètres, 162.
Fathom or braza, Spanish, 142.
" or faden, Prussian, 142.
Favn or fathom, Danish and Norwegian, 142.
Fathom or favn, Danish and Norwegian, 142.
Fathom or famn, Swedish, 142.
" or sashine, Russian, 142.
" or vaden, Dutch, 142.
Fever and ague, treatment of, 306.
Field gun's, crews and officers, operations on shore, 245.
Field guns, operations on shore, 253.
Field ice, 118.
Figures underlined on charts, 141.
Fishing vessels, lights for, 63.
Fits, treatment of, 308.
Fitz-Roy's, Admiral, scales for barometer, 91.
Fixed lights, 132.
Flags of slave traders, 232.
Flags, signal, for British merchant service, 31.
Flashing signals, Colomb's, 23 to 28.
Flashing signals with flags, 27.
Fleet sailing, 21, 22.
Fleet sailing, technical terms, 21, 22.
" " trying rate, 22.
Floating power of spars, 333.
" " tanks and casks, 334.
Flood and ebb tides, 167.
Fog signals, rule of the road at sea, 63.
Foreign charts, notes on, 142, 143, 144.
Foreign sea terms and phrases, 349.
" sails, 349.
" masts and yards, 350.
" miscellaneous, 351.
" phrases, 353.

2 D

FRA

France, money of, 315.
„ weights and measures, 324.
„ system of buoyage, 137.
„ tide signals, 138.
French and English words for the state of the sea, 358.
French, brasse or fathom, 142.
„ mariner's compass, 52.
„ mètre, 142.
„ polish, 371.
„ sea terms and phrases, 349 to 356.
„ words descriptive of weather equivalent for, 356.
Frost-bite or numbness, treatment of, 307.
Froude, Mr., trials in towing, 387.
Frozen limb, treatment of, 307.
Fuel, table to economize, 127.
Fumigate, to, a ship or clothing, 310.

GAS for lighthouses, 134.
Geographical mile, definition of, 201, note.
German mariner's compass, 52.
Germany, money of, 315.
„ weights and measures, 328.
Gibraltar, money of, 315, 316.
„ weights and measures, 324
Gilding, 367.
Glue, marine, 374.
„ cement, 374.
Gray, Mr., on rule of the road at sea, 57.
Greece, money of, 316.
„ weights and measures, 327.
Greek mariner's compass, 53.
Gruel, 378.
Guinea current, 107
„ „ limits of, 105.
„ „ velocity and temperature, 105.
„ „ between C. Verde and C. Palmas, 105.
„ „ in the Harmattan season, 106.
Gulf of Aden currents, 109.
Gulf Stream, 103 to 105.
„ „ its colour, 103.
„ „ warm water limits, 101.

HOR

Gulf Stream, cold wall, 103.
„ „ velocity of, 104.
„ „ average temperature of, 104.
„ „ climate where it prevails, 104.
„ „ taking advantage of, 105.
Gun polish, 373.
Gunboat, centrifugal pump for, 387.

HALL'S, Dr. M., method of treatment of the apparently drowned, 296, 297.
Hammocks for saving life, 289, 290.
Harmattan season, 106.
Harpoon ship's log, Walker's, 161.
Harpoon sounding machine, Walker's, 160.
Havre, tides at, 170.
Health, bill of, ship's papers, 226.
Heights, measuring, 198.
„ roughly, 185.
„ by barometer, 200.
„ table showing number of feet subtending an angle of 1' at a given distance, 206.
„ on charts, how expressed, 141.
Height of masthead, 7, 8.
Heights, table to find when distance is known, 198.
Helmet, diving, 397.
Hindoostanee mariner's compass, 53.
Hints for the guidance of officers having charge of boats, 214.
Hints to bathers, issued by the Royal Humane Society, 291.
Holland, money of, 316.
„ weights and measures, 327.
„ system of buoyage, 139.
Holophotal arrangement of lights, 132.
Hong-Kong, money of, 316.
Horary table, 28.
Horizon, dip of, 198.
Horizon table, 17 to 20.
„ „ Admiral Ryder on, 17, 18.
„ „ method of ascertaining the distance of an enemy's ship, 17 to 19.
„ „ table of masthead angles, 19

HOR

Horizon table, dip table, 20.
  ,,   ,, showing distance of at different elevations, 197.
Horses, hoisting in, 380.
  ,, landing, 266.
Horse-power, formulæ for, 383.
Horsey, Captain de, on rule of the road at sea, 57.
Hospital ship, disembarking troops, 270.
Hot and cold belts of Japan Stream, 114.
Howard's, Dr., method of treatment of the apparantly drowned, 300, 301.
Hull, Staff-Commander, on practical nautical surveying, 151.
Humane Society's hints to bathers, 291.
Hurricanes, cyclones and typhoons, general notes on, 78 to 82.
Hurricanes, fall of barometer in, 94.
  ,, the knowledge of, important, 78.
  ,, direction of, 78.
  ,, increasing violence in centre or vortex. 79.
  ,, warnings of, 79.
  ,, rule for discovering position of, with regard to ship, 80.
  ,, when centre of storm has passed over ship, 81.
  ,, Dove's tables, 83.
Hydrographical abbreviations, 140.
Hydrographic information, 140 to 206.
  ,,   ,, notes on reporting, 145 to 150

ICE in Atlantic Ocean, 118.
  ,, limits of field, 118.
  ,, in high southern routes, 118.
Icebergs, where met with, 118.
  ,, in high southern routes, 118.
  ,, when the greatest number may be met with, 118.
  ,, East of Cape Horn, 118.
  ,, South of Cape Leeuwin, 118.
  ,, indications of thermometer, 119.
  ,, should be passed to windward, 119.
Illuminants for lights, 133, 134.

KNO

Immersion of ship, 348.
Inches, table for converting into millimètres, 98
India, China, and Australia passage tables, 123, 124.
India, China, and Australia to England, passage table, 120.
India, money of, 313, 314.
  ,, weights and measures, 323.
Indian Ocean, currents, 108, 109, 110.
  ,,   ,, cyclones, 78.
  ,,   ,, tides in, 170.
  ,,   ,, winds, 77.
Information for divers, 397.
Injuries, *see Accidents*.
Instruction for divers, 397.
  ,, for dressing divers, 400.
International Code of Signals, 30, 31, 32.
  ,, answering Pendant, 31.
  ,, for British Merchant Service, 31.
  ,, two, three, and four flags, 32.
Invoices, ship's papers, 225.
Irish stew, 376.
Iron lacquer, 374.
Iron tanks and casks, capacity of, 334.
Italian mariner's compass, 52.
Italian metro, 143.
Italy, money of, 316.
  ,, weights and measures, 326.
Italian sea terms and phrases, 349 to 356.

JAPAN. money of, 316.
  ,, weights and measures, 330.
  ,, stream, see *Kuro-Siwo*.
Japanese mariner's compass, 53.
Java, currents on south coast, 405.
  ,, to Amboina, passage, 406.
Jetties, piers, &c., construction of, disembarking troops, 272.
Jib, to find number of square yards in, 339.

KAMCHATKA current, 115.
  Kisbie's life buoy, 283.
Knots, table to convert into statute miles, 201.

Knots, to kill, 368.
Kuro Siwo, or Black or Japan stream, 112 to 115.
,, during the monsoons, 112.
,, description and limits of, 113.
,, velocity of, 113.
,, temperature of, 114.
,, hot and cold belts of, 114.

Lacquering, 373.
Lac varnish, 370.
Lading, bills of, ship's papers, 225.
Ladders for divers, 401.
Landing in a surf, 216.
,, on steep beach, 221.
,, horses, disembarking troops, 266.
,, water, disembarking troops, 265.
Land transport, preparation of disembarking troops, 275.
Launches, steam, *see Steam Launches*.
Law of storms, 78 to 84.
Dove's tables, 83, 84.
Lead lines, 153 and 159.
Lemonade, 379.
Life belts, 286, 287.
Life-boat Institution, rules by, on the management of open rowing boats, beaching them, &c., 217.
Life buoys, 283, 284, 285.
,, the service, 283.
,, Kisbie's, 283.
,, Welch and Bourchier's, 284.
Light woods, to colour, 372.
Lights, 131 to 135.
,, catoptric, 131, 132.
,, dioptric, 131, 133.
,, catadioptric, 131.
,, magneto electric, 131.
,, conditions of superiority, 131.
,, holophotal arrangement, 132.
,, fixed, 132.
,, reflectors, 132.
,, revolving, 132, 133.
,, illuminants or combustibles for, 133, 134.
,, gas for, 134.

Lights, annual consumption of oil, 134.
,, apparent, 134, 135.
,, notes on reporting, 147.
,, rule of the road at sea, 62, 63
Lloyd's classification of timber, 336.
Log book, ship's papers, 225.
Log line, 161.
,, Massey's patent, 161.
,, Walker's harpoon ship, 161.
Logarithms, table of, 394.
Loss of blood, to check, 302.
Lunars, sextant, 194.
Low Archipelago, hurricanes in, 78.

MAD Dog, bite of, treatment for, 308.
Madras, weights and measures, 323.
Magnetic bearing, to determine the real or correct, 40.
Magneto-electric lights, 131.
Malabar coast cyclones, 78.
Malay Archipelago, winds and weather, 404.
Manifest, ship's papers, 226.
March, order of, operations on shore, 249 to 251.
Marine glue, 374.
Marine steam engine, 382.
Mariner's compass of various nations, 52, 53.
Marsh poison, cure for, 308, 309.
Massey's patent log, 161.
Massey's sounding machine, 160.
Masthead angles, 9 to 16, and 19.
,, height of, 7, 8.
Masts, yards, &c., foreign terms, 350.
Mattresses, cork, 287 to 290.
,, ,, Admiral Ryder on, 288.
Mauritius, money of, 316.
,, weights and measures, 324.
,, caution regarding storms of, 81.
McDougall and Calvert's disinfectants, 309.
Measuring distances, 181 to 184, 186, 187. 196, 197, 199.
,, ,, at sea, 181.
,, ,, of a target, 181.
,, ,, Raper's method, 182, 183.

INDEX.

MEA

Measuring distances, log. of cosines of principal points of high land, 183.
,, ,, by sound, 183, 184.
,, ,, on shore, 186.
,, ,, extemporary measurements. 186, 187.
,, ,, pacing, 187.
,, ,, step of track horses, 187.
,, ,, walking horse, 187.
,, ,, meridian, by Chronometers, 188, 193.
,, ,, sextant, 194, 195.
,, ,, table of distances at which objects can be seen at sea, 196.
,, ,, table showing distance of horizon at different elevations, 197.
,, ,, dip of the horizon. 198.
,, ,, to find the distance of an object by two bearings, and distance run between them, 199.
,, ,, table to convert knots into statute miles, 201.
,, ,, table for converting statute miles into knots, 202.
,, ,, trial trip table, 203, 204.

Measuring heights, 185 and 198.
,, ,, by barometer, 200.
,, ,, table showing number of feet subtending an angle of 1', at given distances, 206.

Measured mile, the, 203, 204.
Meldrum on cyclones, 82, *note*.
Men, provisioning for a given number of days, 390.
,, clothing for 90 days, 393.

MON

Mensuration, useful notes in, 331.
Merchant service, signals, 31.
,, Shipping Acts, 67 to 69.
,, vessels, boarding, 223.
Meridians adopted in the construction of foreign charts, 142.
Meridian distances, measuring, 192 and 193
,, ,, notes on, 148.
Mètre, French, 142.
,, the unit of French measures, 324.
Mètres and decimètres, table for converting into English feet and fathoms, 162, 163.
Metro, Italian, Portuguese, Spanish, 143.
Mexican current, 116.
Mexico, Gulf of, tides in, 170.
,, hurricanes of, 78.
,, money of, 317.
Mile, length of, English, 319.
,, ,, Indian, 323.
,, ,, French, 325.
,, ,, Dutch and Grecian, 327.
,, ,, German and Russian, 328.
,, ,, Austrian, Swedish, and Danish, 329.
,, ,, Turkish, Chinese, and Japanese, 330.
Mile, measured, the, 203, 204.
,, nautical and geographical, definitions of, 201, *note*.
Millimètres, table for converting into English inches, 97.
Mixing colours, 366.
Molucca Channels, currents in, 407.
Money, foreign, with its English value 313 to 318.
,, Africa, East Coast of, 314.
,, Africa, West Coast of, 318.
,, America (U.S.), 313.
,, Argentine Republic, 313.
,, Austria, 313.
,, Belgium, 313.
,, Brazil, 313.
,, British India, 313, 314.
,, Burmah, 314.
,, Canada, 314.
,, Cape of Good Hope, 314.
,, Chili, 314.
,, China, 314, and 407.

MON

Money, Denmark, 314.
,, Egypt, 315.
,, Falkland Islands, 315.
,, France, 315.
,, Germany, 315.
,, Gibraltar, 315, 316.
,, Greece, 316.
,, Holland, 316.
,, Hong-Kong, 316.
,, India, 313, 314.
,, Italy, 316.
,, Japan, 316.
,, Mauritius, 316.
,, Mexico, 317.
,, Mozambique, 314.
,, New South Wales, 317.
,, Norway, 317.
,, Portugal, 317.
,, Russia, 317.
,, Siam, 317.
,, Singapore, 317.
,, Spain, 318.
,, St. Helena, 318.
,, Sweden, 318.
,, Turkey, 318.
Monsoons, Indian Ocean, 77.
Monsoons, &c., notes on reporting, 147.
Morning rainbows, 88.
Morse alphabet, signals, 26.
Mortar, 374.
Movements in the field, 242.
Mozambique, money of, 314.
,, Channel, winds, 77.
Muster roll, ship's papers, 226.
Mutton, how to stew, 376.

NATIONAL Life-Boat Institution, rules published by, for management of open rowing boats in a surf, &c., 217, *et seq.*
National Life-Boat Institution, treatment of the apparently drowned, 293.
Naturalization of chronometers, 189.
Natural scale, how obtained, 142.
Natural sines, &c., table of, 396.
Nautical mile, definition of, 201, *note*.

OPE

Nervousness in diving, 399.
New South Wales, money of, 317.
N.E. Monsoon, Indian Ocean, 77.
N.E. Trade, Atlantic, 74.
,, drift, 116.
,, Pacific, 76.
New measurement, tonnage of vessels, 336.
New Guinea, currents in, 407.
,, ,, winds and weather, 404.
Nipper and buoy, Burt's, 160.
Northern hemisphere, thermometer in, 99.
Norway, money of, 317.
,, weights and measures, 329.
Norwegian fathom or favn, 142.
,, mariner's compass, 52.

OBSERVATIONS at night, sextant, 195.
Oceanic rivers, 102.
Officers, value of training, 241.
Official log book, ship's papers, 225.
Oil, annual consumption of, in lighthouses, 134.
,, boiled, 369.
,, baked, 370.
,, polish, 371.
Old measurement, tonnage of vessels, 338.
Omāni (slavers), 232.
Ombay Passage, currents in, 406.
Operations on shore, disembarking troops, 255 to 277.
Operations on shore, 241 to 254.
,, ,, employment of small-arm men when landed for service, 241 *et seq.*
,, ,, value of training officers and men, 241.
,, ,, movements in the field, 242.
,, ,, landing battalion and 4 guns, 243.
,, ,, battalion, consists of, 243.
,, ,, the battery, 244.
,, ,, dress of men, 244.

INDEX.

Operations on shore, what the men will carry, 244, 245.
,, ,, field gun's crews will carry, 245.
,, ,, field gun's officers will carry, 245.
,, ,, provisions, daily scale, 245
,, ,, camp equipment, 246.
,, ,, reserve ammunition, 246.
,, ,, do. carriage for, 247.
,, ,, pioneers, 248.
,, ,, bivouac for night, 248
,, ,, order of march, 249, 250, 251.
,, ,, attack of a position, 252.
,, ,, field guns, 253.
,, ,, defence of posts, 254.
Opium, caution as to using, 306.
Order of march, operations on shore, 249 to 251.
Oregon, tides in, 171.
Oya Siwo current, 115, 116.

PACIFIC Ocean, average limits of trade winds, 75, 76.
,, ,, cyclones, 75.
,, ,, hurricanes, 76.
,, ,, rainy seasons, 75.
,, ,, variable belt, 75.
,, ,, N.E. Trade, 76.
,, ,, S.E. Trade, 76.
,, ,, currents, 112 to 118.
,, ,, ,, Aleutian, 112.
,, ,, ,, Kuro Siwo or Japan Stream, 112 to 115.
,, ,, ,, Kamchatka, 115.
,, ,, ,, Oya Siwo, 115, 116.
,, ,, ,, drift of N.E. Trade, 116

Pacific Ocean Currents, Mexican, 116
,, ,, ,, Counter, 116.
,, ,, ,, Peruvian, 117.
,, ,, ,, Bight of Panama, 117.
,, ,, ,, Equatorial, 117.
,, ,, ,, Australian, 117
,, ,, ,, S.E. Drift, 117
Pacing, 187.
Pacific Ocean, passage table, 125, 126.
Painting ship, 365.
,, on glass, 370.
Paint, mixing, 366.
,, old, to remove, 367.
Panama, Bight of, current, 117.
Passage tables, 119 to 127.
,, ,, Atlantic, 121, 122.
,, ,, England to India, China, and Australia, 120.
,, ,, England to Pacific, 120.
,, ,, Australia to England, 120.
,, ,, India, China and Australia, 123, 124.
,, ,, Pacific, 125, 126.
Passport, ship's papers, 224.
Patent log, Massey's, 161.
Pendant board, 5.
Persia (slavers), 232.
Peruvian current, 117.
Pilots, signals for, by day or night, 68.
Pioneers, operations on shore, 248.
Phrases and sea terms (foreign), 349.
Plata, Rio de la, currents off, 108.
Points of compass, table for converting into degrees, &c., 49, 50.
Poisons and antidotes, 305.
Poisonous fish, effects of, and remedies for, 305.
Polish for guns, 373.
,, for iron, 373.
Polishing, French, 371.
Pork, how to stew, 376.
Portugal, money of, 317.
,, weights and measures, 326.
Portugese fathom or braca, 142.
,, metro, 143.

Position, attack of a, operations on shore, 252.
Posts, defence of, operations on shore, 254.
Practical Nautical Surveying, Staff-Commander Hull on, 150.
Practical recipes, 368.
Presence of mind in diving, 402.
Pressure, barometric, tables of, Atlantic, 95, 96.
Pressure of steam, 384.
Programmes for boat racing, 234, 237.
Provisions, disembarking troops, 262.
,, operations on shore, 245.
,, scale of for a given number of men and days, 390.
,, weight of, 321.
Provisioning and cooking, boat cruising, 227, 228.
Prussian fathom or faden, 142.
Prussia, money of *(see Germany)*, 315.
,, weights and measures of, 328.
Pulling boats, weights of, 213.
Pump, boat cruising, 227.
Purifying water, 379.
Putty, 369.

RACES, rowing, 237.
Racing, boat, *(see Boat Racing)*.
Rainbows, 88.
Rainy seasons, Atlantic 74.
Rain awning for boats, 227.
Rainy seasons, average limits of, 74 to 77.
,, ,, Pacific, 75.
,, ,, Malay Archipelago, 404.
Range of barometer in hurricanes, 94.
Range of tides, 168.
Raper's method of measuring distances, 182, 183.
Rations of meat, to soak and boil, 376.
Reading the barometer, 92, 93.
Rating chronometers, 192 and 194.
Reaumur's thermometer, rules for converting scale of, into Fahrenheit and Centigrade scales, 99, 100.

Reaumur's thermometer, table showing comparison of with Fahrenheit and Centigrade scales, 101.
Receipts for cooking, 375.
Recipes useful and practical, 368.
Red Sea, currents, 109.
Reflectors for lights, 132.
Rendezvous for disembarking troops, 256.
Reserve ammunition, operations on shore, 246.
Re-silvering sextant glasses, 152.
Restoring the apparently drowned, 294, 301.
Revolving lights, 132, 133.
Revolving storms, 78.
,, ,, rate of, 78.
,, ,, seasons of, 78.
,, ,, characteristics of, 78.
,, ,, direction of, 79.
,, ,, vortex of, 79.
,, ,, warnings of, 79.
,, ,, rules for avoiding, 80.
,, ,, caution in using theory of, 81.
,, ,, barometer, 79 and 82.
Rhymes, weather, 89, 90.
Rig, boat cruising, 227.
Rio de la Plata, currents off, 108.
,, ,, tides, 170.
Roasting, receipt for, 375.
Rocket line, 281.
Rocket and mortar apparatus, saving life from shipwreck, 281 to 283.
Rope equal to wire or chain, 362.
Rope making, 360.
Rope, proportionate strength, 361.
Rope stropping, size of, 362.
Rosewood stain, 372.
Rowing regulations, boat racing, 235.
Royal Humane Society's hints to bathers, 291.
Royal National Life-boat Institution, treatment of the apparently drowned, *see Drowned*.
Rules for avoiding revolving storms, 80, 81.
Rules for boat racing, 233, 235.

INDEX.   427

Rules for converting thermometer scales, 99, 100.
Rule for finding number of square yards in sails, 338.
  „ for finding tonnage of vessels, 336.
Rules on management of open rowing boats, beaching them, &c., 216, *et seq.*
Rule of the road at sea, 57 to 66.
  „ Captain de Horsey on, 57.
  „ Mr. Gray's remarks on, 57.
  „ general rule for steam ships meeting and crossing, 58.
  „ how to avoid collision, 59.
  „ aids to memory to avoid collisions, 60.
  „ extract of Board of Trade regulations for preventing collisions, 61.
  „ steering and sailing rules, 61.
  „ lights for steam tugs, 62.
  „   „ for ships being towed, 62.
  „   „ for sailing pilot vessels, 63.
  „   „ for fishing vessels and open boats, 63.
  „ fog signals, 63.
  „ catechism for the use of examiners in seamanship, issued by Board of Trade, 63 to 66.
Rule of the road for boats, 214.
Rules useful in mensuration, 331
Russia, money of, 317.
  „ weights and measures of, 328.
Russian mariner's compass, 53.
  „ sashine or fathom, 142.
Ryder, Admiral, on cork mattresses, 288.
  „   „ on horizon table, 17, 18.

SAILS, area of, table for computing, 33.
  „ dimension of, table for finding, 339.
Sailing, fleet, 21, 22.
  „ pilot vessels, lights for, 63.
  „ regulations, boat racing, 236
  „ trying rate of, 22.
Sail, sounding under, 155.
Sailors, a secret of success to, 195
Sale, bill of, ship's papers, 225.

Salt meat, to prepare, 376.
Sashine or fathom, Russian, 142.
Saving drowning persons, instructions for, 292, 293.
Saving life from Shipwreck, 281, *et seq.*
Scalds, cure for, 304.
Scales for barometer, Fitzroy's, 91.
Scotland, system of buoyage, 137.
Sea letter or brief, ship's papers, 224.
Sea, measuring distances at, 181.
Seaports, notes on reporting, hydrographic information of, 145.
Seasons in which hurricanes, cyclones, and typhoons prevail, 78.
Seasons, rainy, 74 to 77, and 404.
Sea terms and phrases (foreign), 349 to 356.
Semaphore signs, 29.
Service life buoy, 283.
Sextant, 194, 195.
  „ lunars, 194.
  „ pointing artificial horizon, 195.
  „ observations at night, 195.
  „ use of, a secret of success to sailors, 195.
  „ *see Sounding*, 151.
  „ arc of excess, 151.
  „ directions for re-silvering, 152.
S.E. drift, 117.
S.E. trade, Atlantic, 74.
S.E. trade, Pacific, 76.
Shadwell, Admiral, on chronometers, 188.
Ships being towed, lights for, 62.
Ship, dimensions of, 348.
Ship log, Walker's harpoon, 161.
Ship's log, ship's papers, 225.
Ship, to fumigate, 310.
Ship's papers, 224, 225, 226.
  „   „ voucher of nationality, 224.
  „   „ certificate of registry, 224.
  „   „ passport, 224.
  „   „ sea letter or sea brief, 224.
  „   „ charter party, 224.
  „   „ official log book, 225.
  „   „ ship's log, 225.
  „   „ builder's contract, 225.
  „   „ bill of sale, 225.
  „   „ bills of lading, 225.
  „   „ invoices, 225,

INDEX

SHI

Ship's papers, manifest, 226.
,, ,, clearance, 226.
,, ,, muster roll, 226.
,, ,, shipping articles, 226.
,, ,, bill of health, 226.
Shipping articles, ship's papers, 226.
Shipwreck, Boxer's rocket, 281.
,, the rocket line, 281.
,, breeches buoy, 282.
,, on a flat shore, 283.
,, saving life from, 281 *et seq.*
Shoals, sounding when out of sight of land, 155 to 157.
Shock or collapse, *see Accidents*.
Siam, money of, 317.
Sick, cookery, 378.
Siebe and Gorman's diving apparatus, 397.
Signals, 1 to 5.
,, boats', disembarking troops, 261, 262, 267.
,, Colomb's flashing, 23 to 28.
,, flashing, with flags, 27.
,, fog, rule of the road at sea, 63.
,, for pilots (Merchant Shipping Act) 68.
,, ,, in the daytime, 68.
,, ,, at night, 68.
,, international code of, 30, 31, 32.
,, answering pendant, 31.
,, for British Merchant Service, 31.
,, two, three, and four flags, 32.
,, of distress, 69.
,, ,, in daytime, 69.
,, ,, at night, 69.
Signal stations, list of, 30.
Signals, tide, in French ports, 138.
Signs, semaphore, 29.
Signs and abbreviations for charts, 140.
Silvester's, Dr., method of treatment of the apparently drowned, 297, 298.
Singapore, money of, 317.
Size, 369.
Sky, colour of, *see Weather*.
Slavers, capture of, remarks on, 230 to 232.
,, ,, special instructions to boarding officers, 230, 231.
,, ,, territorial waters, 231, 232

SPA

Slave catching, Captain Colomb on, 231.
,, traders, flags of, 232.
,, trading, ships engaged in, where likely to be met with, 232.
Small-arm men, employment of, when landed for service, 241, *et seq*.
Smith's straight line course table, 46, 47.
Sokótra, currents off, 109.
Sounding line, deep sea, 160.
,, machine, Massey's, 160.
,, ,, Walker's harpoon, 160.
Sound, measuring distances by, 183, 184.
Sounding, notes on, 150 to 163.
,, Staff-Commander Hull on, 150.
,, the sextant, 151.
,, ,, arc of excess, 151.
,, ,, directions for re-silvering. 152.
,, the lead line, 153.
,, noting, 153.
,, time to be noted, 154.
,, to ensure the lines sounded being straight lines, 154.
,, under sail, 155.
,, Bareca for beacon, 155.
,, boats' anchor, 155.
,, gaining local information, 155.
,, shoals out of sight of land, 155, 156, 157.
,, table showing time occupied in, 158.
,, lead lines, 159.
,, ,, for surveying, *Note*, 159.
,, deep sea lead line, 160.
,, Massey's sounding machine, 160.
,, Walker's harpoon sounding machine, 160.
,, Burt's nipper and buoy, 160.
Soundings, notes on reporting, 148.
,, on charts, how expressed, 141, 142.
,, on foreign charts, 142, 143.
,, table for converting, 162, 163.
Southern hemisphere, thermometer in ,99.
Spain, money of, 318.
,, weights and measures, 326.

INDEX. 429

Spanish fathom or braza, 142.
,, mariner's compass, 52.
,, metro, 143.
,, sea terms and phrases, 349 to 356.
Spars, varnish, to make, 370.
,, floating power of, 333.
Specific gravities, 335.
Speed table, 6.
Speed, to find co-efficient of, 389.
,, do. in paddle-wheel vessels, 389.
Spheroidal tables, Carrington's, 164 to 166
Spirits of wine, 371.
Sprains, cure for, 304.
Spring and neap tides, 167.
Squalls, 89.
Staining, 371.
Standard chronometer, 191, 192.
Standard compass, 37 to 53.
,, ,, description of and instruction for using, 37, 38.
,, ,, cards and pivots, 38.
,, ,, caution as to concussion 39.
,, ,, when used as Azimuth compass, 38.
,, ,, do. or steering compass, 38, 39.
,, ,, where it should be placed, 39.
,, ,, process by bearing of a distant object, 40.
,, ,, to determine the real or correct magnetic bearing, 40.
,, ,, process by Azimuth's and Amplitudes of the sun, 41.
,, ,, process by reciprocal, bearings, 41.
,, ,, deviation in iron vessels affected by heeling of ship, 42.
,, ,, affected by the funnel, 42.
, ,, ship's course should only be directed by, 43.

Standard compass, deviation table, 44.
,, ,, steering table, 45.
,, ,, Archibald Smith's straight line course table, 46.
,, ,, do. explanation of, 47.
,, ,, forms for registering the observations for determining the deviation of the, 48.
,, ,, when deviation is to be ascertained, 40.
Stations, signal, list of, 30.
Statute mile, table for converting into Admiralty knot, 202.
Steam launches, instructions for working the engines of, 210 to 212.
,, ,, to get up steam, 210.
,, ,, the boiler, 210, 211.
,, ,, starting the engine, 211.
,, ,, running, 212.
,, ,, sea water for feed, 212.
,, ,, stoking, 212.
Steam ships, meeting or crossing, see *Rule of the road at sea.*
Steam tugs, lights for, 62.
Steering table, 45.
Stewing, 376.
St. Helena, money of, 318.
Sting from wasps, &c., 306.
Storms, revolving, see *Revolving storms.*
Storm tables, Dove's, 83, 84.
Stowage of coal, 332.
Stowing chronometers, notes on, 188, 189.
Straight line, Smith's, course table, 46, 47.
Stratus clouds, description of, 87.
Stream currents, 102.
Streams, tidal, 172.
S.W. monsoon, Indian Ocean, 77.
Success, secret of, to sailors, 195.
Sun stroke, remedy for, 304.
Sun's true bearings, or Azimuth tables, 41.
Sunset, to find time of, 381.
Surf, management of boats in, 217 *et seq.*
Surveying, Staff-Commander Hull on, 150.
Sweden, money of, 318.
,, weights and measures of, 329.

SWE

Swedish fathom or favn, 142.
" mariner's compass, 52.

TABLE, A. Smith's straight line course, 46, 47.
Table, deviation, 44.
" dip, 20 and 198.
Tables, Dove's wind, 83, 84.
Table for ascertaining velocity of tide, 205.
" for computing area of sails, 33.
" for converting English inches into millimètres, 98.
" for converting mètres and decimètres into English feet and fathoms, 162, 163.
" for converting millimètres into English inches, 97.
" for converting points of the compass into degrees, minutes, &c., 49, 50.
" to find the distance of an object by two bearings, and the distance run between them, 199.
Tables for reducing soundings, 175, 176.
Table, horary, 28.
" horizon, 17 to 20.
Tables of barometric pressure, Atlantic, 95, 96.
Table of distances at which objects can be seen at sea, 196.
" of masthead angles, 9 to 16 and 19.
" passage, 120 to 126.
" for measuring heights by barometer, 200.
" of specific gravities, 335.
" of mariner's compass of foreign nations, 52, 53.
" showing comparisons of Fahrenheit's, Centigrade, and Reaumur's thermometers, 101.
" showing distances of horizon at different elevations, 197.
" showing number of feet subtending an angle of 1' at a given distance, 206.
" showing time occupied in sounding, 158.

TID

Table showing the size chain or wire rope which is used as a substitute for hempen rope, 362.
" speed, 6.
Tables, spheroidal, Carrington's, 164 to 166.
Table, steering, 45.
" to correct heights for curvature, 198.
" to convert knots into statute miles, 201.
" to convert statute miles into knots, 202.
" to economise fuel, 127.
" trial trip, 203, 204.
" of natural sines, &c., 396.
" of logarithms, 394.
" pressure in diving, 398.
Tanks, size and capacity, 321.
Target at sea, finding distance of, 181.
Tarpaulins, paint for, 368.
Tea, how to make, 377.
Technicalities of winds abroad, 159.
Tensile strain of chain cables, 363.
" " anchors, 364.
Territorial waters (slavers), 232.
" " Egypt, 232.
" " Omani, 232.
" " Persia, 232.
" " Turkey, 232.
" " Zanzibar, 232.
Thermometer, in the northern hemisphere, 99.
" in the southern hemisphere, 99.
" rules for converting the various scales, 99, 100.
" table showing comparisons of Fahrenheit, Centigrade and Reaumur's, 101.
" indications of, in approaching ice, 119.
Tide pole, instructions for the use of, 173 to 175.
Tides, notes on, 167 to 176.
" how expressed on Admiralty Charts, 167.
" flood and ebb, 167.
" spring and neap, 167.

Tides, range of, 168.
," establishment of Port, 168.
," diurnal inequality, 168, 169.
," in the English Channel, 169.
," at Havre, 170.
," Rio de la Plata, 170.
," Gulf of Mexico, 170.
," Indian Ocean, 170.
," China Seas, 170, 171.
," Australia, 171.
," California and Oregon, 171.
," wind affecting the, 172.
," tidal streams, 172.
," and half tides, 172.
," up deep bays, rivers, &c., 173.
," instructions for use of tide pole, 173 to 175.
," tables for reducing soundings, 175, 176.
," diagram to explain terms, 176.
," notes on reporting, 146.
," signals in French ports, 138.
," velocity of, table for ascertaining, 205.
," velocity of, how expressed on charts, 141.
," how measured, 141.
Timber, classification of, 336.
Time to fire daylight gun, 381.
Tonnage of vessels, rule for finding, 336.
Towing boats, 216.
Towing trials, 387.
Trade winds, Pacific, 75, 76.
," Atlantic, 74.
," average limits of, 74, 75.
," &c., notes on reporting, 147.
Treatment of the apparently drowned, 293 to 301.
Treaty, weights by, China, 409.
Trial trip table, 203, 204.
Trinity House, corporation of, buoy systems, 136.
Troops, disembarking, *see Disembarking troops*.
True bearing, notes on, 149.
Trysail, to find number of square yards in, 339.
Turkish mariner's compass, 53.
Turkey, money of, 318.

Turkey, weights and measures, 330.
," (slavers), 232.
Typhoons, China Sea, 77, 78.
," hurricanes, and cyclones, general notes on, 78 to 82.

UNITED States, weights and measures 327.
," ," money, 313.
," ," buoyage, 139.
Useful notes on the marine steam engine. 382.

VADEN or fathom, Dutch, 142.
Variable belt, breadth of, in Atlantic, 74
," ," ," Pacific, 75.
Variation of compass, notes on reporting, 148.
Varnish, 370.
Veal, how to stew, 376.
Velocity of tide, table for ascertaining, 205.
Vigias, notes on reporting, 147, 148.
Voucher of nationality, ship's papers, 224.

WALKER'S harpoon ship log, 161.
," ," sounding machine, 160.
Warm and cold bands, Agulhas current, 112.
Warps, 216.
Warrior (ironclad) to lighten one foot, 348.
Waste heat, causes of, 382.
Water, to filter, 379.
," ballast for boats, 215.
," weight of, 321.
Waterproofing boots, 375.
," cloth, 375.
Weather, 85 to 89.
," Beaufort notation, 85.
," colour of sky, 85.
," clouds, 86.
," French equivalents, 356.
," scale adopted for noting amount of cloud, 86.

Weather, what the various clouds indicate, 86.
" description of clouds, 87, 88.
" rainbows, 88.
" squalls, 89.
" squalls, how preceded and accompanied, 89.
" rhymes, 89.
" the barometer, 90 to 98.
Weights and measures, Argentine Republic, 327.
" " Austria, 329.
" " Belgium, 326.
" " Bengal, 323.
" " Bombay, 323.
" " Brazil, 327.
" " British, 318 to 322.
" " British West Indies, 324.
" " Canada, 324.
" " Cape of Good Hope, 324.
" " China, 330, 407, 408.
" " Denmark, 329.
" " Egypt, 330.
" " France, 324 to 326.
" " Germany, 328.
" " Gibraltar, 324.
" " Great Britain, 318 to 322.
" " Greece, 327.
" " Holland, 327.
" " India, 323.
" " Italy, 326.
" " Japan, 330.
" " Madras, 323.
" " Mauritius, 324.
" " Norway, 329.
" " Portugal, 326.
" " Prussia, 328.
" " Russia, 328.
" " Spain, 326.
" " Sweden, 329.
" " Turkey, 330.
" " United States, 327.
" " West Indies, 324.

Weight, to find the, of a known substance of given dimensions, 332.
Weight to immerse ship one inch, 348.
Welch and Bourchier's life buoy, 284, 285.
West Coast of Africa, money of, 318.
White squall, West Indies, 89.
Wilson, Commander A. K., on finding distances at sea, 181.
Wind, 73 to 82.
" Beaufort's notation, 73.
" rate and pressure, 73.
" average limits of the regions of trade winds and monsoons, cyclonic storms and rainy seasons, 74, 75.
" Atlantic Ocean, 74.
" Pacific Ocean, 75, 76.
" Indian Ocean, 77.
" Arabian Sea, 77.
" Bay of Bengal, 77.
" China Sea, 77.
" East Coast of Africa, 77.
" Mozambique Channel, 77.
" between the Equator and the parallel of 10° S., 77.
" between the parallels of 10° and 30° S., 77.
Winding chronometers, notes on, 190, 191.
Winds abroad, terms for, 359.
Winds and weather, Malay Archipelago, 404.
" " New Guinea, 404.
Wind and current charts, note on, 76.
" charts, 74 and 76.
Wine, spirits of, to test, 371.
Wire rope or chain equal to hempen rope, 362.
Wreck, boarding a, 222.
" on a flat shore, 283.

YELLOW, to dye, receipt for, 372.

ZANZIBAR (slavers), 232.

GRIFFIN & CO., Publishers, The Hard, Portsmouth; and 15, Cockspur St., S.W.

Griffin & Co., Publishers, The Hard, Portsea,

ACTIVE LIST OF FLAG OFFICERS,

Captains, Commanders, and Lieutenants of the Royal Navy, with particulars exhibiting the progress, &c., of Officers, from their entry into the Service to July 1st, 1871. By Captain WILLIAM ARTHUR, R.N. Demy 8vo. Price 3s. 6d.

ACTIVE LIST OF THE FUTURE.

By the Author of the "Active List." Price 1s.

NAUTICAL SURVEYING.

Reprinted from Alston's Seamanship. With Charts. 2s. 6d.

"It is, however, the treatise on nautical surveying which more particularly distinguishes this work, and which we can earnestly recommend to the officers of the service generally, as it is well calculated to instruct them on a subject very little understood outside the surveying service. This treatise was originally written by Staff-Commander RICHARDS, now in command of H.M. surveying vessel 'Lightning,' an officer of very large and varied experience, and well known to be one of the best practical surveyors in the Royal Navy. The treatise is simple and clear in arrangement, and written with the especial object of instructing the officers of the naval service in general, and only deals with the use of such instruments as are found on board of every man-of-war. We have never met with any treatise on nautical surveying by any means so well calculated to answer the purpose for which it was written.."
Naval Science.

Engineer Officers, Watch, Station, Quarter, and Fire Bills

By William J. J. Spry, R.N. Price 3s. 6d. A complete *vade mecum* for Engineer Officers in Her Majesty's Navy.

and Cockspur Street, Pall Mall, London, S.W

Stanford University Library
Stanford, California

In order that others may use this book, please return it as soon as possible, but not later than the date due.